The Pictorial History of Steam Power

CASE

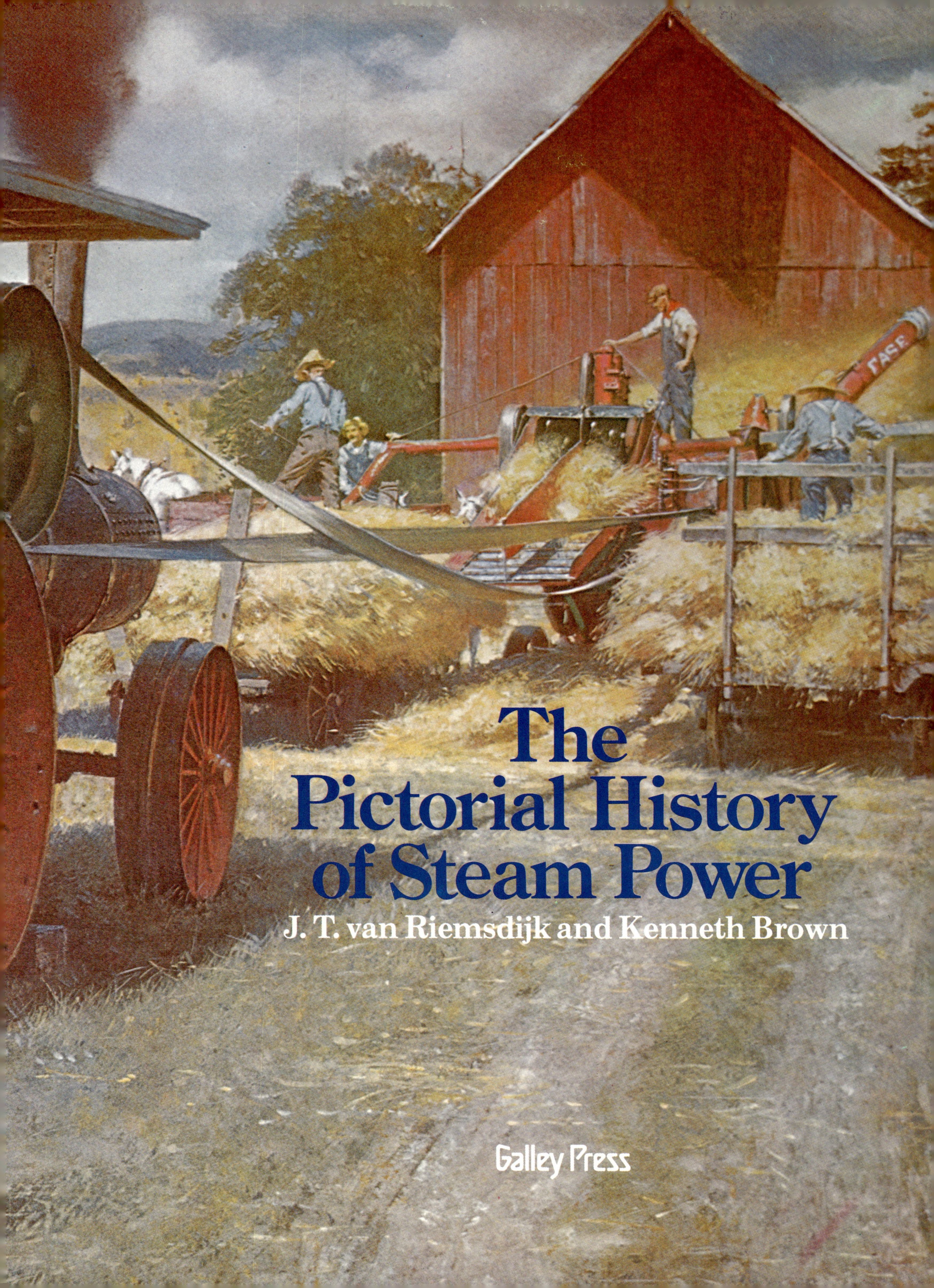

The Pictorial History of Steam Power

J. T. van Riemsdijk and Kenneth Brown

Galley Press

Contents

First published in 1980 by
Octopus Books Limited
59 Grosvenor Street
London W1

This edition published in 1984 by Galley Press, an imprint of W.H. Smith and Son Ltd. Registered No. 237811 England, trading as WHS Distributors, St. John's House, East Street, Leicester LE1 6NE

ISBN 0 86136 852 5

Printed in Hong Kong

Page 1: *an inside-cylinder 0-6-0 goods locomotive of the early 20th century in the classic British style, seen traversing the Punjab plain of Pakistan.*

Pages 2–3: *Metcalf's famous painting of a threshing scene on a North American farm at the turn of the century. J. I. Case of Racine, Wisconsin, manufactured traction engines and tackle and is a thriving concern today.*

This page: *a steamboat on the Mississippi, at New Orleans, USA.*

Introduction

Facing page: *steam spectacular, a geyser in Yellowstone Park, USA. The heat within the earth offers a vast resource of virtually untapped energy, but it does not usually offer itself in so readily exploitable a form.*

If you put ½ litre (1 pint) of water in a kettle and let it boil away, it will produce more than 900 litres (200 gal) of steam. If you confine the steam to a smaller space it will exert pressure upon its container: the more it is confined the higher the pressure will be. If it is confined within the kettle, and the kettle does not burst, the steam will press upon the surface of the water and thereby allow its temperature to rise above boiling point. If the pressure is reduced, some of the water will immediately turn into steam until a new equilibrium is reached between the liquid and the vapour.

These physical facts, together with the natural abundance of water, account for the pre-eminent importance of steam as the link between heat and power: even today steam engines produce by far the greater part of the world's electricity (even in stations fuelled by atomic energy) and propel the largest ships. And there is no sign that the pre-eminence of steam is near its end.

However, the way in which steam is used has changed enormously. Until the late 19th century the reciprocating steam engine, with pistons working in cylinders, predominated; but it was decisively challenged in 1897 by the multi-stage steam turbine of Charles Parsons. At Queen Victoria's diamond jubilee review of the fleet, conducted for her by the Prince of Wales but watched by Her Majesty through a telescope from Osborne House,

The fountains of Italy were supplied by aqueducts and streams from the hills, as here at the Villa d'Este at Tivoli. Attempts to emulate their splendour in northern Europe led to some inventions important in the pre-history of the steam engine.

Previous page: *steam drifting from the cooling towers of a power station. It comes from the water used to cool the condensers, and represents a great loss of energy. In urban situations this energy can be applied to district heating systems, as was done at Battersea in London.*

Parsons made an unauthorized appearance in his 30 metre (100 ft) turbine yacht *Turbinia*. Before the eyes of the Royal Family, of the Royal Navy and of many distinguished overseas guests, he demonstrated that nothing could equal the speed of his strange little craft, and started the train of events which resulted in the total supersession of the reciprocating steam engine for all really high powers. Turbines, basically of Parsons' pattern, are the steam engines of today.

Steam engines, and other gas or vapour pressure machines, existed, in experimental form or as primitive toys and curious novelties, even before the piston and cylinder arrangement was devised. Their history goes back to classical antiquity. The natural phenomena exemplified in some of these ancient devices were the subject of much speculative science, interest in which was revived during the Renaissance. The new scepticism of the 16th century, and the rise of experimental science together with the development of scientific apparatus in the 17th, provided a basis of theory and experience upon which such experimenters as Papin and Savery could build, but success eluded both of them because their devices could only be made in a small size.

When a practical steam engine was at last produced, by Thomas Newcomen in 1712, it soon became an object of scientific attention, and the existence of such machines led to great advances in theoretical physics for the next century or longer. The piston-in-cylinder steam engine eventually transformed the developed world, and this is well documented by technical and economic historians, but what has never been fully researched is the interaction between the engineers and the scientists during its evolution. Equally important is the fact that the steam engine gave rise to the first large machine tools, and indeed to the basic workshop machinery of the present day. The other important source of techniques for manufacturing in metal was the clock trade, but its secrets were still jealously guarded at the beginning of the 18th century, and its products were so small compared with the early steam engines as to represent a different order of ideas altogether. All the same, the fusion of the two techniques in the 19th century was to give us precision engineering.

The steam engine, then, is the central machine of the industrial revolution, because it was the great educator, the great proving ground, the great ancestor as well as an increasingly important provider of the power to drive the machines. Its first industrial rôle was to pump water from deep mines and this has a strange echo about it, an echo of the purposes for which some early classical experiments were made and, indeed, of the motivation of a 17th-century experimenter whose work was important in the prehistory of the Newcomen engine: Otto von Guericke. His object was the creation of fountains.

The sight and sound of gushing water has always had a strange fascination, and natural springs have been endowed with supernatural attributes in the imaginations of many different peoples. Major and minor deities have been thought to inhabit or protect these places, and the therapeutic qualities of their waters have been held to be magical. The recreation of the natural spring within a temple or a rich man's courtyard was a classical ambition, but though heat engines may have provided the magic in a temple it was usually gravity that supplied the fountains in the gardens of the rich. At least, that was the case in Italy, where tremendous aqueducts brought water from the hills to Rome, and dammed streams fed the gardens of the villas of Tuscany.

When the ideas of the Renaissance spread from Italy to less naturally favoured places, splendid fountains required splendid pumps. There never was so imposing, so vast, or so unreliable a pumping plant as the machine at Marly which used water wheels to drive a long train of pumps supplying the water of the Seine to the magnificent gardens of Versailles. And it was the troubles von Guericke encountered with his pumps, when he wished to adorn his gardens and impress his friends in Magdeburg with fountains worthy of Italy, that led him to the invention of the vacuum pump and the celebrated experiment with the 'Magdeburg hemispheres' which demonstrated so clearly how the force might be obtained which Newcomen was to put to use in his atmospheric steam engine. And, before Newcomen, Savery's engine, intended for the draining of deep mines, proved suitable only for working a fountain.

The piston-in-cylinder steam engine became a familiar object on the railways. Here everybody could see it in action, at close quarters, and even understand something of its operation. It was a permanent popular demonstration of mechanical principles such as the conversion of reciprocating to rotary motion by means of connecting rod and crank, and a vivid illustration of the fact that heat could be turned into power by means of water and steam. Its exciting presence made it a popular subject for toys and models, and those toys that really worked by steam possessed some of the glamour, and probably more of the educational value, of the real things.

Beauty was there, too. The beauty of the fountain was often emulated, in sight and sound, by a white jet of steam climbing into a blue sky on a cold sunny day—when a safety valve blew off—or by the vigorous white cloud of exhaust steam tumbling over a train at speed. In the landscape, from far away, one could trace the presence of an invisible railway line by following the white trail of steam from an unheard train, just as today one can trace the path of an unheard jet aircraft by its vapour trails high in the sky. For the aircraft leaves steam in its wake as well.

The fact is that all engines are versions of steam engines. The internal combustion engine was developed on the basis of steam engine technology, and it was the boiler that was suppressed. The gas turbine which powers the jet aircraft is the result of a similar transformation of the steam turbine. In both cases the fuel is burned in compressed air, either in the cylinder or in the combustion chamber of the gas turbine, and the products of combustion—the working fluid—are largely steam and oxides of carbon. The steam liquefies into droplets when it meets cold air, so the visible trail of locomotive and aircraft are both the same thing.

The association of steam with the romance of travel is one obvious reason for the devotion of many non-technically minded people but, long before steam

Steam-driven toys have provided an introduction to science and engineering for many fortunate children. This is a steam-driven toy battleship of German manufacture, dating from around 1900.

engines could move themselves about, the French *savant* B. F. de Belidor remarked upon the animal-like qualities of these strange new machines, and suggested a reason for human beings identifying themselves with them. In 1737 he wrote 'This is the most marvellous of machines—there is no other with a mechanism more akin to that of animals. Warmth is the origin of its movement, it creates in its various pipes a circulation like that of the blood in the veins, with valves which open and close as required; it breathes in and out regularly and automatically and by its own work provides all it requires to keep going'. The last is no doubt a reference to the invariable association, at the time, of the steam engine with mines. But it was the steam locomotive that impressed this personality upon the general public with its warmth, its lithe steel limbs and the noisy breathing, amounting at times to positive shouting, that accompanied its progress.

The engines of steam-driven vessels were less in evidence and are today quite invisible. At first they were essentially stationary engines, but soon evolved to special types adapted for driving paddle wheels while keeping the main weight low down in the hull. Because of the need to condense the steam—at first because the engines worked almost at atmospheric pressure, and later to conserve fresh water for the boilers—the progress of a steamer was more associated with smoke than with steam, as is well depicted in numerous paintings and prints. However, on entering and leaving harbour, which was when steamers were seen by most people, the steam whistle occasionally produced a graceful plume, and the writer's sole recollection of the magnificent *Normandie* is of just such a graceful salutation—like a plume of white feathers against the sky, and somehow a typically French gesture.

At the other end of the scale, one can to this day enjoy the engines of paddle steamers, and nowhere better than on the Swiss lakes, where the manoeuvring of the engines at the frequent stops and rapid starts is not unlike the driving of locomotives. The engines are generously exposed to view, and usually attract almost as many admirers as the scenery, because the spectacle of a large reciprocating engine at work can be enjoyed in few places nowadays, but is surely one of the most splendid spectacles that engineering has to offer.

The largest and grandest of marine reciprocating engines were seldom seen by the public. These were the screw engines of the largest liners of the late 19th and very early 20th centuries, such ships as the *Titanic* and its luckier, long-lived sister, the *Olympic*. As tall

The vapour trails of aircraft are largely made by steam. The burning of fuel in jet or piston engines produces large quantities of steam as the hydrogen in the fuel combines with the oxygen in the air.

PUBLIC
SCHOOL

THROUGH LINE
NEW YORK

Previous page:
the States of America were united by the railway, and the pioneer railroad, opening up the vast landscapes of the interior, is part of American folklore. This is one of many prints published by Currier and Ives showing diverse aspects of 19th-century American life.

Steam vessels on the Bosphorus, a stretch of deep water in Turkey which separates Europe from Asia. None of the passenger and vehicle ferries shown is more than 25 years old.

A typical tramp steamer of the early 1900s, coal-fired and with triple-expansion machinery. A large tonnage of vessels of this type lies on the sea bed as a result of two world wars.

as a three-storey house and longer than they were tall, such engines used the steam in three or four stages and had four, five, or six cylinders. The largest might develop 25,000 horsepower. The *Titanic* had two engines, plus a turbine for its third propeller. Multiple expansion made long sea voyages possible by steam, because the high efficiency of such engines reduced the amount of coal needed. In cargo service, even the smallest space not needed for coal could increase the payload of the vessel, and for more than half a century virtually all the world's sea freight was carried in ships powered by three-crank triple expansion engines, about the size of a double-decker 'bus.

Even in smaller vessels, the steam engine at one time was solely in charge, despite the heavy demands on limited space made by the storage of coal. To sustain steam trawlers for up to three weeks in the North Sea fishing grounds, for instance, it was the practice to set off with coal filling the fish holds. As the coal was consumed, the holds would be washed down ready to receive the catch. It was a hard life with steam: conditions in the stokehold of a small vessel in a rough sea had to be experienced to be believed; the heat, the physical effort, the cramped conditions and the constant fight to keep one's balance all beggared description.

It is possible even today to get a taste of life on a steamer fired by coal, aboard the passenger vessels which ply the Bosphorus. With the magnificent minarets of Istanbul as the backdrop, one may enjoy a glass of Turkish 'chai' and watch the clinker from the firebars being dumped overboard as soon as one is some way from the shore. Also at Istanbul, men with wicker baskets on their heads may be seen carrying coal from a dumb barge and tipping it down unobtrusive circular hatches in the ship's deck. When the bunkers are filled and the hatch replaced, the deck is scrubbed down to remove any trace of grime with that attention to detail which was a hallmark of the steam age. If one is lucky enough to find an engineer who speaks English, one may even be invited below, to experience at first hand the sight, sound, and smell of a reciprocating steam engine working in a confined space when the external temperature is in excess of 38°C (100°F).

The steam engine fitted happily into a totally different setting on the village green. Here, once or twice a year, a vivid splash of colour signalled the arrival of the local funfair, and this was a place of romance, especially after dark. Mighty showman's engines, resplendent in maroon and gold paint and ornamented with polished brasswork, were there, rocking gently back and forth against their wheel chocks as their spinning crankshafts and flywheels gave out light and power. Steam-driven mechanical organs trumpeted out the latest hit tunes, their music books slowly unfolding and refolding into fascinating cardboard squares. There were richly carved roundabouts, with steam engines at their centres, round which one swung dizzily to one of several competing tunes as the view of nocturnal revelry rotated about one; and everywhere there were crowds of people, young and romantic, old and raucous, hesitant or adventurous. The smell of trampled grass, hot oil and coal smoke can now only be experienced at traction engine rallies and steam fairs for enthusiasts: it is no longer a part of the calendar.

The steam engine was part of the landscape, which it enhanced or despoiled according to circumstance. The curve of a railway line could be elegant, a civilized compromise with nature, neither subservient nor dominating. In Cornwall, during the heyday of tin and copper mining in the last century, the engine houses, with their great beams projecting and nodding up and down, lent interest to a rather barren landscape as the mines were drained, the ore raised and crushed and all the activities of the industry proceeded. The arrival of a horse team dragging a 40 tonne (40 ton) beam for a new engine was a spectacle in itself, as was the raising of the beam to its bearings on the lofty 'bob wall'. There was rejoicing too, for a new engine signalled more work and a brief respite from poverty for some of the families in the village.

Today, many of these engine houses have been standing for 100 years, monuments to the engineering past and tourist attractions for the present. Few still contain engines, but three have been taken over complete by preservation societies in Britain.

Today, the strangest chapter of all in the history of the steam engine is unfolding. This is the preservation of engines and their operation, not to do any useful work but because an engine in motion is a thing of interest and beauty. They have become art objects. A movement towards live steam preservation which began with narrow-gauge railway locomotives and steam road vehicles has now spread to large stationary pumping and mill engines. For all their working lives these gentle giants were hidden from public gaze in their own stately homes. Now, on most weekends, somewhere in the country, a tall smoking stack and the clink of a shovel beckon the visitor to part with a modest sum and come inside to be moved by the grandeur of a steam engine in motion. He will find a devoted team of men and women who have spent countless hours in filthy toil to remove rust, scale, and the other evidence of disuse; to retube and restay boilers, remetal bearings, reglaze windows and paint coat upon coat to restore the glory to a piece of Victorian design. And the main lines of British Railways now provide a racing ground for locomotives once sold for scrap, but now lovingly restored to service by amateurs.

Perhaps the tide of history has been reversed, but perhaps it was always a bit like that really: those who looked after steam engines were often devoted to them and the love felt for them was not a matter of engineering at all. They were always art objects to some, human or animal to others, to be loved or hated, cursed or grieved over, for such is the nature of man. As Belidor said, 'this is the most marvellous of machines'.

Industrial archaeology: the abandoned pumping engine house of an old Cornish mine.

Steam still rises from this South Wales colliery. Steam engines were used for winding the cages up the shafts, for pumping and ventilation, for working conveyors and sifting the coal to separate sizes, and for numerous other jobs around the pit-head.

All the excitement of a steam-driven merry-go-round. English examples revolve clockwise in contrast to Continental and American machines.

A line-up of steam traction engines at a rally in southern England. Such events are popular in Britain, USA, Australia and New Zealand where many machines have been lovingly restored to their former glory.

Prelude: Fountains and Fire-Machines

Otto von Guericke

HERO OF ALEXANDRIA

Experiments with the effects of heat on gases and vapours, and with the pressure of steam that could be generated by boiling water, were made and recorded in classical antiquity: they no doubt had an unrecorded prehistory. The one ancient device that is well known is the 'aeolipile' of Hero of Alexandria, which is believed to have been first made around the year AD 50. This would now be called a reaction turbine. It consisted of a boiler placed over a wood fire, and the steam from the boiler passed into a pivoted hollow sphere of copper or bronze. From the sphere two pipes emerged on opposite sides, and were bent round at right angles at their tips, and provided with constricted nozzles. When steam issued from these, the sphere spun round.

Many modern reconstructions of this device have been made, and it has been sold as a toy. It works because, although all the internal pressures balance out in the sphere and most of the pipes, there is a tiny area in the bend of each pipe which comes opposite the nozzle, and the pressure there is not balanced. Consequently it succeeds in pushing against this area, which makes the aeolipile spin.

Hero made other experiments which have been recorded. One was a heat engine which simulated the supernatural. In a temple, a hollow altar was provided for sacrificial fire. Behind it, a shrine containing an image of a god was hidden by a pair of hinged doors. When the fire was kindled, the air within the altar expanded, and its pressure was conveyed to the surface of water in a closed vessel which had a pipe rising from the bottom and turning over into the bottom of an open vessel suspended on ropes. The heat thus caused the open vessel to fill with water, and its weight on the ropes opened the doors of the shrine, normally kept shut by a counterbalance. The open vessel did not descend far enough to expose the end of its filling pipe to the air, with the result that when the fire went out and the altar cooled, the water was sucked back, the counterweight took charge, and the doors closed.

The same principles could be applied to a fountain which played automatically when the sun shone upon it, and recharged itself when sunlight was succeeded by shadow. Installed in a south-facing colonnade, such a fountain could go through its cycle of operations several times every day, as the sunlight moved behind each column in turn. The term 'Hero's Fountain' was also applied to a device in which air pressure, not heat, played a part. In this, a very small fountain was set at the top of a frame inside which was a curious dumbell-shaped vessel, which could be turned so that either enlarged end could be at the top. In operation, water in the upper end trickled to the lower, displacing air which was piped up to another chamber in the upper part, in which its pressure forced another supply of water up and through the fountain. This would go on until the main vessel in the upper part of the dumbell was empty, when all that was needed was to rotate the dumbell and bring the water back to the top. The actual motive power was human, and it worked because water is heavier than air, but it is of interest to us because gas pressure provided the motive power link.

SEVENTEENTH-CENTURY DEVELOPMENTS

Leaving the many fascinating devices of the remote past, of which tantalizingly incomplete accounts survive as invitations to conjecture, we find many of the same ideas revived, with other classical learning, at the Renaissance. Not only was the spirit of experiment and scientific enquiry reborn in that period, but also many of the notions of what constituted civilization, culture and pleasure. It is not then surprising that in addition to Florentine versions of ancient Roman villas, we find a new interest in gardens, and especially those making use of the decorative and soothing effect of water. Of the elaborate water gardens of the Italian Renaissance much remains, but of the various mechanisms associated with them almost nothing has survived, although where gravity fed the fountains, they still play.

There can be little doubt that the desire to play has been a powerful force for technological advance, and the plaything—aeolipile, magic altar or musical fountain—has been the proving ground for many ideas which have since been of prime importance in the workaday world. The prehistory of the steam engine contains many such ideas, and in the 17th century attempts were made to apply them in a useful way, but doubts as to the real motivation linger. There were many attempts to make a machine to raise water by fire, to make what was in those days called a fire engine, but though there was already a need to drain mines by some such means, there was an apparently equally urgent desire to operate fountains.

Torricelli demonstrated the possibility of a vacuum in 1642, in the process of inventing the mercury barometer. Otto von Guericke began to work on the possibilities of harnessing the force of atmospheric pressure (which was to operate Newcomen's steam engine) about ten years later, and actually made the first vacuum pumps, but his interest, as already stated, derived from his desire to equip himself with pump-operated fountains. There was a touch of showmanship about von Guericke, best revealed by his experiment with the 'Magdeburg hemispheres', in which the strength of 16 horses was pitted against that of atmospheric pressure, before an invited audience. He has another claim to our attention, and gave another example of his showmanship, as the man who erected a large air thermoscope on his house in Magdeburg. This was a type of primitive thermometer, depending on the fluctuating pressure of a quantity of air trapped in a copper sphere. The air displaced water in a large U-tube, with a float in the outer leg. The float was connected to a gilded angel, by means of a chain and pulley, and the angel pointed to a written description of the weather, on a vertical scale.

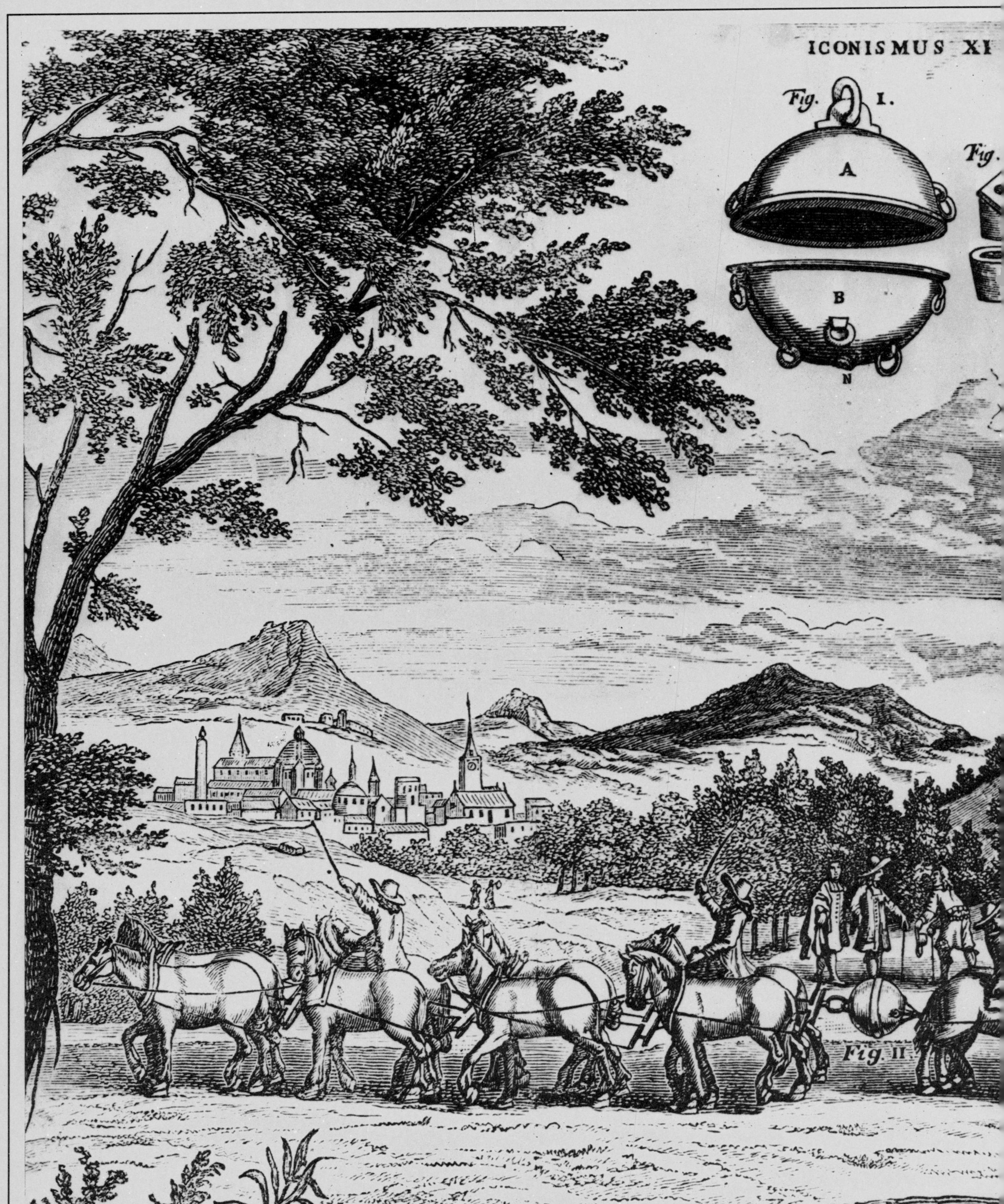
ICONISMUS XI
Fig. I.
A
B
N
Fig.
Fig. II

Von Guericke's experiment with the Magdeburg hemispheres. The hemispheres were lightly held together by grease, and the air pumped out from inside them by means of Von Guericke's newly invented air pump. Sixteen horses could not then pull them apart, but opening the air tap to break the vacuum released them at once.

The possibilities of the vacuum were realized in the first practical steam engine of Thomas Newcomen, which used steam to create the vacuum, in 1712. Without Von Guericke's demonstrations, made over half a century earlier, the steam engine might not have appeared when it did. But Von Guericke's original motivation was in fact the creation of fountains in the Italian style, for which pumps were needed.

This, then, was another gas pressure device.

Expanding air by heat, to displace water, was vital to the operation of Amontons' 'fire wheel' of 1699. The heat had the effect of lifting water within the wheel, to keep it out of balance and therefore in rotation. It depended on only a small part of the wheel being heated at a time, and its rate of rotation would have been small, had it worked at all, because though heating could be rapid, the necessary cooling would have been slow.

The impulse turbine made an appearance in the middle of the 17th century, probably not for the first time. Giovanni Branca proposed driving a horizontal scoop wheel by means of a steam jet from a boiler rather like that supposed to have supplied Hero's aeolipile. This would certainly have worked though, as any real power would have depended on high pressure, it would equally certainly have exploded if required to be useful. It was illustrated connected to a crushing mill operated by drop stamps, and so pointed to the growing awareness that such things might have a serious purpose.

At the end of the 17th and beginning of the 18th century, Savery attempted to apply the accumulated knowledge of a century of experiments in a device intended specifically for raising water from deep mines. 'The Miner's Friend' was a device without moving parts, the principle of which is still with us in the Pulsometer pump. It proved impossible to make it robust enough to work reliably in a mine, but it operated fountains very well, and as the only near approach to success before Newcomen, it deserves to be described.

Savery's pump centred upon a pressure vessel provided with automatic check (i.e. non-return) valves in its connected piping, so that it could go through the following cycle of operation: first, it was filled with steam from the boiler; then steam was shut off and the outside of the vessel was cooled so that the steam condensed (this was done by pouring cold water over it from a header tank); the vacuum so created drew water from a lower level, and when the vessel was filled the steam, at such pressure as was necessary, was readmitted, and because the suction valve at the bottom had closed, it forced the water upwards through a discharge pipe, replacing it with steam in readiness for repetition of the cycle.

The trouble was that the pressure required to lift the water was unavoidably a function of the height to which the water had to be lifted. There was no 'leverage' or multiplication factor such as was introduced by Newcomen when he balanced a piston against a pump bucket. And it was simply not possible to make large boilers or pressure vessels able to withstand the required pressures. However, the idea was an excellent one, and the French scientist Denis Papin devised something similar which he proposed to apply to the less onerous job of pumping water round a waterwheel, but this remained at the level of a curiosity, because the real need was not for rotative power at this time, but for pumping; and it seemed cumbersome to introduce an intermediate rotation stage in the process.

It was Papin who, in 1690, made a miniature piston-in-cylinder engine. Again, it could not have been made very large, given the technology of the time. The cylinder was upright and open-topped. The piston was connected to a counterpoise by a cord and pulleys, but it was airtight and could not rise from the bottom of the cylinder unless the vacuum beneath it was broken. In fact, there was a little water beneath it. Heat applied to the bottom of the cylinder boiled the water, and the piston rose. The heat was removed, the cylinder cooled and the pressure of the atmosphere pushed the piston down, lifting the weight. This was the Newcomen cycle in miniature, but Newcomen provided a separate boiler, and cooled the steam itself, rather than the cylinder, by an internal water jet. He did this 20 years later, and gave motive power to the Industrial Revolution.

The Italian renaissance water garden was the inspiration of many later and more northerly water gardens, in which, as in this fantasy, the baroque degenerated into the absurd.

Pumping and Early Rotative Machines

Denis Papin

THOMAS NEWCOMEN

It was Thomas Newcomen, a relatively obscure Dartmouth ironmonger, who astonished the world by producing a successful piston-in-cylinder engine. The first recorded example was erected at a colliery near Dudley Castle, Staffordshire, in 1712 when he was 49.

Like the earlier contrivances of Savery and Papin, Newcomen's 'atmospheric' engines, as they became known, were conceived and used solely as a means of raising water. Although Newcomen engines could be used to pump water to a waterwheel and hence produce rotary motion, the world had to wait until the 1780s and James Watt for engines which turned a crankshaft and produced rotary motion directly.

When we look back upon Newcomen's primitive machines, it is hard to realize what a breakthrough they represented at the time. L. T. C. Rolt, in his biography published in 1963, said: 'in the whole history of technology it would be difficult to find a greater single advance than this and certainly not one more pregnant with significance for all humanity'. Every prime mover of today, even a spacecraft, owes something to the atmospheric engine.

The term 'atmospheric' was used because it was the pressure of the atmosphere acting on the piston which caused it to move. Again, it was not until after the turn of the century that boilers could be produced which were capable of withstanding steam pressures substantially higher than that of the atmosphere, enabling more work to be obtained from smaller cylinders and drastic economies in fuel consumption to be effected.

The reciprocating action of a Newcomen engine was ideal for marrying up to existing types of deep-well and mine pumps which hitherto had been operated by water or animal power with severe limitation in output. Little is known about Newcomen's early experiments, nor is it by any means certain that he was aware of the state-of-the-art in applying the use of steam to pumping water at the time.

What is known is that he had an able assistant, a plumber called John Calley, and that he was engaged in the same line of research as Denis Papin in France. In about 1690, Papin succeeded in producing an experimental engine with a vertical open-topped cylinder, which worked by generating steam in the space beneath the piston. By allowing the cylinder to cool until the steam condensed, a vacuum was formed beneath the piston, allowing atmospheric pressure to push the piston down. The cycle was then repeated.

The action of such an engine must have been intolerably slow and it is reasonable to suppose that one of Newcomen's most serious problems was finding a more rapid way of cooling the steam in the cylinder. According to Desaguliers and other writers, Newcomen tried to cool the cylinder walls externally, using a flow of cold water, probably in a lead jacket surrounding the cylinder.

The other significant advance in Newcomen's thinking was in generating the steam in a separate boiler beneath the cylinder with a shut-off cock which was worked by hand to control the stroking of the engine. Although Savery had used a separate boiler in his apparatus, of which there were examples at work in Cornwall, it is not certain whether Newcomen was aware of this when he began his experiments.

Newcomen's cylinder, in view of the simple tools at his command, was probably brass and of the order of 178 mm (7 in) bore, common for pump barrels at the time. To get the cylinder and piston reasonably true meant painstaking work using hand tools. In addition to the use of soft leather seal round the piston, Newcomen placed a small quantity of water on top of the piston moving up and down with it, to prevent air from entering the cylinder and weakening or destroying the vacuum. The system for supplying the water seal and making good the loss by evaporation became a normal feature of the atmospheric engine, as did the overhead wooden rocking beam, with chains passing over arch-heads at the ends, to connect it to the piston and to the pump rod in the shaft.

How Newcomen discovered that introducing a jet of cold water into the cylinder produced the much more rapid condensation needed to speed up the action of the engine is uncertain. Martin Triewald, the learned Swede who came to know Newcomen after the atmospheric engine had become established, records that water leaked into the cylinder from the jacket with a rush, through the melting of some tin-solder which had been used to mend an imperfection in the cylinder wall. This created 'such a vacuum that the weight attached to the little beam, which was supposed to represent the weight of water in the pumps, proved to be so insufficient that the air, which pressed with a tremendous pressure on the piston, caused its chain to break and the piston to crush the bottom of the cylinder as well as the lid of the small boiler'.

Newcomen's biographer, Rolt, expressed himself strongly inclined to accept this report, apart from observing that 'steam at no more than atmospheric pressure would not have been hot enough to melt the solder, but that the flux had not "taken" in the blowhole is much more likely'.

Though this discovery of the condensing water jet by fortunate accident must have given Newcomen and Calley a great fillip after ten years of struggling in their primitive workshop, it still left other problems to solve. How to rid the cylinder of accumulating hot condensate and the air entering with the steam, and how to make the steam and admission valves self-acting, had still to be worked out. The fact that both had been solved by the time the 1712 engine went to work suggests that the condensing water jet discovery must have taken place well before 1710. Indeed, stories are told of Newcomen engines with hand-operated valves which must almost certainly have been built before 1712, but none of them has ever been substantiated.

The exact site of the Dudley Castle engine, incidentally, has never been proved, and many other mysteries still surround the inception of the Newcomen engine.

Facing page: *a typical small rotative beam engine as used for a wide variety of industrial drives early in the 19th century. This one was built in 1827 and used to drive pumps at Holyhead docks. The beam is supported on an A-frame to make the engine a self-contained unit. Watt's famous parallel motion may be seen, as can the valve chest at the back of the cylinder and, at the bottom, the disengaging gear so the valve could be worked by hand at starting instead of by eccentric.*

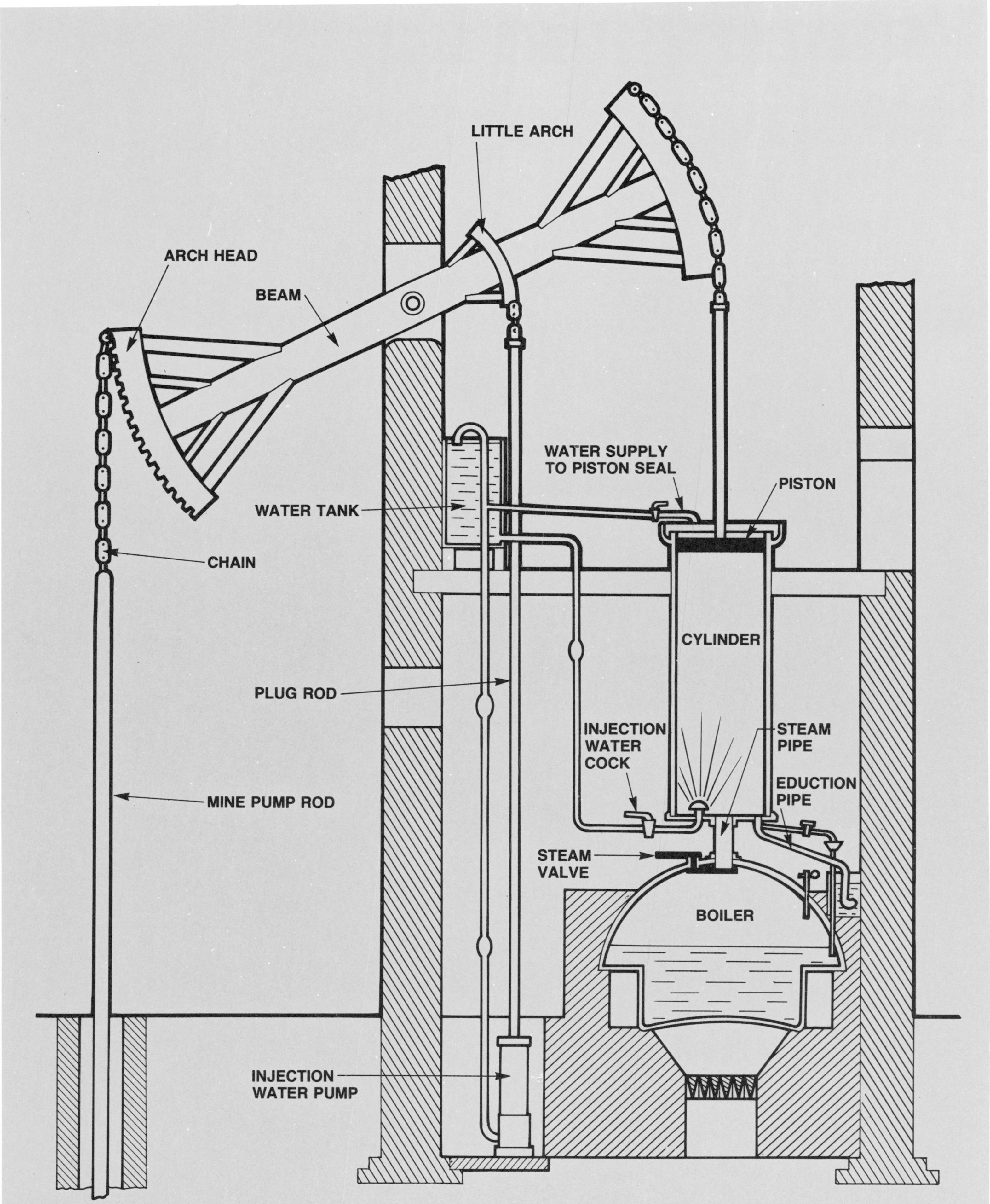
LITTLE ARCH
ARCH HEAD
BEAM
WATER SUPPLY
TO PISTON SEAL
PISTON
WATER TANK
CHAIN
CYLINDER
PLUG ROD
INJECTION
WATER
COCK
STEAM
PIPE
MINE PUMP ROD
EDUCTION
PIPE
STEAM
VALVE
BOILER
INJECTION
WATER PUMP

The working of the Dudley Castle engine will be appreciated from the diagram. An upright cylinder, open at the top, was supplied with steam from a boiler resembling a brewer's copper beneath it. The piston, lined with leather and with a water seal on top, was hung by a chain from the arch-head on the 'indoors' end of the overhead rocking beam. The pump rods were similarly suspended by chain from the 'outdoors' end. When steam was admitted into the cylinder via the steam valve on top of the boiler, the piston was drawn up by the weight of the pump rods, and any air or water blown out of the cylinder through water-sealed non-return valves.

When the piston reached the top of its stroke, the steam valve was closed and the injection water cock opened, causing rapid condensation of the steam inside the cylinder when it met the jet of cold water. Atmospheric pressure then drove the piston down, raising the pump rods and hence making the pumping stroke.

The cycle then repeated itself, the steam valve and injection cock being opened and closed by means of leather-faced tappets fixed to the plug rod hung from the beam and engaging pivoted, curved levers. (The valve gear is omitted from the diagram for clarity.) There was no engine framing as such, support and anchorage of the various parts being from the masonry and timber house.

The Dudley Castle engine had a cylinder of 533 mm (21 in) bore and the piston travelled 1.8 m (6 ft). At each stroke, its bucket pump raised 45 litres (10 gal) of water through a distance of 47 m (51 yd), and at the normal working speed of 12 strokes per minute the engine developed about 5½ horsepower. Although engines increased greatly in size, the basic design proved reliable and remained virtually unaltered for 60 years.

To appreciate more fully the scale of his achievement in launching upon the world a machine of considerable complexity and refinement compared with anything which preceded it requires some knowledge of the circumstances prevailing at the time. It was a feat which required much more than merely inventing the working principles.

Thomas Newcomen was not a man of letters but a practical mechanic, albeit one with great vision and skill. Since he had his own business to run, the time he could devote to his experimental work must have been limited. How much he knew about the work of the well-known scientists of his day is open to doubt, especially in view of Dartmouth's remoteness from London. For instance, it is unlikely that Newcomen saw the account of Papin's model cylinder and piston experiment until long after it was published by the Royal Society in 1697.

However, Dartmouth did have one asset; it was close to the tin and copper mining areas of Devon and Cornwall where minerals had been worked from Roman times. By the end of the 17th century, mines were reaching depths which posed considerable problems due to ingress of water. The limited power of waterwheel-driven pumps to keep the workings dry was becoming a growing hindrance to mining operations.

Newcomen is known to have visited Cornwall: supplying tools and equipment to the mines doubtless accounted for a fair proportion of his business. There he would have become familiar with the bucket, or lifting, pump driven by waterwheel and he might also have seen a Savery engine with its separate boiler. Ultimately he had to come to an arrangement with Savery before he could exploit his invention because the latter held the master patent for raising water by the use of fire, the actual means not being specified. In fact, all early steam engines were known as 'fire-engines', a term which persisted for well over a century.

Even today, inventors face enormous problems in getting their brainchild 'off the ground', so imagine what it must have been like for Newcomen in his day of poor communications and strong prejudices. The price of a Newcomen engine was of the order of £1,000, a very substantial sum, and he must have faced difficulty in obtaining sufficient backing. On Savery's death in 1715, a syndicate of London businessmen was formed to exploit his patents which of course included the atmospheric engine. Newcomen died in 1729.

On the technical side, Newcomen offset the problems of his engine's novelty and complication to some extent by building his engines of parts which had already proved satisfactory in practice. Gunmetal castings, for instance, were in common use in cannon. Wrought-iron was obtainable but it was expensive, so its use was limited: the working gear for operating the valves, for example, was composed largely of wood and leather. Brass was also used for such items as bearings and stuffing boxes which could be machined because there were primitive wheel lathes in existence which could turn softer metals up to about 50 mm (2 in) diameter. Cast-iron was generally of poor quality and as much as possible of the boiler and static parts of the engine were made of bricks or masonry. Timber was used for the rocking beam and engine-house framing.

Newcomen's biggest constructional problem was undoubtedly making the cylinder. His early engines used brass with the interior rubbed smooth. With the invention of the boring bar still 60-odd years away, he had to rely on the skill of the brass founders of the day to produce castings which were dimensionally accurate. They also had to be free from flaws.

Cast-iron cylinders did not come into general use until long after Newcomen's death. The ironmasters of the time steadily improved their casting techniques because, as time went on, cylinders for atmospheric engines reached 1,778 mm (70 in) and more in diameter.

Another mystery which surrounds the atmospheric engine is why, in view of Newcomen's West Country association, the first engines were not erected in Cornwall. There is a report of a steam engine at Wheal Vor mine which worked from 1710–5 but unfortunately this has never been substantiated. The healthy appetite for fuel of a Newcomen engine may be one of the reasons why its use was mainly at collieries in its early years.

The first definite record of a Newcomen engine in Cornwall was one erected at Wheal Fortune near Penzance in 1720, and in the same year the first Newcomen waterworks plant was erected, at York Buildings in London. Two years later saw the first engine erected abroad, in Hungary, and by the time of Newcomen's death his engines were in use in Western Europe.

The 'duty' of these early engines, expressed in the units of the time, was of 3–4 millions, that is, million pounds of water lifted 1 foot high by the consumption of each bushel (94 lb) of coal (or 454,000 kg of water lifted 305 mm high by the consumption of 43 kg of coal).

Once the atmospheric engine had proved its efficacy, the number soon ran into hundreds. Apart from increase in size, few improvements were made until John Smeaton made model studies to determine better proportions for the various parts, enabling higher duties to be obtained. He also extended the use of iron in construction of engines. By 1775, the duty of the best engine had risen to 10–12 millions but this still represented an overall thermal efficiency of no more than one per cent. There were then 60 engines at work in Cornwall and 100 in the Tyne Basin alone.

JAMES WATT

By the 1760s, the Newcomen engine was ready to be revitalized by what has been described as the greatest single improvement ever made in the steam engine—James Watt's separate condenser. Watt was a maker of mathematical instruments at Glasgow University, and

Facing page: *a typical Newcomen atmospheric engine of about 1712. Arch-heads with chains passing over them were used to keep the piston and pump rods in line; use of a bucket pump kept the pump rod in tension. The cock for admitting steam beneath the piston to permit its up stroke was situated between the boiler and cylinder.*

he became attracted to the Newcomen engine when, in 1764, he was given a scale model (owned by the university) to repair. The work done, he tried it and found it would only run erratically because the boiler quickly ran short of steam.

In endeavouring to account for the high steam consumption, he eventually reached the conclusion that it was due to the high heat loss at each stroke caused by alternately heating the cylinder wall by live steam and then cooling it again. Early in 1765, Watt conceived the use of a separate vessel in which to condense the steam, in communication with the cylinder via an exhaust valve. This vessel, the condenser, was kept clear of water from the condensed steam and injection water by scavenging with a pump worked off the engine. Since this pump also removed any air which leaked into the system, it was termed the air pump.

As a further measure to minimize heat loss, Watt introduced a steam jacket around the cylinder, the jacket itself being lagged by a non-heat-conductive material. He also put a top cover on the cylinder and introduced steam at atmospheric pressure above the piston to push it down, instead of air, using the same steam again for condensing. Further model experiments fully validated Watt's theories and in 1769 he took out a patent for 'a new method of lessening the consumption of steam and fuel in fire engines'.

As time went on, Watt increased the steam pressure acting above the piston to about 900 g (2 lb) above atmospheric to increase work output from the engine. The working cycle of the engine was now as follows: steam at low pressure was admitted above the piston, pushing it down against the weight of the pump rod in the shaft. When the piston reached the bottom of its travel, the equilibrium valve opened, and this allowed the expanded steam above the piston to pass to the underside without change in volume, the piston being raised by the weight of the rods. On the next steam stroke, opening of the exhaust valve allowed the steam below the piston to be drawn rapidly into the condenser which was already under vacuum from previous strokes. Inside the condenser it met the injection water jet, causing it to condense and settle at the bottom. Connected to the base of the condenser via a non-return (foot) valve was the air pump bucket worked from the engine beam, which drew air and water out of the condenser on its up strokes.

The historic meeting with Matthew Boulton which led to the famous partnership between dour Scots mechanic and astute Birmingham businessman took place in 1773. It was to have enormous repercussions. Boulton established the famous Soho manufactory in Birmingham in 1774, and Watt moved there from Glasgow to continue his model experiments full-time. A year later they went into full partnership, the name Boulton & Watt soon becoming famous for both inventiveness and insistence on quality workmanship.

The first two pumping engines on Watt's improved plan were produced in 1776, a 1,270 mm (50 in) cylinder pumping engine for a colliery at Tipton and a 965 mm (38 in) blowing engine—that is, an engine which forced air rather than pumped water—for John Wilkinson's blast furnaces in Shropshire. The connection with Wilkinson is important because, a year earlier, he had invented the boring mill for machining smooth the interior surfaces of cylinders.

Clearly, with the Watt engine it was no longer possible to seal the piston with water, so Wilkinson's invention was essential to development of the species. To maintain the seal in a bored cylinder while minimizing friction, Watt put a peripheral groove round the piston and packed it with hemp soaked in tallow for lubrication. By means of a 'junk ring' bolted down on top, the packing was compressed outwards so that it pressed lightly against the cylinder wall. Though it meant raising the cylinder cover every few weeks for men to enter the cylinder and tighten the bolts or to repack the piston, the arrangement worked perfectly well with low steam pressures and persisted for many years.

The two 1776 engines proved an instant success. They burnt only a third of the fuel used by atmospheric engines and resulted in a flood of inquiries. In Cornwall, where the price of coal was inflated by high transport costs, every atmospheric engine was replaced within a few years. By the early 1800s, the Boulton & Watt pumping engine, with its cast-iron (instead of wood) beam, Watt's elegant parallel motion, finely moulded valve gear and other metal parts finished bright, was very far removed from the relatively crude atmospheric engine.

Quality of workmanship and finish was a hallmark of

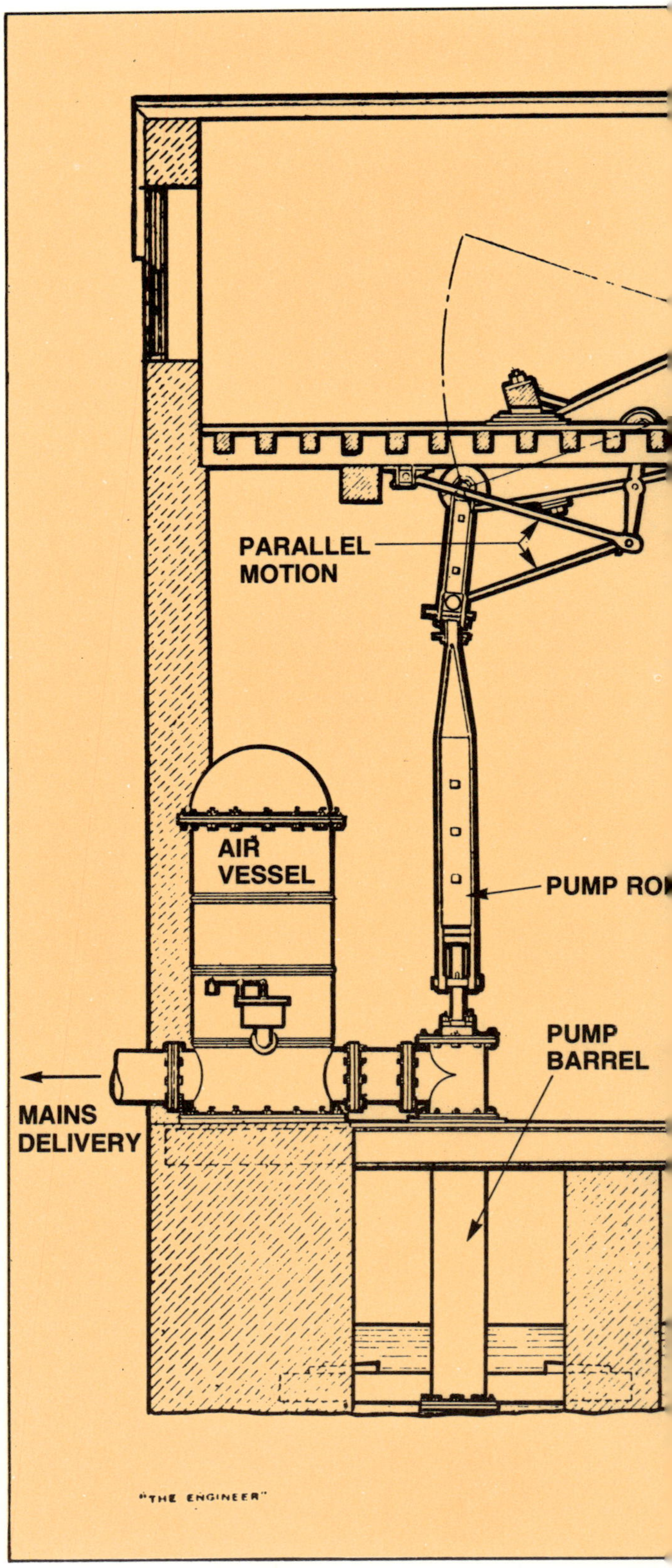

Boulton & Watt engines from the start. It created serious problems for the partners, especially in the early days when the bulk of the engine parts came from outside suppliers. Impelled by Watt's attention to detail and insistence on the best materials and workmanship, Britain's engineering industry gradually prepared itself for the Industrial Revolution which was to come. At this time and for many years after, Britain was in the forefront of development, not only of the steam engine itself but also of the growing number of new technologies which followed in its wake.

The steam engine's biggest limitation at the start of the Boulton & Watt partnership was that it could still only pump water; the piston was free in the cylinder rather than being tethered to a crank. Not surprisingly, Watt interested himself in ways of producing rotary motion by steam power but, instead of adapting his beam engine to turn a crank, his first experiment was with a 'steam wheel' as he called it. Built at the Soho Works in 1774, it consisted of an annular chamber 1.8 m (6 ft) in diameter, mounted on a horizontal axis. The chamber was divided into three parts by flap valves at 120 degrees. The bottom was filled with mercury so that steam admitted between the liquid and the flap valves caused the chamber to rotate.

The result of the first of many attempts in history to produce rotary motion directly from fluid pressure, the steam wheel suffered the usual problem in rotary steam engines, namely leakage. Watt, somewhat disheartened, would happily have concentrated his efforts on pump-

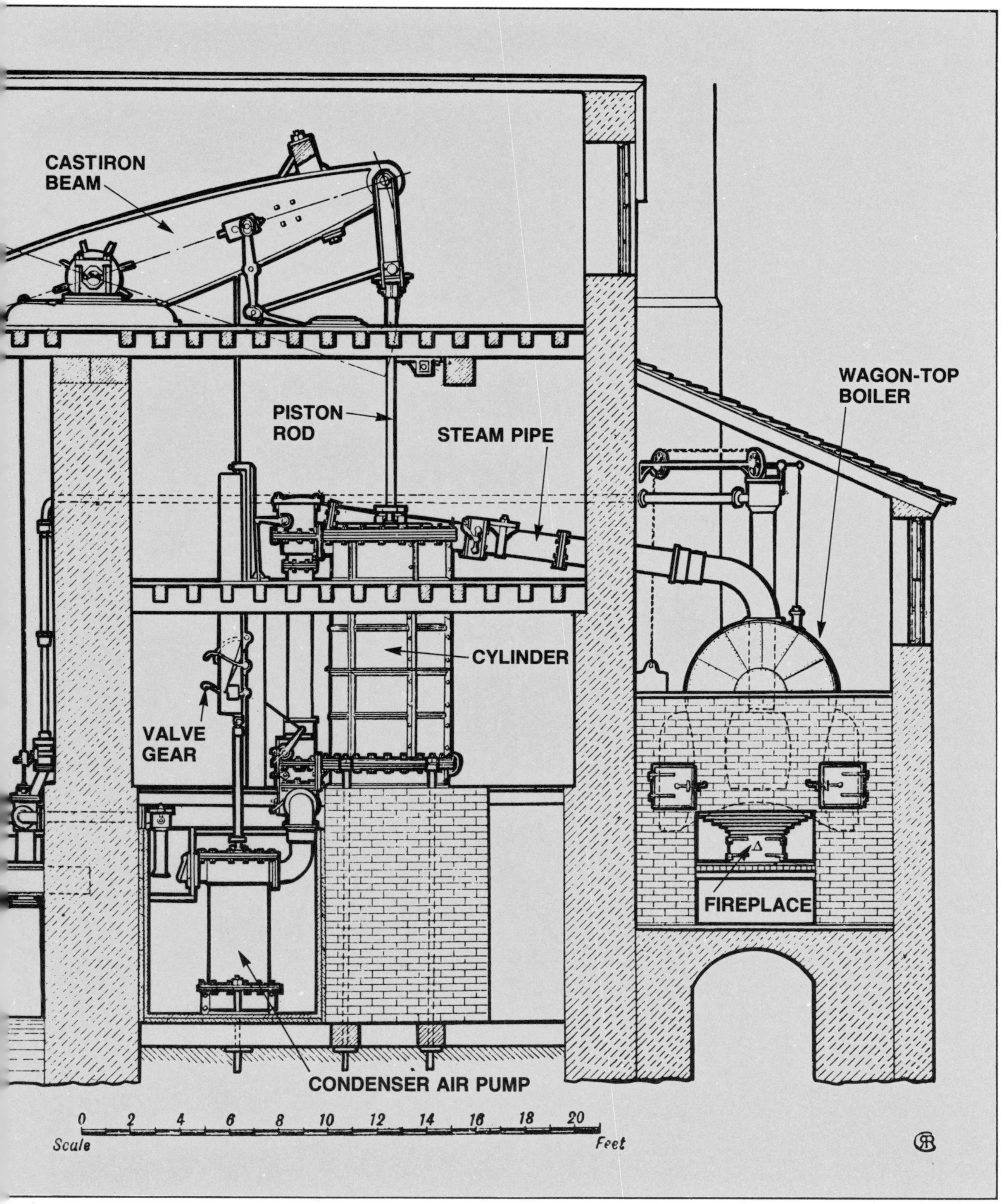

A typical Watt pumping engine installation of 1803 at London's Chelsea waterworks, Pimlico. On the left is the main bucket pump; on the right is one of Watt's 'wagon-top' boilers which supplied steam at 14–20 kN per sq m (2–3 lb per sq in) above atmosphere. The jet condenser is situated beneath the driver's floor, between the cylinder foundation and the centre wall which supports the beam. The condenser air pump is worked by the 'plug rod' which also works the valve gear and is hung from the beam just in front of the parallel motion. The small pump just in front of the wall circulates cooling water to the condenser. The domed vessel on the left is an air vessel to protect the mains from shock due to rushes of water as the main pump valves open and close. This engine had a 1,219 mm (48 in) diameter cylinder and 2.4 m (8 ft) stroke and replaced an atmospheric engine of 1742 on the same site.

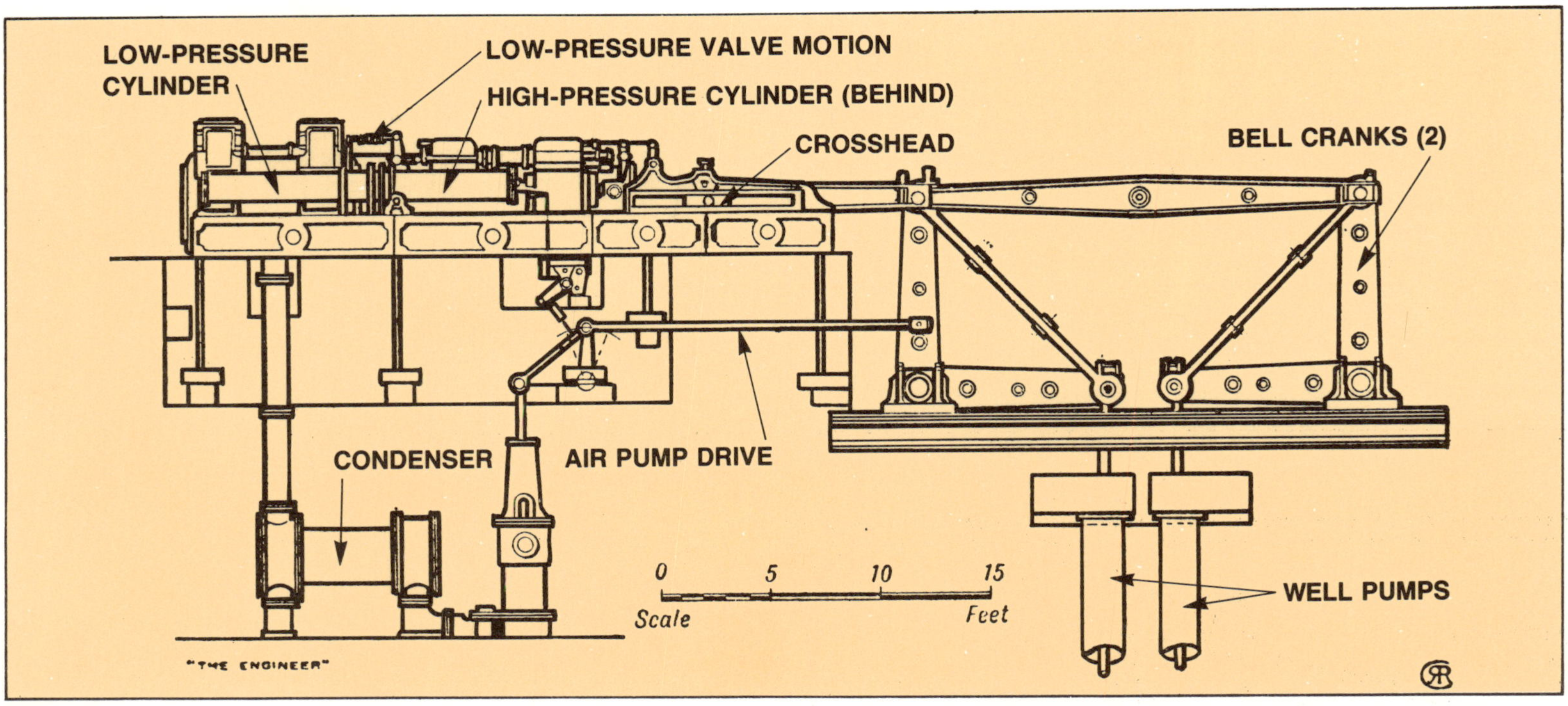

Popular around the turn of the century for mine and waterworks pumping was the horizontal non-rotative double-acting engine of the Davey type, said to give an efficiency comparable with a Cornish engine but requiring a much smaller engine-house. This example had compound cylinders and was supplied in 1891 to Copper Mills waterworks in East London, where it worked for many years.

ing engines, but Boulton was more far-sighted. He could see a vast potential market for rotative engines and he succeeded in persuading Watt to adopt the simple expedient of adapting his reciprocating engine to produce rotary motion.

The first was put up at Soho in 1782 and, from this point on, the development of pumping and rotative engines began to diverge. They did not come together again until after halfway through the 19th century when it became commonplace to use rotative forms of engine for pumping duties.

Watt's first rotative engines were single-acting beam engines, generally a smaller version of a pumping engine but with a connecting rod instead of a pump rod which turned the flywheel shaft through a 'sun-and-planet' gear. The use of a crank seems obvious now, but Watt did not appear to realize how a flywheel in combination with a crank would regulate the stroke of the piston and carry the crank over the dead centres. The matter was settled in 1780 by James Pickard, who applied a crank to a Newcomen engine in Birmingham and immediately patented it. Watt responded by patenting a number of alternative solutions of which the sun-and-planet had the most to commend it. It did have one advantage, which was that it doubled the speed of the flywheel.

Sun-and-planet engines were produced until 1802, crank engines being made from 1794 onwards, the year Pickard's patent expired.

Next in Watt's stream of inventions was the double-acting engine, patented in 1782. In it, steam acted alternately on the two sides of the piston, giving double the power from a single cylinder and a more even motion. It meant providing a rigid connection between the top of the piston rod and the beam in place of the chains which were only effective in tension.

Watt's first solution to this problem was a toothed rack attached to the piston rod and engaging a toothed sector on the end of the beam. This contraption was noisy and not very reliable, so Watt substituted a simple pivoted link. To guide the top of the piston rod in a straight line while minimizing friction, he devised his famous parallel motion with fixed anchorages attached to the engine (or engine-house) framing. He patented this beautifully symmetrical device in 1784; it is said that this mechanical invention was the one of which he was most proud.

Watt's next step, made in 1788, was to fix a conical pendulum governor to his rotative engines. This governor was linked to a butterfly valve in the steam pipe to the engine and arranged so that, as the engine speeded up, the governor balls would move outwards under centrifugal force and rotate the butterfly valve towards the closed position. The engine would thus stop accelerating or slow down; in other words, its speed would automatically adjust itself, within limits, to the load on the engine. Watt did not invent the centrifugal governor since it was already in use on windmills, but he adapted the principle to introduce an element of automatic control to his engines, more and more of which were being built to drive machinery where the load was not constant. The governor was also a safety device since it prevented the engine from 'running away' if it lost its load due to, say, the breakage of a drive shaft, which could otherwise do catastrophic damage to the engine.

Engine valves at that time were of the simple mushroom type, usually made of bronze. This simple form of drop valve was attached to a rod working through a gland by which the valve was lifted off its circular seat to open and lowered back to close. A double-acting engine required four such valves. It fell to William Murdock, an employee and later partner to Boulton & Watt, to design and patent a sliding, rectangular valve worked by a single eccentric on the crankshaft in place of the four valves. Thereafter the slide valve was adopted for all but the largest slow-speed engines and of course the pumping engines, which continued to use separate valves.

Watt is also credited with devising the term 'horsepower' which is still used today (or its metric equivalent) to define the rate at which an engine is capable of doing work. One horsepower is defined as being equal to 0.745 kW, or 33,000 ft-lb per minute. In other words, it is a function of both the force or torque which the engine is capable of producing, and the speed. Watt measured the work capability of the average horse and then multiplied it by 1.5 before rounding it off to the above figure, so that there was no possibility of a customer making a direct comparison and being disappointed. The firm charged customers for their engines on a *per horsepower* basis.

The demand for rotative engines soon overtook that for pumping engines, popular though these were. By 1800, rotative engines up to 50 horsepower had accounted for more than 60 per cent of the 500 or so engines produced at Soho during the first 25 years of the firm's existence.

It might reasonably be asked, what were other engineering firms doing about engine building during this period? The simple answer is that, until Watt's patent embracing the separate condenser expired in 1800, the Birmingham firm exercised a virtual monopoly. This was because the patent was worded more as an expression of principles than of details. However, there were some blatant breaches of the patent, particularly in Cornwall, and in the 1790s Boulton & Watt became deeply embroiled in litigation with the result that considerable royalties were eventually recovered. Fortunately for the development of technology, an application to extend the patent was refused so that, after 1800, other engineers were at last free to exploit their ideas.

Recognizing the extent of the competition they would face once the patent lapsed, Boulton & Watt built a new factory at Soho in 1795, laid out and equipped solely for building steam engines. After 1800 there was a rapid growth in engine manufacturing, ranging from small country works building engines for local customers, to large concerns such as Matthew Murray of Leeds, whose workmanship and reputation soon excelled. Engine builders in the rest of Europe also began showing the world that they, too, could build steam engines and make them do effective work. In the USA, where engineering technology lagged behind Britain, a Boulton & Watt engine was used to power the first commercial steamboat in 1807, after unsuccessful attempts to commercialize the use of high-pressure steam.

Thus the steam engine took the first steps in ridding itself of the shackles of being tailored to the particular purpose it had to perform. Nothing could now stop it developing into an all-purpose prime mover, like the diesel engine and gas-turbine of today.

WATT IN CORNWALL

Before following the course of the rotative steam engine in industry in a later chapter, we must return to the Boulton & Watt 'honeymoon' of 1775–1800. During this period, as we have seen, the firm produced almost as many non-rotative pumping engines as rotative engines.

Despite the dramatic savings in fuel resulting from the replacement of the atmospheric pumping engines in Cornwall by Watt's improved engines, engineers in the West and elsewhere believed that even more dramatic savings were possible. But because of the irksome patent, there was little they could do about it.

Pumping engines which Boulton & Watt supplied to Cornwall followed the firm's usual practice whereby the customer paid for all materials and found the labour for erection, while the firm provided drawings and an engineer to supervise erection. Some important parts were made at Soho, the rest were made by outside firms to Watt's exacting specification. All cylinders, for example, had to come from Wilkinson. As payment, the firm claimed one-third of the saving in coal over the atmospheric engines.

To register the number of strokes from which premium payments were calculated, a counter worked by a pendulum was fixed to the engine beam in a locked box so the client could not 'fiddle' the readings. (This was a forerunner of household gas and electricity meters of today.)

Though Watt's wagon-top boiler was an improvement on earlier steam-raising vessels, he stuck doggedly to steam pressures around 15–20 kilonewtons per sq metre (2–3 lb per sq in), maintaining that any higher pressure was dangerous. Consequently, though his 1782 patent included expansive working whereby the steam valve was shut part way through the stroke and expansion of the steam used to propel the piston the rest of the way, the full economy was not realized because the initial steam pressure was far too low.

Watt's engines on the mines inherited a number of details from the atmospheric engines, including the use of a 'cataract' for regulating the length of the pause between strokes.

The cataract consisted of a water-filled cylinder with a weighted piston whose rate of descent in the cylinder could be controlled by varying the opening of an escape cock. Usually placed below the driver's floor, the cataract was connected by linkage to trips which caused the exhaust and steam valves to open to initiate the steam, or 'indoor', stroke.

A balance box or boxes were used to counteract the weight of the timber pump rod which became considerable in a mine being pursued in depth. The balance box consisted of a large wooden box carried on one end of a simple pivoted beam, the other end being linked to the pump rod. Loading the box with stone or scrap iron enabled the balance of an engine to be nicely adjusted for smooth operation and maximum efficiency of working.

The pumps were usually of the bucket type, the column of water being lifted on the engine's indoor stroke. When Watt engines were used in waterworks—as they came to be in large towns and cities which demanded large quantities of water—the bucket pump often had to force water up to a level above that of the engine and was known as the jack-head type. Both types of pump had been used with atmospheric engines long before the Watt era.

The incentive to economize on fuel was nowhere stronger than in Cornwall, and it was here that the first major infringement of a Watt patent occurred. In 1781 Jonathan Hornblower patented a two-cylinder engine in which steam used normally in one cylinder passed into a second where it did further work before being passed to the condenser. In an attempt to equalize the work done in each cylinder, the second or low-pressure cylinder was larger than the first, since the steam entered at much lower pressure.

Hornblower's first engine was put to work at Radstock colliery in Somerset in 1782. Its cylinders were 482 and 609 mm (19 and 24 in) in diameter and the piston rods were connected to the beam at different points so that the stroke of the low-pressure piston was 2.4 m (8 ft), while that of the high-pressure, being nearer the fulcrum, was 1.8 m (6 ft). Between 1784 and 1791 he put up others in Cornwall on which the owners eventually paid royalties to Boulton & Watt to escape litigation. These engines proved no more efficient than Watt single-cylinder engines, again because the pressure was too low. After 1800 another Cornishman, Arthur Woolf, revived the idea of compounding using high-pressure steam, with more success. But due to the dramatic improvements made to the single-cylinder engines after expiry of the Watt patent, his compounds, though made in some numbers, still did not offer sufficient advantage to outweigh their extra cost and complication. The term 'Woolf compound' was subsequently applied to double-acting rotative compound beam engines with cylinders arranged on the same plan, and these became popular later in the 19th century.

Another thorn in Watt's flesh was Richard Trevithick, an impulsive and brilliant Cornishman who is mentioned in most chapters of this book because of the diversity of the contributions he made to development of the steam engine. He and a friend, Edward Bull, introduced pumping engines without a beam, which had the cylinder inverted over the shaft. The piston rod was coupled directly to the pump rod. The first engine on this plan was put up at Balcoath mine in 1792 and, because the upper part of the cylinder was open to the condenser, it infringed Watt's patent. Subsequently

Richard Trevithick

Trevithick put an engine up at Ding Dong mine, near Penzance, which he arranged to exhaust to atmosphere in the face of threats from Boulton & Watt. The Bull engine, as it became known, came into limited favour in later years, principally for waterworks pumping.

Trevithick's greatest claim to fame, however, was the successful transformation from concept to reality of the high-pressure engine—an engine which used steam at a sufficiently high pressure that it could exhaust to atmosphere without reliance on the vacuum formed in a condenser to enable it to perform useful work.

The idea was not new—it had been postulated by Jacob Leupold of Leipzig as far back as 1725—but to make it work meant constructing boilers to stand much higher pressures than those used by Watt. Trevithick's first high-pressure engine was a beam winding engine built for Cooks Kitchen mine in 1800. This engine had a condenser but a subsequent one at Dolcoath exhausted to atmosphere and was known as 'the Valley puffer'.

Trevithick began using boilers which had a cylindrical cast-iron shell and an internal return flue of wrought-iron. The fire-grate was at one end of the flue and the chimney at the other. Pressures were usually in the range of 170–350 kN per sq m (25–50 lb per sq in).

A contemporary engineer in America, Oliver Evans, was also using high-pressure steam at this time, going up to 690 kN per sq m (100 lb per sq in) and more, with a boiler similar to Trevithick's. Evans was hampered by less advanced technology and was forced to revert to an external fireplace as used in earlier boilers. His high-pressure engine of 1804 was a single-cylinder, double-acting overcrank vertical, separate from the boiler.

Despite a setback, when one of Trevithick's boilers exploded in 1803 (because the safety valve had been tampered with)—there was enough prejudice against the dangers of using high-pressure steam without a practical demonstration of its pent-up force—the Cornishman persisted with his ideas. His first objective, namely making the steam engine self-moving by eliminating the cumbersome condenser and its appurtenances, was achieved with resounding success in 1801–4, with his first road and railway locomotives. Now the stage was set for applying high-pressure steam to a Watt-type condensing pumping engine, with results that were to startle the world.

THE CORNISH PUMPING ENGINE

In 1806 Trevithick planned to replace the wagon boilers of a Boulton & Watt pumping engine on the famous Dolcoath copper mine by his cylindrical boilers, and to work the engine with steam at 170 kN per sq m (25 lb per sq in). The steam was to be cut off early in the stroke so that the piston would complete its travel under the expansion of steam already in the cylinder; this was what Watt had tried to do with low-pressure steam.

Nothing actually happened until 1812, in which year Trevithick also tried his ideas in a small pumping engine built new for Wheal Prosper mine near Gwithian. This was a single-acting condensing engine with a 609 mm (24 in) diameter cylinder and a stroke of 1.8 m (6 ft). It was supplied with steam at 275 kN per sq m (40 lb per sq in) by a cylindrical boiler 1.8 m (6 ft) in diameter and 7.3 m (24 ft) long with a straight internal fire tube running from end to end. This became known as the Cornish boiler whose derivative, the twin-flue Lancashire boiler, is still in use today.

When worked at one-ninth to one-tenth cut off, the Wheal Prosper engine was credited, unofficially, with a duty in excess of 30 millions.

About this time Trevithick began using the plunger pole instead of the bucket pump. This was a hollow cast-iron tube closed at the end, which worked up and down in the 'pole-case' through a gland, water being constrained to keep moving up the rising main by simple flap-type non-return valves.

The engine now pumped on the equilibrium stroke, or down stroke of the pump rod, under its own weight, the steam on the piston being used to raise the rod and the poles fixed to it. (On a deep mine the pump rod carried a series of plunger lifts.) The plunger was easier to maintain than a bucket and came into widespread use.

Trevithick also built a few high-pressure 'pole engines' which had a trunk piston like a plunger pole. High-pressure steam was applied beneath the piston to raise the pump rods and either passed to a condenser or exhausted to atmosphere on the down stroke. It is said that non-condensing engines of this type could be heard working several miles away!

Development of the Cornish engine received a big stimulus from production of Captain Joel Lean's monthly duty report of engines in everyday use on Cornish mines. In 1813, the best engine reported (Watt's engine at Herland) was doing 26.4 millions while the average for the 29 engines reported was 19.3 millions. From then on duties gradually climbed as rival engineers applied lagging to boilers, cylinder and steam pipes, tried different valve arrangements and perfected the balance of their engines. In 1828, for example, the average duty of the 57 engines reported was 37.1 millions. The highest figure, of 76.8 millions, was achieved by Samuel Grose with one of his two 2,032 mm (80 in) engines at Wheal Towan.

The best duty ever recorded was in 1834 when William West's brand new 2,032 mm (80 in) engine at Fowey Consols attained 125 millions in the course of a 24-hour trial. Admittedly conditions were probably ideal and the engine was cutting off at only one-tenth of the stroke, but even so it was an astonishing achievement and represented an efficiency of some 11 per cent.

News of this performance quickly spread and resulted, in 1838, in a London engineer called Thomas Wicksteed ordering a sister engine and erecting it at Old Ford Works of the East London Waterworks Company. There it met its guaranteed duty of 90 millions which was 2½ times that of the best Boulton & Watt engine. After this, Cornish engines became standard plant for large waterworks and remained so for 30 years. At one time there were 70 engines pumping water for London alone, some built new, others conversions of Watt engines.

The majority of Cornish engines were built for the mines during the heyday of Cornish mining, which occurred during the first half of the 19th century. Many other engines went to collieries or to important mining areas abroad: Australia, Mexico, South Africa. Large cylinders of up to 2,286 mm (90 in) diameter were common from the start; these and cast-iron beams weighing up to 50 tonnes (50 tons) in charge of horse teams were familiar sights in Cornwall when engines were moved from site to site as mining fortunes rose and fell. Three famous engine builders became established in Cornwall, the most prolific producer being Harvey & Company of Hayle Foundry, who built their last engine, a 1,778 mm (70 in) for Rio Tinto mines in Spain, as late as 1899.

The biggest Cornish engines of all were three annular compound engines with an inside cylinder of 2,134 mm (84 in) bore and an annular piston in an outer cylinder of 3,657 mm (144 in) which went to Holland for draining the Haarlem Meer in 1845–9. One had eleven beams protruding radially from a massive circular engine-house, the others had eight, and the pump rod depending from each carried a large bucket pump. Most of the main parts were made in Cornwall but the beams, of 'hollow-work' pattern, were made locally. Happily, one of these giants is preserved; it last worked in 1933 and there is a short piece of film of that occasion.

Rotative beam engines were also used extensively in

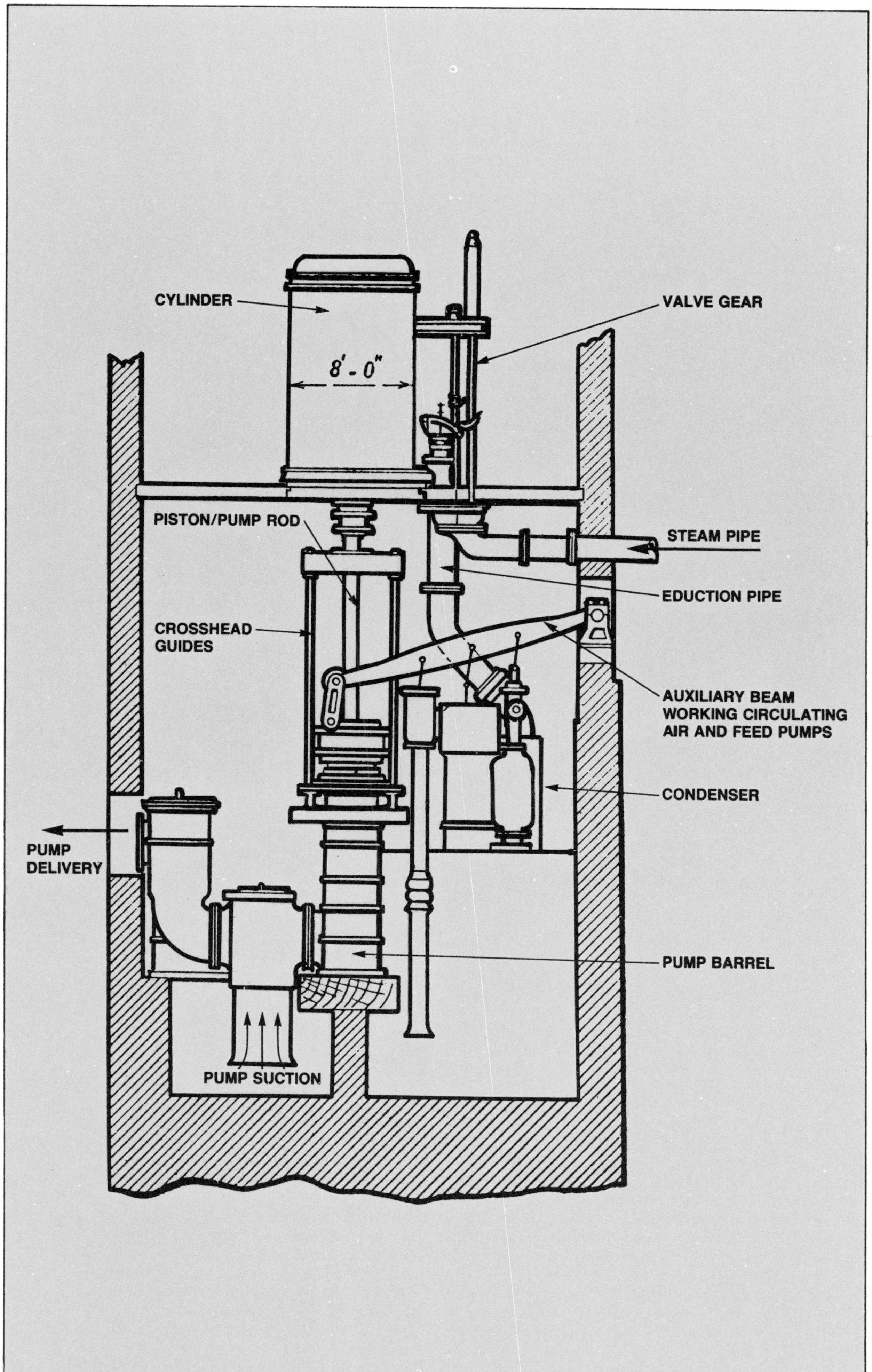

One of a pair of 'Bull' Cornish engines supplied by Harvey & Co. of Hayle to London's Hampton waterworks of the Southwark & Vauxhall Company in 1855. The cylinder, top, *was 1,676 mm (66 in) in diameter and the diameter of the pump, directly beneath it, was 990 mm (39 in); the common stroke was 3 m (10 ft). The Bull arrangement of inverting the steam cylinder over the pump and dispensing with the beam was to save space but in practice such engines were less efficient than the true Cornish beam engine. The Hampton engines worked at 6–10 strokes per minute and lasted until 1931; a similar engine with a 1,778 mm (70 in) cylinder is preserved at Kew Bridge Museum.*

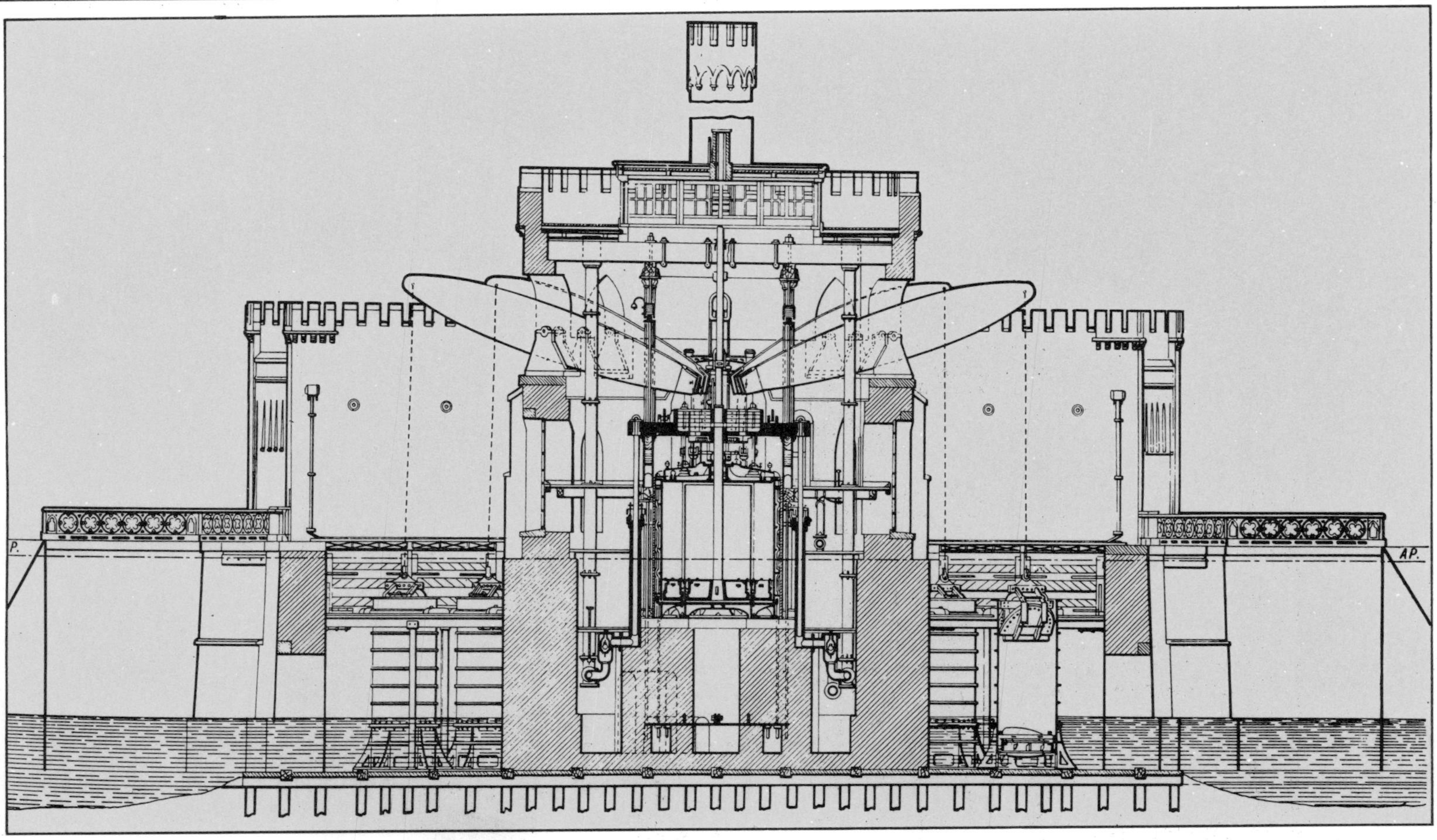

Above: *the three biggest pumping engines the world has ever seen were annular compound single-acting beam engines designed and manufactured in Cornwall to drain the Haarlem Meer in Holland. The Cruquius engine depicted has been preserved. It has eight 'hollowork' iron beams protruding radially from a circular engine-house and each coupled to a large low-lift bucket pump at its outdoor end. It was built by Harvey & Co., Hayle, in 1849.*

Facing page: *a typical, medium-sized Cornish engine for draining water from a deep mine, built in 1871. The massive wall supporting the beam is actually the front wall of the engine-house which keeps the main machinery weight away from the shaft collar. The condenser and boiler feed pump are in the open between the wall and the shaft, an improvement on the Watt arrangement since it kept the engine-house free from vapour.*

Cornwall. They drove batteries of ore stamps, wound ore from the shaft and sometimes, by means of bell-cranks and flat-rods, combined pumping with other work. Early examples were single-acting engines working on the Cornish cycle, a heavy connecting rod acting in place of the pump rod to perform the outdoors stroke. Later ones were made double-acting with drop valves. For economy, they invariably had a condenser.

Rotative engines were also used to work 'man-engines', first used in Germany to save miners having to climb ladders in shafts deeper than about 200 m (650 ft). Michael Loam introduced them to Cornwall in 1843 and, though never very numerous, man-engines soon spread to Belgium, South Africa and the USA. Their usual form was a timber rod moved up and down by a crank at surface through a stroke of up to 3.7 m (12 ft). At the top and bottom of its travel, small platforms fixed to the rod paused level with fixed platforms in the shaft so that as men stepped nimbly on and off, they were carried either up or down. Drive from the steam engine was usually geared down to give a speed of 3–6 strokes a minute in the shaft. The last man-engine worked at Levant mine in Cornwall until 1919 when a breakage with the rod full of men cost 31 lives.

WATERWORKS PUMPING

The decline of Cornish mining during the latter half of the 19th century, coupled with the rapid population growth in towns and cities, swung the pumping emphasis from mining to water supply. Though the Cornish engine was deservedly popular where large quantities had to be pumped—the Harvey 2,540 mm (100 in) engine at Kew, for example, moved 3,228 litres (710 gal) of water per stroke against a head of 53 m (175 ft)—country waterworks required something smaller. Even large urban waterworks often had to think twice about the cost of putting up an engine-house big enough to shelter a large Cornish engine.

One solution to this problem was the inverted Bull engine which required a much smaller house, but since it still worked on the Cornish cycle it was at its most efficient in large sizes. The largest Bull engine built for waterworks was a 2,286 mm (90 in) cylinder example by Harvey, put up at Campden Hill in London in about 1870. (An even bigger example, built locally, was a 3,048 mm (120 in) engine on a colliery in Fifeshire.)

The non-rotative single-cylinder engine is also liable to wreck itself instantly should it go out of control, for example by breakage of the piston rod. Since reliability of supply is all-important to water engineers, they turned to the Woolf compound engine, which was one with the two cylinders in line at the same end of the beam, but in rotative form. In about 1845, James Simpson, then engineer to the Chelsea Waterworks Company, brought out the double-acting compound beam rotative engine, with a flywheel to take control. Simpson later set up his own works at Pimlico to make these engines, and they had considerable success. This type of engine was made in a wide range of sizes to suit a range of jobs in addition to pumping, and by many firms.

Meanwhile, from 1825 onwards, engine-builders had begun adopting a horizontal position for the cylinders of stationary engines, just as Trevithick had done with small engines at the turn of the century. The drawback was the fear of undue wear of the bottom of the cylinder due to the weight of the piston. This was to some extent overcome by the use of a piston tailrod with a second gland and support slide. As described later on, the horizontal engine with single, compound or duplex cylinders was in common use during the latter part of the century on a wide range of duties, of which waterworks pumping was just one.

But the purpose-built non-rotative pumping engine was not dead yet. An American engineer, Henry R. Worthington, built his first engine for Savannah waterworks, Georgia, in 1854. Two years later he invented the horizontal duplex pump which, 70 years later, could still be found all over the world in sizes ranging from small

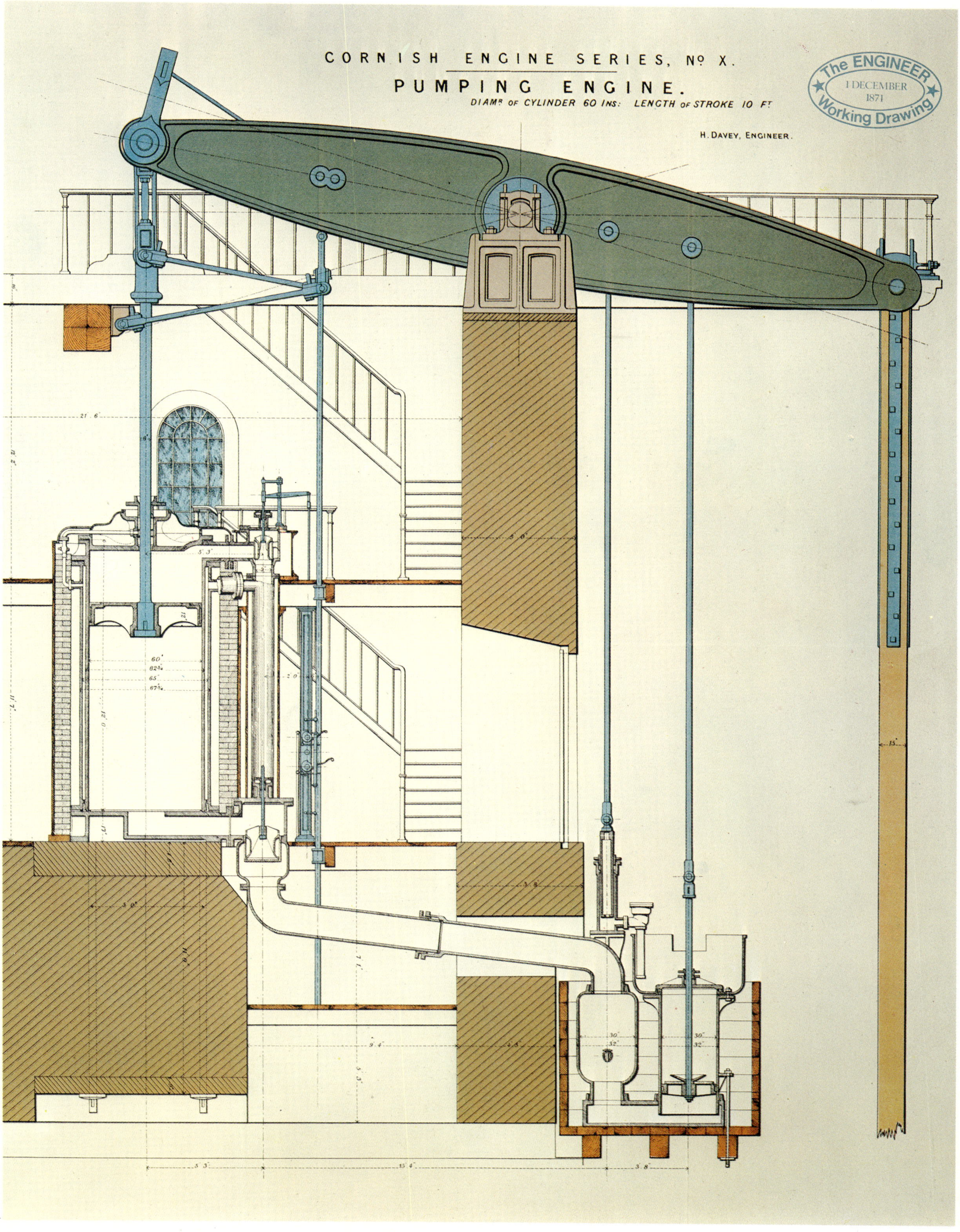
CORNISH ENGINE SERIES, No X.
PUMPING ENGINE.
DIAMR OF CYLINDER 60 INS: LENGTH OF STROKE 10 FT
H. DAVEY, ENGINEER.
The ENGINEER
1 DECEMBER
1871
Working Drawing

Right: *valve gear of the Grand Junction 90-inch Cornish pumping engine of 1846, preserved at Kew in West London. Since a Cornish engine has no rotating parts, the valves are worked by a dead-weight and trip system which is fascinating to watch.*

Below: *also at Kew is this splendid example of a Woolf compound rotative beam engine built by Easton & Amos in 1867. It has the two cylinders arranged one behind the other and a Porter-Allen governor.*

boiler feed pumps to large triple-expansion duplex pumping engines. The duplex pump had steam cylinder and pump in line on a common rod, two sets being arranged side by side so that the valve on one side was actuated by the motion of the piston rod on the other. The advantages were that the engine did not need a cumbersome flywheel and the speed accommodated itself to the load. In 1879 another American, John D. Davies, patented the steam compensator which stored up energy at the beginning of the stroke and gave it out towards the end, thus enabling a stroke to be completed at a very low speed indeed. This was adapted to the duplex pump in 1885, making feasible pumps of large horsepower, compound or triple expansion. By that date no less than 45 per cent of all waterworks plant in North America was of the duplex type.

About the same time, the first Worthington waterworks pump came to England. The firm of James Simpson entered into agreement with the American company for exclusive manufacturing rights in Great Britain. London got its first in 1886 with an engine at Chelsea capable of delivering 23 million litres (5 million gal) a day.

Besides needing a house only the fraction of the size required for a rotative engine, there was a noticeable absence of large, heavy parts in a Worthington engine. A few engines were arranged vertically where space considerations dictated, for instance they could sit happily in a space vacated by a beam engine. A list of 170-odd engines built on both sides of the Atlantic in

One of the most popular types of pumping engine was invented by Henry Worthington in the USA, a horizontal non-rotative duplex machine where the steam and pump cylinders are arranged in line on the same rod. This is an early example dating from 1889 and now preserved at Upper Cherry Gardens Waterworks, Folkestone. It has two close-spaced piston-cum-pump rods, triple-expansion steam cylinders with wooden lagging, and slide valves above the cylinders.

the period 1886–92 shows examples going to Russia, South Africa, India, Australia, Mexico, Holland and Portugal. Truly the Worthington engine transcended man-made barriers! By 1900 steam pressures in excess of 690 kN per sq m (100 lb per sq in) were being used with them. Happily, two early examples supplied to Cherry Gardens Waterworks, Folkestone, have been preserved.

By this time the jet condenser had given way to the surface condenser where the steam and cooling water were kept separate by circulating the cooling water through a bank of tubes. Part of the pumping discharge normally provided circulation.

Worthington engines and many rotative horizontal engines were coupled to a double-acting bucket pump working horizontally in tandem with the steam cylinders. This type of bucket was, in reality, a plain piston fitting in a machined bore which delivered on both strokes via non-return valves connected to both ends of the pump cylinder, or 'chamber'.

Another non-rotative pumping engine was Henry Davey's. He devised a differential valve gear actuated by a master piston and governed by a cataract which could be used with a horizontal engine and do away with the crank and flywheel. He patented his differential gear in 1871. At first he used a single cylinder which drove single acting pumps through a bell crank, and later set up his own firm, Hathorn Davey of Leeds, to manufacture them. Later on his engines were also made compound or triple expansion with the cylinders arranged in line, and sometimes in place of a bell-crank he used a large oscillating disc. A typical compound Davey differential engine was supplied to Copper Mills pumping station in East London in 1896. It had cylinders 610 mm and 1,067 mm (24 and 44 in) in diameter and a 1.8 m (6 ft) stroke, and drove through a bell-crank two 584 mm (22½ in) diameter well pumps with the same stroke. Small Davey-type and rotative pumping engines were sometimes arranged diagonally instead of horizontally with the pump at right angles, to save space. A triple-expansion example was later added to the same station, steam being supplied at 690 kN per sq m (100 lb per sq in) by marine-type return-tube boilers.

The crucial need for fuel economy, which prompted higher steam pressures and passing the steam through more stages, and the availability of a ready supply of cold water from the source, provided a common link with the marine steam engine where similar considerations applied. Hence the last and highly successful type of reciprocating engine applied to waterworks duty was almost a carbon copy of a marine engine, namely the inverted vertical surface condensing triple-expansion engine.

In 1886 an American, Edward P. Allis, first applied this type to waterworks. The only major departures from marine practice were eliminating the reversing gear and providing an overspeed governor. England quickly followed and the pumping arrangement became standard, namely three plungers below the crankshaft driven up and down by rods depending from the engine crossheads. Drop valves or Corliss valves (to be described later) became standard and instead of one huge flywheel most triple-expansion engines had two, either arranged between the cranks or at the ends of the crankshaft. One of the triples supplied by the Lilleshall Company to Cricklewood pumping station in 1902 was used to simulate one of the *Titanic's* engines in the film classic *A Night to Remember*. This engine had three cylinders 533, 864 and 1,321 mm (21, 34 and 52 in) in diameter and a 1.3 m (4 ft) stroke, driving three single-acting plunger pumps 610 mm (24 in) diameter. It ran at 24 revolutions

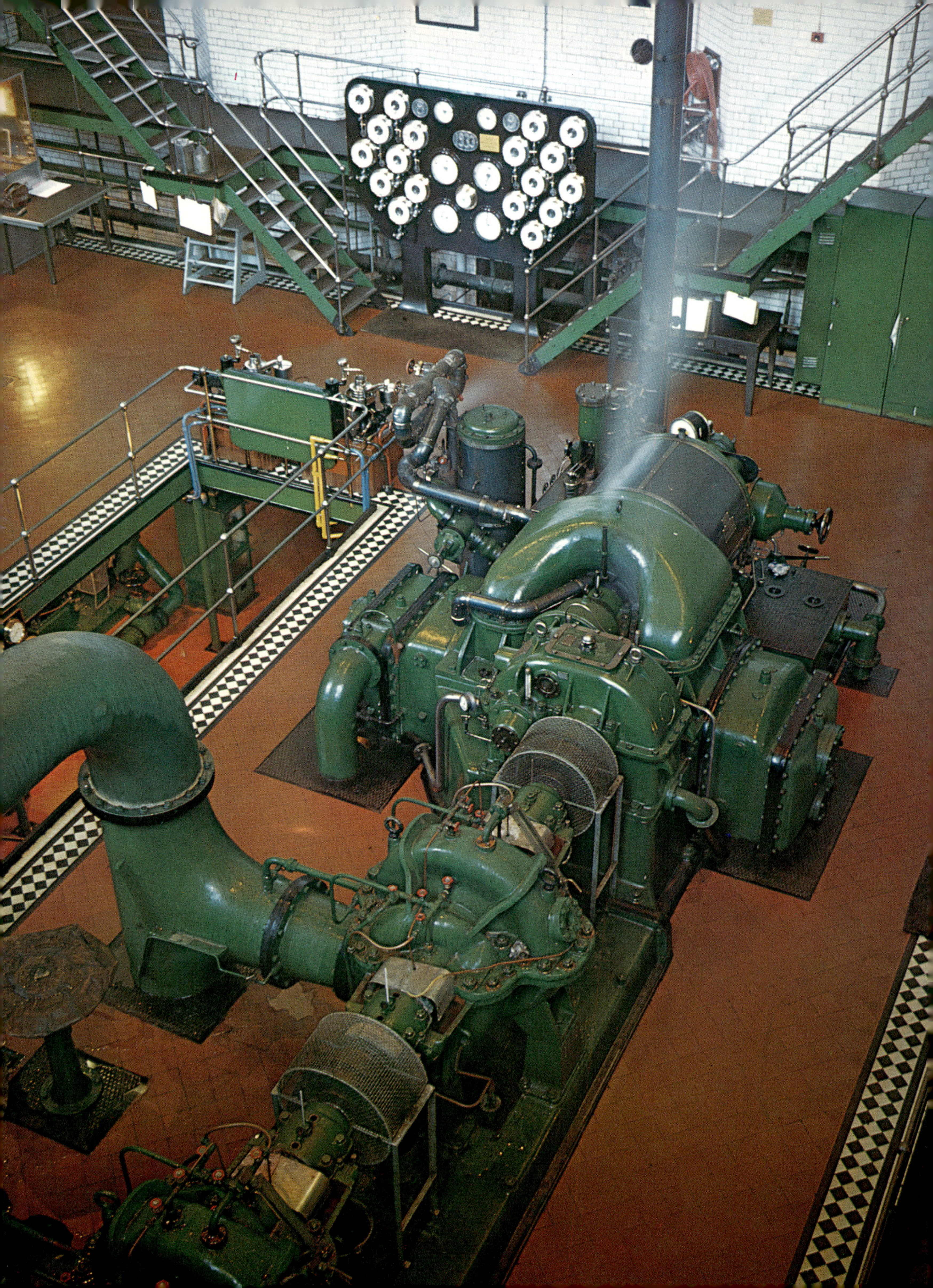

a minute, rather slower than its marine counterparts would have done.

For efficiency and reliability this type was hard to beat, though the initial cost including the building was rather high. The last examples were made in about 1940 and at the time of writing, two magnificent engines are still working at Kempton Park, the very last steam reciprocating waterworks engines in daily use in Britain.

Waterworks engines were seldom dull. Sometimes, instead of being built to an orthodox design, established principles were combined to produce a 'one off' to suit particular conditions or the whim of the waterworks engineer. One of the biggest and best-known examples was the double inverted compound beam engine built by James Watt & Company (Boulton & Watt's descendants) for the Whitacre station in Birmingham.

This was, in effect, a pair of compound non-rotative engines with cylinders 838 mm and 1425 mm (33 and 60 in) in diameter, mounted on pillars above a system of rocking beams working in a recess in the driving floor. The beams were so arranged that all the pistons and the 660 mm (26 in) plunger pumps had a common stroke of 3 m (10 ft), a link arrangement between the two engines keeping their strokes in phase to give an even water flow. Control was by Davey differential gear, commonly referred to as a 'Davey steam man'. When doing 8 strokes a minute under steam at 552 kN per sq m (80 lb per sq in), the engine(s) delivered 23 million litres (5 million gal) a day against a 24 m (80 ft) head.

The speed of a large reciprocating pumping engine was normally in the range 8–24 rpm, equal to and governed by the stroking rate of the pumps. Smaller works often employed horizontal engines running at a relatively high speed of 80–100 rpm and driving the pumps through reduction gearing.

OTHER PUMPING DUTIES

Sewage pumping was a requirement akin to waterworks pumping. It likewise resulted in some fine engines being built and housed in palace-like engine-houses. Four fine rotative Woolf compound beam engines built by Gimson of Leicester as late as 1891 have been preserved in that city. These engines pumped on both strokes by the simple expedient of having two single-acting plungers, one driven off the flywheel end of the beam and the other from a piston tailrod below the high-pressure cylinder.

Despite widely fluctuating load conditions, hydraulic pumping engines which supplied water under high pressure to docks, swing-bridges and other installations which used hydraulic machinery tended to follow later waterworks practice.

The invention of the steam turbine by Charles Parsons in 1884 will be discussed in detail later, but it provides a fitting steam finale to this chapter on pumping engines. Since the multi-stage centrifugal pump is a high-speed machine like the turbine, the two were frequently supplied in combination for a wide range of pumping duties. The centrifugal pump itself owes its origin to Osborn Reynolds who introduced it for low lifts in 1850. Later research produced pumps capable of the highest lifts required in water supply, not only above ground but also in deep well pumping. The moves towards automation and hence electrification in water supply, especially since the Second World War, gave

Facing page: *a turbine installation for waterworks pumping. This unit was supplied to Kempton Park pumping station by Worthington-Simpson of Newark in 1933 and comprises a Fraser & Chalmers two-stage impulse turbine single-reduction geared to a centrifugal pump. It delivers 55 million litres (12 million gal) a day to North London.*

Below: *Horizontal compound rotative pumping engine built by Hathorn Davey & Co. of Leeds in 1926 and now preserved at Millmeece in Staffordshire. The well pumps are worked by horizontal rods and a bell crank arrangement, and force pumps from the tailrod from the low-pressure piston, on the right of the photograph.*

the turbo-driven centrifugal pumping set a shorter life than its high efficiency and low space requirement deserved.

In mining, the use of steam power for pumping produced machines no different in principle from those used for waterworks. In fact, owing to physical limitations, steam had virtually disappeared from mine pumping by the 1930s except for a handful of Cornish engines still active in their native county. At one time, collieries were extensive users of Cornish engines, just as they had been of atmospheric engines. Thereafter steam pumping plants installed were almost invariably horizontal installations, both rotative and non-rotative, with pump rods driven from the surface via bell-cranks. In some remote mining regions of the USA where electric power was not readily available, it became the practice to use Cornish-type pitwork (plunger pumps fixed to poles fixed to moving pump rods) driven by a very large rotative engine of established form at surface. One such installation, now preserved, is illustrated.

Draining low-lying areas such as the Dutch polders and the British Fens was done traditionally by windmill. When steam power arrived in the 1830s, it commonly took the form of a rotative beam engine driving a scoop wheel through reduction gearing. Some of these engines were quite small and had the beam supported on iron framing independent of the engine-house. The scoop wheel was purpose designed for lifting a very large quantity of water against a low head and worked rather like an undershot waterwheel in reverse. Such wheels, with a series of curved buckets, went up to diameters as great as 15 m (50 ft) and ran at speeds of between four and eight revolutions per minute.

From 1850 onwards, the high-speed centrifugal pump, driven by a small horizontal or vertical engine and, later, by turbine, gradually took over drainage duties. In some parts of the world, notably in Egypt and parts of Europe, Archimedian screws driven by a conventional steam engine were used for large-scale low-lift pumping for drainage or irrigation. One such installation is preserved in Denmark.

Many fine steam pumping plants made by the principal British manufacturers were supplied abroad during the latter part of the 19th and early 20th centuries, particularly to the colonies. There still stands at Buenos Aires a pumping station built on the grand scale common in the Victorian era, which once housed rotative beam engines supplied by James Watt & Co in about 1880.

Great Britain was, of course, the principal steam engine building centre in the world. The massive yet simple and reliable reciprocating engine, ousted by electric power in its country of origin, is still doing duty in some far-flung corners of the world.

In this all-too-brief chapter on pumping engines and early rotative engines, the lap of honour must certainly go to the Cornish beam pumping engine. This highly refined machine—the direct descendant of the early, creaking, atmospheric engine—was built over a period of no less than 100 years. When the last engine ceased operation at the Severn Tunnel in 1962, exactly 250 years after Newcomen's Dudley engine made its first tentative strokes, it ended one of the brightest chapters in the whole history of technology.

By the greatest good fortune, a handful of large Cornish engines survive at Kew Bridge and they are being restored to run as a public spectacle. Future generations will no longer risk being deprived of the awesome sight of a huge beam rocking back and forth with that strange, intermittent, motion of an engine with no crank, nor of the polished cranks and levers of the plug-handle valve gear lifting and tumbling as if worked by an unseen hand. But, more important, the beam pumping engine taught man to build engines and machines. In doing so, it exerted a profound change on the way of life of every citizen in the civilized world.

Approximate coal consumption of steam pumping engines per water horsepower per hour

		kg	lb
1725	Newcomen	14.5	32
1775	Smeaton	7.7	17
1800	Watt	4.0	9
1840	Cornish	1.4	3
1870	Horizontal compound	.9	2
1885	Vertical triple-expansion	.7	$1\frac{1}{2}$
1900	Steam turbine	.45	1

Top left: *this giant mine pumping engine, built in 1890 by the E. P. Allis Co., Milwaukee, is preserved in the USA at Iron Mountain, Michigan. The cylinders are in a steeple tandem compound arrangement, the high-pressure cylinder being on top; the whole engine is 16 m (54 ft) tall. Crankshaft and walking-beam are sunk in a pit. This engine is accredited with an average delivery of 8,737 litres (1,922 gal) a minute from a depth of 461 m (1,513 ft).*

Top right: *the Cruquius pumping engine in Holland last worked in 1934, and is now preserved. Some of its eight 'hollowork' iron beams may be seen protruding outside the house.*

Left: *one of two splendid inverted vertical triple-expansion rotative pumping engines of 1926–8 still serving London at Kempton Park Waterworks. The resemblance to a marine engine is very striking, the twin flywheels and lack of reversing gear being the only significant differences. The three-throw pumps are beneath the floor, driven by vertical rods from the cross-heads. All the engine valves are of the Corliss rotary type which, like this type of pumping engine, originated in the USA.*

Steam on the Railway

Facing page: *this painting shows a bridge of the Stockton and Darlington Railway with the inaugural train of 1825, 'Locomotion'. A full-size working copy of 'Locomotion' was built for the 150th anniversary of the opening of the railway, in 1975. (See page 49.)*

THE FIRST LOCOMOTIVES

The locomotive appeared early in the development of the steam engine. It was as well that it did, because it evolved into something quite unlike other steam engines, and indeed unlike almost any other machine. From the beginning it was not a stationary engine adapted to move itself on rails, because the stationary engines of the period could not be adapted in that way: most of them relied heavily on brick or stone for their structures and those that did not were built like Elizabethan timber-framed manor houses.

The steam locomotive very rapidly evolved into its familiar form, which possesses several important characteristics which entitle it to be regarded as unique among machines. One of these is its physical flexibility: the whole structure is not and cannot be rigid, though it must be strong. The crankshaft or driving axle, like the other axles, is connected to the frame by springs which permit movement at right angles to the direction of drive, and even in the direction of drive there is more play than should be found on stationary engines, to allow proper freedom of the up and down movement. Then the whole thing is a vehicle, and proper provision has to be made for it to run around curves and negotiate junctions, and the way in which it does this affects the working of the engine mechanism, just as the working of the mechanism affects the quality of the locomotive's 'ride'. This 'ride', moreover, affects the steaming of the boiler, in ways which are well recognized but not altogether understood. It is a curious fact that locomotives tested on stationary test plants never give as great a maximum power output as when tested on the railway, and this may be attributed to the vibration of the fire and the movement of water in the boiler. The boiler circulation within is assisted, with a consequent increase in heat transfer rate and the liberation of steam bubbles clinging to evaporative surfaces.

However, the most important of these almost biological interactions between the components of a steam locomotive is undoubtedly the effect of the exhausted steam upon the fire, which it stimulates by producing a partial vacuum within the smokebox as it passes from blast pipe to chimney. It is in this part of the locomotive, commonly known as the front end, that redesign has, from time to time, produced enormous improvements in performance at exceedingly low cost, but the design of this part has to be related to the design of the boiler itself. The unusual nature of the locomotive might be summed up by saying that it is different from

A model representing Trevithick's Pen-y-Darren locomotive of 1804 in the Science Museum, London. (See page 42.)

'Puffing Billy' of 1813. This photograph was probably taken around 1860, when this locomotive was taken out of service owing to a change of the track gauge of the Wylam railway, and shows the condition in which it is now preserved in the Science Museum, London.

the sum of its parts, which cannot be designed and tested separately and then brought together with any assurance of a predicted result. It is therefore hardly surprising that many locomotive designers seem to have only partly understood what they were doing, and the most successful, at least in the 19th century, were great empiricists, able to learn from their own and other people's experiences. It should be added that only André Chapelon seems to have been able to grasp all these interactions and to handle them in a truly scientific manner.

COALBROOKDALE AND PEN-Y-DARREN

The first steam locomotive for use on a railway was built in England by Richard Trevithick, apparently in 1802. Trevithick had decided to use 'strong steam', i.e. steam at a pressure well above that of the atmosphere. For this he was roundly condemned by James Watt, who continued to prefer steam at a little above atmospheric pressure—kettle steam, in fact, just able to rattle the lid—all his life. Steam at atmospheric pressure had been a necessity for Newcomen early in the 18th century, because a boiler could not then be made to withstand any appreciable internal pressure; but it was no longer a necessity 90 years later, and Trevithick saw two advantages in using a higher pressure. One was that the cylinder could be smaller and yet produce the same power, and the other was that a condenser might be dispensed with. With atmospheric steam on one side of the piston, a vacuum was essential on the other, to produce any piston thrust. With high-pressure steam on one side, atmospheric pressure on the other could be overcome and the thrust but little reduced by its resistance. Trevithick favoured a pressure of 345 kN per sq metre (50 lb per sq in) above that of the atmosphere, and with this, without a condenser, he could expect about four times the thrust from a piston of given size, as compared with the Watt system.

The next stage in Trevithick's thinking was the idea of a very strong cast-iron boiler which could also serve as the main structural member of a complete engine-and-boiler unit. The cylinder was partly sunk into this boiler, and the crankshaft bearings were bolted to it. He very quickly made one or two self-propelled road vehicles based on this unit, but the driven wheels had no springs (at least as far as one can make out, for none of these machines has survived) and in the state of the roads as they then were there was nothing to suggest that mechanically propelled road vehicles of this or any type had a future. A far smoother track was offered by the railways which already existed, operated by horses, ropes or gravity, in and around some industrial sites. These private routes were also a better place to try out the first non-condensing steam engines with high-pressure boilers, because they were extremely noisy and on a public road might well cause a horse to bolt with its rider. This was because the steam exhausted to atmosphere was practically at full boiler pressure—expansive working resulting in softer puffs had not yet been attempted in these engines—and silencers had not yet been fitted.

It was at Coalbrookdale in Shropshire that this first locomotive was tried. The place was appropriate, for it was there that the foundations of the modern iron and steel industry had been laid by Abraham Darby the first, when, in 1709, he pioneered the smelting of iron ore with coke. The railway at Coalbrookdale was almost certainly the first to be equipped with iron rails, and it was on these that Trevithick's locomotive was to run.

In the event, it did not run for long. There was an accident, followed by an enquiry, and the experiment was abandoned. Mystery surrounds all this and whatever happened seems to have been hushed up. The locomotive became a stationary engine.

In February 1804 Trevithick's second and most famous locomotive set out on a brief but successful career at the Pen-y-Darren ironworks near Dowlais in South Wales. Its designed duty was the haulage of

The 'Rocket' of 1829, as exhibited in the Science Museum. Compared with its original condition, the cylinders have been lowered and a smokebox has been fitted, while many parts had been removed before this famous locomotive was rescued from a scrapyard in the 1860s.

10 tonne (10 ton) loads of iron over 14 km (9 miles) of railway to Abercynon, where the iron was loaded onto barges. It was also designed to win a wager of £500 for ironworks owner Samuel Homfray, who had backed the power of steam against that of the horse with a neighbouring ironmaster, Anthony Hill. Expectations were fulfilled in both respects. The locomotive took 25 tonne (25 ton) loads, ran at 8 km (5 miles) per hour, climbed gradients as steep as 1 in 36, and managed a round trip without any feeding of the boiler. This last is perhaps the most surprising feature of the whole performance, as it involved nearly four hours of steaming time, but the return journey was probably without a load. This machine was not the first steam locomotive, but it was the first one to pull a train to any purpose.

In a letter dated 20th February 1804, Trevithick records the crucial observation which was to give the steam locomotive its most characteristic feature. He writes: 'the fire burns much better when the steam goes up the chimney than when the engine is idle'. So already

A model of Stephenson's 'Planet' locomotive, the first successor to the 'Rocket' type, in the Science Museum.

in 1804 one of the biological interactions is present and observed, and perhaps the most important one. How fortunate it was that the fire burned better when the demand for steam was greater! Trevithick had turned the pipe upwards in the chimney, apparently purely by instinct, but it could have been natural enough just to terminate it in a side inlet, to muffle the blasts of the exhaust. Had he done that, the fire would have burned less brightly when the engine was working and the whole machine would have been a failure.

Much more will be said about the action of the exhaust steam on the fire in a locomotive boiler, but here it will suffice to emphasize that this action was responsible for the fact that the vast majority of steam locomotives puffed as they went, thereby providing the magnificent spectacle of billowing white steam above the train which is one of the images of the railway.

This engine weighed 5 tonnes (5 tons). From evidence which is not wholly conclusive but which has never been challenged, we can deduce that it had a horizontal cast-iron boiler, with a horizontal cylinder set into one end. This end was removable, and the U-shaped flue was attached to it, so if it was unbolted the cylinder, and the flue, could come out with it, leaving the inside of the boiler clear for inspection. The outside of the flue was also thus exposed. To one end of the flue (which was somewhat tapered to be wider at the furnace end) was attached an elbow from which the chimney rose vertically. The other end contained the fire-grate and was provided with fire- and ash-doors.

The piston rod drove a wide crosshead carried on slide bars, the outer ends of which were supported on long brackets from the boiler. The ends of the crosshead were attached to the connecting rods which ran past the boiler on each side and worked upon a crankshaft carried in bearings fixed on the boiler at the far end. On one side of this shaft was a large flywheel, which carried one crankpin; the other side had a flycrank and a gearwheel, which meshed with a larger one mounted upon a stub axle also bracketed to the boiler. This larger gear was in mesh with gears fixed to the two wheels on this side of the engine, and the overall gear ratio was not much different from 1:1. The carrying wheels rotated upon fixed axles, again firmly attached to the boiler, so the drive was to the wheels on one side only. There could be no springs and the whole structure was built round the boiler—an admirable feature of Trevithick's stationary engines because it made them compact and dispensed with the engine-house as a structural component of the engine, but fatal to the long-term success of Trevithick's locomotives because this unyielding mass broke the rails beneath it. At Pen-y-Darren these took the form of tram plates, flat plates with a flange along one side, able to guide the flangeless wheels of wagons suitable for alternative use on any hard surface without rails.

The valve gearing was of the simplest: an arm on the crosshead hit a collar on the valve rod at the end of each stroke, so knocking the valve backward and forward. This allowed admission of live steam for almost the full stroke and if there was any expansive working it could only be the results of restricted valve opening not admitting steam fast enough to keep up with the piston.

'CATCH ME WHO CAN'

After a very short spell of use, the Pen-y-Darren locomotive was taken out of service and probably converted into a stationary engine. The tram plates were not strong enough for it. Had the railway been equipped with edge rails the story might have been different. In any case it was far more powerful than it needed to be, and Trevithick planned a much lighter version. This, of which drawings survive, was built for the Wylam colliery. Its design was as described for the Pen-y-Darren locomotive and it probably also broke rails, because after trials it was not accepted and was again converted for stationary use. Trevithick's last locomotive, probably his fourth, was his London locomotive, also known

A model of Stephenson's 'long boiler' locomotive in the Science Museum. The capacious overhanging firebox and the overhanging outside cylinders are characteristic.

Bury 0-4-0 of the Furness Railway, in the National Railway Museum at York. Though it resembles the long boiler type in some ways, the cylinders are inside and the frames are made of bars, not plates.

'Columbine' in the National Railway Museum at York. This small Crewe-built 'single driver' was designed by William Buddicom and Alexander Allan, and is sole survivor of a numerous class of small engines which once worked the West Coast route to Scotland.

as 'Catch me who Can'. This was built in an attempt to arouse public interest, and it operated on a circular track near where Euston station now stands. It drew a four-wheeled carriage—an elegant open affair of the kind one associates with Ascot—and for the first time in history people paid to ride behind a steam locomotive. Tickets cost two shillings, the equivalent of several pounds today. Thomas Rowlandson painted a lively watercolour of the scene and, from this, and the picture printed on the tickets (some which have survived), it is apparent that this locomotive closely resembled a type of stationary engine for agricultural use which Trevithick had developed and which has been well illustrated in contemporary literature. The main difference from the earlier locomotives was the placing of the cylinder vertically at the other, closed, end of the boiler. The piston rod worked upward to a transverse crosshead, and the connecting rods worked down on each side to the wheels on the rear axle. There was no drive to the front axle, which supported the front of the boiler where firehole and chimney were placed together as before. 'Catch me who Can' was a 2-2-0 in the Whyte notation, though this is not immediately obvious from the pictures because the wheels are all the same size. The earlier engines with one-sided drive simply do not fit the Whyte system.

In the Science Museum in London there is a stationary engine of the type of which 'Catch me who Can' was clearly a version. This stationary engine-cum-boiler seems to be the only one existing of this type, which suggests that, in spite of being illustrated and described, it was not made in more than a very few examples. The Science Museum engine was rescued a century ago by F. W. Webb, the eminent engineer of the London and North Western Railway, who had it brought from Shropshire to Crewe and put together again, a few missing parts being supplied and broken ones repaired. There are many features of this engine which bear testimony to more than one reconstruction, and the reason for various holes and cuts offers great scope for conjecture. However, there are two aspects of it in particular which may mean that this was once 'Catch me who Can'. One is that there is no provision for taking off any drive to another machine, which would not be necessary if the large flywheel were replaced by a smaller track wheel and the flycrank also replaced by a similar wheel. The other point is that the connecting rods are cranked outward to clear the flywheel on the one side, but similarly cranked on the other side to clear nothing, though an offset is needed to line up with the flycrank. The drawings of these stationary engines do not show anything but straight connecting rods, and the idea that these rods were cranked to clear two track wheels is irresistible. What seems possible is that this started as a stationary engine, was converted to run on rails, reconverted to stationary use and finally reconstructed in its present form, in which traces of these earlier states are still there to be detected and analysed.

Trevithick had no greater success with 'Catch me who Can' than with his other self-propelled steam engines. He turned his attention elsewhere, and so the inventor of the steam locomotive and the discoverer of the blast pipe (as the exhaust pipe, so arranged as to create a draught upon the fire, is usually termed) leaves the history of railways.

THE MIDDLETON RAILWAY

The first commercially successful locomotives were built by Matthew Murray for John Blenkinsop and used on the Middleton Railway near Leeds. They ran on four wheels, but these were not driven by the engine. Drive was by a cogwheel on one side, which engaged teeth cast on the side of the rail. This then was a rack railway, the sort of thing which was to become familiar on mountain-sides in Switzerland and elsewhere.

The engine bore a resemblance to 'Catch me who Can' in that the cylinders were set upright in the boiler, with piston rods connected to overhead crossheads, from which depended connecting rods on both sides of the boiler. However, this time there were two cylinders, and

Matthew Kirtley's double-framed 2-4-0 of the Midland Railway, as modified by S. W. Johnson. Some of these engines of the 1860s and 1870s were still working in the 1930s.

the cranks upon which they acted were maintained at right angles to ensure self starting in all positions and a relatively smooth action, as one cylinder was in mid-stroke when the other was at a dead centre. This, of course, was the commonest arrangement throughout the history of the steam locomotive, the vast majority of engines built having two cylinders and their cranks at right angles, so the Middleton engines established a very fundamental feature.

The cylinders of these engines were placed one ahead of the other, and each drove its own crankshaft, which was geared to the cogwheel axle between the two crankshafts. The engines thus had five cross shafts: two carrying axles at the ends, the cogwheel shaft in the middle, and the two crankshafts symmetrically placed on each side of it. The framing consisted of two longitudinal baulks of timber carrying bearings for the shafts and, though there were no springs, the whole thing was more flexible than Trevithick's machines.

The boiler had a single, straight-through flue, with the fire in one end and the chimney at the other. Originally the exhaust passed directly to the atmosphere via a silencing box on top of the boiler, but later the Trevithick exhaust arrangement was adopted, to judge by later pictures of scenes on the railway. Even with the stimulus of blast action, these boilers could hardly have made much steam, because they presented little heating surface, yet the engines could haul 100 tonnes (100 tons) on the level at walking pace. The track must have been a good one.

This type started working in 1812 and the last of them exploded in 1834, so they provided a continuing example and encouragement to other experimenters during the whole experimental period and beyond. Not least of their merits in the eyes of engineers and industrialists was the fact that four of them replaced 50 horses and thus put 200 men out of work. By 1834 it had been amply demonstrated that a rack railway was only needed when exceptional gradients, of 1 in 12 or steeper, had to be surmounted, and the rack locomotives were replaced by more orthodox machines.

'PUFFING BILLY'

The idea that iron wheels had little grip upon iron rails was common in 1812 and for some time after. Accordingly there was a proposal by William Chapman for a locomotive which hauled itself along by means of a chain wrapped round a drum underneath, and one or more machines were built by William Brunton in which the cylinders operated steam legs which pushed against the ground in the manner associated in more recent years with stranded motorists facing backwards as they tried to push their recalcitrant vehicles. However, the matter was settled in an admirably scientific way by William Hedley, viewer or superintendent of the Wylam colliery. He was, in his own words 'forcibly impressed with the idea that the weight of an engine was sufficient for the purpose of enabling it to draw a train of loaded wagons. To determine this important point', Hedley continued, 'I had a carriage constructed . . . loaded with different parcels of iron, the weight of which had previously been ascertained; two, four, six etc. loaded coal wagons were attached to it, the carriage itself was moved by the application of men at the four handles and in order that the men might not touch the ground, a stage was suspended from the carriage at each handle for them to stand upon. I ascertained the proportion between the weight of the experimental carriage and the coal wagons at that point when the wheels of the carriage would surge or turn round without advancing it. . . This experiment, which was on a large scale, was decisive of the fact that the friction of the wheels of an engine carriage upon the rails was sufficient to enable it to draw a train of loaded coal wagons.'

So Hedley was the first to measure the factor of adhesion, a quantity which varies greatly according to the composition of wheel tyre and rail, and even more according to weather and the state of the rails. He made a model of his test carriage, which is now in the Science Museum. The carriage itself was turned into a locomotive, which was not a success, but was immediately

Stephenson Patentee type 2-2-2 locomotives inaugurated many European railways and, though few have survived, many have been reconstructed in connection with centenary celebrations, films, etc. This is 'Bayard', which inaugurated the Italian railways on the Naples–Portici line.

William Stroudley's 'Gladstone', now preserved in the National Railway Museum and shown here decorated for working a Royal train, was one of an unusual but successful express type with coupled wheels leading.

followed, in 1813, by 'Puffing Billy' which is also now in the Science Museum. This is the oldest surviving steam locomotive in the world, and ranks as the first adhesion engine to be a commercial success. It ran for 48 years along an 8 km (5 mile) railway joining the Wylam colliery with coal staithes on the Tyne. Its usual load was 50 tonnes (50 tons) of coal, and its speed was such that the driver often chose to walk alongside. There were two others of the type: 'Wylam Dilly', now in the Royal Scottish Museum, and 'Lady Mary' which has unfortunately disappeared.

'Puffing Billy' probably incorporates parts of its predecessor and indeed of the experimental carriage. It now has four flanged wheels, though it started life with four flangeless ones, because the Wylam line was a plateway at the time. The old trouble of breakage of the tram plates caused it to be altered to an eight wheeler, but it reverted to its four-wheeled form when the line was relaid with edge rails. The four wheels are driven by gearing from a central cross-shaft which has a crank at each end (very suggestive of the experimental carriage) and the cranks are driven by connecting rods hanging from beams which run fore-and-aft on each side of the engine. At the rear, each beam is connected to the piston rod of an upright cylinder, so the whole machine might be called a double 'grasshopper' beam engine. The valves are operated by tappets on the valve operating rods hanging from the beam.

The boiler followed Trevithick also, with the chimney and fire-door side by side. The shell, however, was made of wrought-iron plates rather beautifully riveted together and the cylinders, because they were at the sides of the engine and could not be sunk into the boiler, were provided with steam jackets.

It is strange to reflect that this locomotive was still working years after such elegant and celebrated machines as Gooch's broad-gauge 4-2-2 express locomotives, or the Northern Railway of France Cramptons, had entered service. In motion, 'Puffing Billy' must have appeared quite a different species from its main line descendants, advancing at walking pace, with the beams waving up and down in that sequential motion which results from cranks set at 90 degrees, with the clanking and clicking of connecting and valve rods and the grinding of gears hidden between the timber frames.

GEORGE STEPHENSON

In 1813, then, the locomotive had reached a practical if ungainly form. The next stage was its evolution through further experiments to a form in which all its parts had assumed what was to be their final basic design, and were in essentially their correct juxtaposition (though enormous improvements in detail design, materials and, above all, in sheer size and power would follow this evolution). The man almost solely responsible for this next stage was George Stephenson.

Stephenson was enginewright at Killingworth colliery, near Wylam, and, in addition to working on stationary engines, built four or five locomotives for its railway and, as his reputation grew, built for other lines as well. His first engine, the 'Blücher', completed at the end of July 1814, was much like the Middleton engines, with vertical cylinders set in the boiler working on to cranks at the ends of two cross-shafts, but these were geared to the track wheels, not a cogwheel. To increase adhesion, the front tender wheels were driven from the locomotive itself by a chain, but this was soon discarded as unnecessary. The chain, however, was an important feature of his next type, in which it coupled the two driven axles together and maintained the phasing of the cranks at 90 degrees. The crankpins were now in the wheels themselves and there was no gearing: in fact the arrangement was that of 'Catch me who Can', duplicated. He also contemplated using coupling rods inside the frames, which would have necessitated two cranked axles.

The final form of this sort of engine was represented by 'Locomotion', the first locomotive of the Stockton and Darlington Railway and the one which inaugurated it in 1825. It was a large example of the type, but its only innovation was the use of outside coupling rods for the first time. Because these had to be phased at 90 degrees to function properly, and each cylinder required cranks similarly phased on each side of the engine (though the front cylinder driving the front wheels had to be phased at 90 degrees to the rear cylinder driving the back wheels) the drive arrangements were not quite straightforward and extra flycranks were needed outside the main crankpin on one wheel on each side. These do not seem to have given trouble and 'Locomotion' had a long career, being finally preserved for posterity.

All these Stephenson engines had simple boilers with straight-through flues, and their steaming rate was limited accordingly. But they were not designed to run fast. Moreover, their vertical cylinders driving directly on to the wheels (except in the case of 'Blücher') precluded the use of springs, though some attempts were made in this direction. In spite of their shortcomings they established the reliability and the practicality of the steam locomotive for industrial haulage, and taught their designer a great deal; but George Stephenson was increasingly being required to attend to the surveying and civil engineering of railways, and it was largely left to his son Robert to devise a machine capable of replacing the horse as the motive power of a passenger train.

THE 'ROCKET' AND ITS SUCCESSORS

The 'Rocket', of which the rather pathetic remains stand in the Science Museum overshadowed by a full-sized replica of itself in its gorgeous youth, fully deserves its fame as the best known of all early locomotives. It should also be regarded as the first modern type, the one in which for the first time the parts fell into logical places and the details of design established a pattern for the long-term future. To list these things sounds exceedingly elementary but it must be emphasized that what appears now to be extremely simple only emerged after a quarter of a century of experiment. In engineering, the complicated solution to a problem is usually the first one arrived at, and simplicity is a matter of refinement.

The logical features of the 'Rocket', then, were: a boiler with a large heating surface consisting of 22 small tubes in place of one large one, and, because a fire-grate could not be arranged in 22 tubes, a separate firebox to contain the grate and the burning fuel, which was water jacketed and, of course, stayed because it formed part of the pressure vessel and water circulated through it and into the main boiler; cylinders which, though inclined, were nearer to the horizontal than the vertical, and drove directly on to crankpins in the wheels, an arrangement which permitted the driving axle to be provided with springs; valves driven by eccentrics instead of the tappet arrangements of the earliest locomotives; and improved exhaust arrangements able to ensure a good draught upon the fire in spite of the greater draught resistance of the multitubular boiler, as compared with one having a single large flue. It so happens that none of these features was completely new, but they all came together for the first time in the 'Rocket'.

As this is meant to be an account of the reasoning behind the development of the locomotive, rather than a plain recital of often repeated history, it is worth pointing out that the interaction of boiler and blast action achieved in the 'Rocket' was a matter of some

George Stephenson

Robert Stephenson

Great Northern Railway no. 990, 'Henry Oakley' of 1898, was the first British Atlantic, a 4-4-2 locomotive. The designer was H. A. Ivatt, and this engine, with Ivatt's enlarged version of 1902, is preserved in running order in the National Railway Museum at York.

Another NRM exhibit is the small tank engine 'Aerolite' of the North-Eastern Railway. Much rebuilt, and with parts possibly dating from the 1850s, this engine is now the sole surviving English two-cylinder compound.

subtlety. Timothy Hackworth, who had assisted in the construction of 'Puffing Billy' and had, on Stephenson's recommendation, assumed charge of the locomotives of the Stockton and Darlington Railway after the opening, had set himself up as a rival to the Stephensons and in his 'Sans Pareil' challenged the 'Rocket' for the £500 prize offered at the Rainhill trials in 1829. Hackworth had constricted the upturned exhaust pipe in the chimney of his engine, thereby sharpening the blast and increasing the draught greatly. However, his boiler had a single flue, and the effect of his modification was excessive: for lack of draught resistance the exhaust puffs pulled half the fire up the chimney. Neither was there enough heating surface to absorb the extra heat. Even if there had been, 'Sans Pareil' could scarcely have used the steam generated, as it was totally unsuited to running at speed, having four coupled wheels very close together, a very high centre of gravity, and no springs (its cylinders being vertical). One might sum it up by saying that the quasi-biological interactions were all awry. To do Hackworth justice, his engine was at least second best, and was bought by the Liverpool and Manchester Railway in spite of not meeting the conditions, but it never did any useful work.

Greater success attended the use of a multitubular boiler without blast action of the exhaust. This was

The first of the famous Midland Railway three-cylinder compounds, built in 1902, also preserved in working order in the NRM.

done by Marc Séguin of the St Etienne–Lyons Railway in France. Séguin arrived independently and simultaneously at the idea of a multitubular boiler, and appreciated that it would need a stronger draught than the single flue type, but instead of sharpening the blast of the exhaust, he forced air into the ashpan beneath the grate. This worked very well but it was cumbersome because the air was blown by fans driven off the tender wheels, and flexible leather pipes were needed to convey it to the engine itself. The engine, too, was of an earlier Stephenson type and almost as unsuitable for fast running as the 'Sans Pareil'. So, original and important as Séguin's idea was, it too lacked that harmonious conjunction of parts which distinguished the 'Rocket'.

Hackworth was a great mechanic, and Séguin a distinguished scientist and engineer, but even a combination of the best ideas incorporated in their locomotives could not have produced anything remotely as good as what the Stephensons produced. In fact, the idea of the boiler was suggested by Henry Booth, secretary of the Liverpool and Manchester company, but the way that idea was carried out, and the entire design of the locomotive in which it first appeared, can only be described as a work of genius.

The story of the Rainhill trials is well known, but their historical significance necessitates a brief account of them here. The point was that until 1829 no steam locomotive had ever run fast. On the Stockton and Darlington the passenger service was provided by horse-drawn carriages on flanged wheels. Locomotives were for freight, and ran at walking pace.

When the Liverpool and Manchester Railway was being built, its motive power was by no means decided: some directors favoured horses; some, fixed engines and cable haulage; and some, locomotives. It was to test the potential of steam locomotives for passenger as well as goods trains that the Rainhill trials were arranged, and locomotive builders invited to compete for a £500 prize. The conditions to be met were, in brief, that the locomotive should consume its smoke; that it should pull three times its own weight at 16 km (10 miles) per hour; that it should weigh no more than 6 tonnes (6 tons) if on six wheels, or $4\frac{1}{2}$ if on four; and the boiler pressure should not exceed 345 kN per sq metre (50 lb per sq in). The site was $2\frac{1}{2}$ km ($1\frac{1}{2}$ miles) of level track with a grandstand in the middle, the trials lasted for several days, and there were more than 10,000 spectators.

The only engine to fulfil all the conditions was the 'Rocket', which showed itself more powerful than re-

The first practical compound, or double-expansion, locomotives were built to the designs of Anatole Mallet for the little Bayonne–Biarritz railway in Southern France, in 1876. 'Anglet' was named after an intermediate station.

quired, was able to reach 48 km (30 miles) per hour without a load, and above all proved reliable, as it covered something like 160 km (100 miles) without mishap. There were three other steam locomotives in the trials: the 'Perseverance' of T. Burstall, which was an early casualty; the 'Sans Pareil'; and the 'Novelty' of Braithwaite and Ericsson. This last was a very light machine with a vertical boiler, based on the makers' fire engines. It was unreliable at the trials, because it had been made in a great hurry, but was in some ways a brilliant design and ran extremely fast. The Liverpool and Manchester Railway offered to buy it, but the deal fell through.

Six more locomotives of the 'Rocket' type were built and five assisted at the grand opening of the L & MR in 1830. They were followed by the 'Northumbrian' type, which, apart from some general increase in dimensions, differed only by having the firebox enlarged to become part of the boiler shell proper, instead of being a separate unit connected by circulating pipes.

After this, the Stephensons changed the style of their locomotives completely, although the vital features of the 'Rocket' were mostly preserved. The cylinders now went to the front and 'inside' (actually in the bottom of the smokebox) to keep warm. There was sense in this, but it necessitated a crank axle, and because of the possible unreliability of this awkward component, owing to the state of metallurgy in the early 19th century, this had six bearings in six frames. Quite gone was the mechanical simplicity of the 'Rocket', and instead there appeared a type of locomotive not much favoured outside Britain, except as a British export. Builders overseas did not imitate this design for long. All the same, this was the type that inaugurated many of the railways of Europe, and which has been preserved or reconstructed in many transport museums.

The first of these were four-wheeled engines, with the firebox overhanging at the rear, behind the crank axle, but it was the six-wheeled version, and especially the 2-2-2 type of which the first was named 'Patentee', which was the great export success of the late 1830s and early 1840s. The six-wheelers had the third axle behind the firebox, and so inaugurated the classic basic configuration of so many British locomotives, in which the space between the wheels is divided into two by the crank axle: the front part contains the cylinders and mechanism and the rear part the bottom of the firebox and the ashpan. Six-wheeled locomotives on this plan were built for 100 years, and some of them were large and powerful, but they were very awkward to work on for servicing or repair.

The multiple framing became less common as the 19th century approached its end and steel generally replaced iron in all parts of the mechanism. Its greatest exponent was Matthew Kirtley of the Midland Railway, whose robust machines were still a common sight between the wars, though none was built later than 1875. The Great Western continued building multiple-framed locomotives into the 20th century.

The six frames were not always fitted. In most recent examples of the multi-framed locomotive, four sufficed: just outside and just inside the wheels. In earlier days the outside frames had usually been double in themselves, having a structure of iron plates on each side of a wooden centre—'sandwich' frames, in fact. Some designs used five frames, and some used only three, which made it possible to use the driving wheels themselves to provide one side of each crank. However, four, with separate crank webs, was the usual arrangement.

EARLY AMERICAN LOCOMOTIVES

Right at the beginning, the British multi-framed locomotive proved unsuitable for North America. It was not flexible enough for the uneven and sometimes sharply curved tracks of American pioneer railroads. After some experiments in the 1830s, in which rotating four-wheeled trucks were applied to the front ends of locomotives which in some respects followed the Stephenson 'Patentee' type, American practice went back in essence to the design of the 'Rocket', with just two frames made of bars, outside cylinders and direct drive to the crank-pins fixed in the wheels. The two important differences were that the cylinders were at the front instead of the back (and the driving wheels were accordingly at the back, actually just in front of the firebox), and the front end was carried by a swivelling four-wheeled truck. The

An Austrian two-cylinder compound express engine of 1905, designed by the great Karl Gölsdorf.

Far left: *a four-cylinder compound 4-6-0 of the Swiss Jura-Simplon railway. The Swiss Federal Railway continued building these until more than 100 were in service. Compound expansion was preferred because it produced a smoother pull up the gradients, and because it was important to economize on the imported coal.*

Left: *the most powerful steam locomotive outside America, André Chapelon's three-cylinder compound 4-8-4. This engine developed 5,500 horsepower, and was intended as the prototype for the SNCF's new range of 6,000 horsepower steam locomotives, which were never built because of the electrification policy.*

most celebrated of these machines were built by Norris of Philadelphia. So powerful were they that as early as 1838 some were exported to Europe, and the Birmingham and Gloucester Railway started using them in 1840 for working the steep incline outside Birmingham known as the Lickey bank, with a gradient of 1 in 37.

The Norris locomotives owed something to Bury, Curtis and Kennedy, Liverpool locomotive builders. For most of his life Edward Bury was a stalwart believer in four-wheeled locomotives, and his engines, largely designed by Kennedy, were excellent for their size. They had inside cylinders, like the Stephenson four-wheeled 'Planet' or six-wheeled 'Patentee' types, but they had bar framing, inside the wheels only, and fireboxes almost circular in plan, with high domed tops. The Bury boiler was completely copied in the Norris locomotives. There were other American single-wheelers, and coupled engines in some variety appeared very early, but the Norris 4-2-0, more than any other type, set the style of the North American locomotive. It evolved into a 4-4-0 in the early 1840s, by the simple addition of another axle behind the firebox, coupled to the one ahead of it. This first type of American 4-4-0 still showed its derivation from the 'Rocket' in several obvious ways, of which the most conspicuous was the steeply inclined position of the cylinders. The second type, which began to become common in the 1850s, was largely created by the Rogers locomotive works, though the great firm of Baldwin's built far more than any other manufacturer and accordingly had a major share in its development. This was the true 'American' type, with horizontal cylinders over a lengthened leading truck, and with a sharply tapered or 'wagon top' boiler, fat at the firebox end and thin at the front. The virtue of this was that it increased weight on the coupled wheels and provided good water circulation round the firebox tubeplate, and at the same time reduced the weight of the front end, which tended to be weighed down by the cylinders on all 4-4-0 locomotives.

The only other point to be made about the emergence of the characteristic American locomotive is one of detail, but none the less significant. It became the usual practice at an early date to equalize the coupled wheel suspension by means of balance beams centrally pivoted and linking the inner ends of the wheel springs. This ensured ready compliance of the driven wheels with the then uneven track, and made the most of the adhesion weight. When the tracks became better, it still was beneficial because, as springs settled unevenly in service, the axleloadings were unaffected, while the tell-tale tilt of the equalizing beam drew mechanics' attention to the need to adjust the spring hangers.

LONG BOILER LOCOMOTIVES

A problem with the small British six-wheeled locomotive was that the boiler barrel was rather short: in

A Nord Atlantic. These engines first appeared in 1899, designed by Gaston du Bousquet of the Nord and built by Alfred de Glehn of the Alsatian Locomotive Works. They were four-cylinder compounds, and they set completely new standards of performance, noted all over the world. They ended their days on fast Pullman trains in the 1930s. This one is now preserved at Mulhouse.

early days it might be only some 2 m (7 ft) long. This resulted in much heat being wasted up the chimney, and sometimes in a red-hot smokebox door (faithfully if uncomprehendingly depicted by J. M. W. Turner in his famous painting 'Rain, Steam and Speed'). This led Robert Stephenson & Co. to devise a new patent locomotive, the 'long boiler' type. The main point of this was the long barrel, but the opportunity was also taken to alter almost everything else. In the original long boiler locomotives the cylinders were placed outside, at the front, of course (the stage of trying different locations was over by this time), giving a considerable overhang and encouraging a forward position for the front tube plate of the boiler barrel. All three axles were beneath this barrel, the central one carrying the eccentrics for the drive to the valves, which were inside the frames. The firebox was behind the trailing axle. Most importantly, there were no frames outside the wheels, because there was no delicate crank axle.

The long boiler locomotive proved a most important development. It was at first made as a 2-2-2, and at what passed for high speed in 1841, when it first appeared (i.e. about 64 km (40 miles) per hour), it was very unstable because of the overhanging weight at the ends. This damned it in Britain, but on the European continent it became extremely popular and was greatly developed by continental builders. As a 0-6-0 it lasted a good hundred years, because its short wheelbase made it ideal for general short-haul freight work necessitating shunting in yards with sharp curves. As a 2-4-0 it became a popular express machine, but as speeds rose the old instability trouble manifested itself and in France it was adapted to the 2-4-2 wheel arrangement. In fact this was first done in 1860 by the Stephensons themselves, to the orders of Robert Sinclair, in some engines for the Grand Luxembourg Railway. But the type was long associated with France and hundreds of such locomotives hauled the passenger trains of the Paris, Lyons and Mediterranean Railway and the Paris–Orleans Railway. The engines of the type designed for the latter railway by Victor Forquenot were amazing machines, still working when 80 years old.

In 1855, John Haswell of the Austrian railways (one of the many distinguished British expatriate engineers who developed railway systems in Europe in the 19th century) produced an eight-wheeled long boiler freight engine, the 'Wien Raab'. It was exhibited in the Paris Exhibition of 1855 and rather surprisingly never returned to Austria, being bought by the French Midi Railway. This was the prototype of another extremely widespread European freight design. As with the 0-6-0 (commonly known as the 'Bourbonnais' type), the 0-8-0 could negotiate sharp curves, and had an easily accessible ashpan, overhanging at the rear, and outside motion; both important features for ease of servicing.

INTRODUCTION OF THE BOGIE

It is fairly clear that the three distinct types which led to three distinct families of locomotive are all due to Robert Stephenson. The 'Rocket', apart from being in some sense the true ancestor of all locomotives, was more specifically the origin of the American, outside cylindered, bar-framed, basically very simple pioneer 4-4-0, from which all later and larger American locomotives were derived; the 'Patentee' established the British inside-cylindered locomotive, even though multiple framing eventually gave way to two simple plate frames inside the wheels; and the long boiler type was the forerunner of the continental plate-framed locomotive with outside cylinders, even though the overhang was eventually reduced by the addition of carrying wheels, as cylinders and fireboxes became heavier. Thus the 2-4-0 became the 2-4-2, and the 0-8-0 became the 2-8-0. The last major contribution to the basic morphology of the modern locomotive was the bogie.

We have seen that Norris fitted a swivelling four-wheeled truck to his six-wheeled locomotives and, far earlier, 'Puffing Billy' was mounted temporarily on two

four-wheeled trucks that foreshadowed the bogie. The early American 4-4-0 likewise had a swivelling front truck, but with a four-coupled locomotive the geometrical action on a curve of a bogie which was merely pivoted could not be correct. The bogie as we knew it today, with a central sliding device to allow lateral movement, and some form of centralizing force, is due to William Adams who applied it to some 4-4-0 tank engines on the North London Railway from 1864 onwards. He also became a firm advocate of equalized suspension and ranks as one of the first engineers to give proper appreciation to the importance of considering the behaviour of the locomotive as a vehicle. He remained for many years a brilliant exponent of the 4-4-0 type with outside cylinders, and so was well outside the British tradition of his time.

This part of the story of locomotive development may be concluded by taking the 'Atlantic' or 4-4-2 type to illustrate the way in which the various schools of design eventually converged, even if they never quite fused. In America, the 'Atlantic' proper arose out of the American 4-4-0, as a result of the need to support a large firebox. On the Continent, it was a development of the 2-4-2, to support a heavier front end and improve the riding qualities. In Britain, it first appeared as an enlargement of a 4-2-2 design on the Great Northern

The 4-6-0 version of the Nord Atlantic was built in greater numbers. Capable of developing 2,000 hp at only 71 tonnes (70 tons) without tender, these engines were du Bousquet's masterpiece. One is preserved at Mulhouse, but this one is in service on the Nene Valley Railway in England.

Railway, providing greater adhesive weight and a larger boiler. One might say that the Americans enlarged the rear, the Europeans the front, and the British the middle, and all arrived at a similar result.

EFFICIENT USE OF STEAM

Of all steam engines, the locomotive is the only one which is required to produce a continuously and widely varying output, and even to run at maximum revolutions while not producing any measurable power at all. For this reason the valve gearing of locomotives has been more developed than that of other steam engines.

It has only been in recent years, and in comparatively few locomotives, that the valves controlling the inlet of steam to the cylinders, and its release to exhaust, have been worked by rotating cams. By far the greater number of locomotives have had a valve for each cylinder, which, moving backwards and forwards alongside it, has opened each end of the cylinder in turn to the live steam chest and to the exhaust passages. This valve moves in something like simple harmonic motion, and its movement therefore corresponds, whatever the setting of the valve gear, to a movement derived from a single eccentric of suitable eccentricity or 'throw', and angular setting. The valve gears applied to steam locomotives, and they have been of many different designs, have almost all been devised to produce the effect of a single eccentric which is of variable throw and angular setting.

The first valve gearing with which we need to concern ourselves is the type applied to the 'Rocket' and some of its contemporaries. This was a single eccentric gear, i.e. there was one eccentric for each valve, and this one was 'loose' and could be made to 'slip'. It was driven by stops on the driving axle, and if the engine reversed after a spell of forward running, the eccentric stood still while the axle turned within it for rather less than half a turn, until the reverse direction stop engaged it, after which it would drive the valve in the correct manner for backward running. The engine was started on its backward course by manipulating the valves by means of handles convenient to the driver, and to make use of these the drive from the eccentrics had to be disengaged first. In fact, each driving rod simply hooked over a pin in the actuating mechanism, and so could be lifted off. So in this gear the angular setting could be varied as between two positions, but the valve travel could not.

Some skill was required to reverse an engine with loose eccentrics, and the next step was to put the disengaging device out of reach of the driver, but to give him a simple lever with two positions, one for forward and one for backward running. This was done by providing two fixed eccentrics for each valve, each with its drive rod ending in an open jaw or 'gab' set sideways. These gabs had widely flared entries, and the driver's lever pushed one or the other into engagement with the valve actuating mechanism, the flared entries pushing the valve into the right position as the lever went over. This was a safer gear from the driver's, the passenger's and the railway company's point of view, but as far as the engine went it produced exactly the same effect as loose eccentrics.

The disadvantage of constant valve travel in a locomotive is that it entails a constant degree of admission of steam. This is usually expressed as a percentage of the piston stroke. In a two-cylinder locomotive, this percentage must be over 60 at the moment of starting, otherwise there may be no steam in the cylinder best placed to start the engine (the other cylinder being at or near dead centre). With constant valve travel this percentage has to remain the same whatever the speed, and the only control available is that of the steam valve

Far left: *a US (Baldwin)-built Vauclain compound 4-4-0 for the French State Railway, c.1900. The superposed cylinders and single piston valve for high- and low-pressure cylinders are clearly visible.*

Left: *in 1903, this Brooks' tandem compound 2-10-2 for the Santa Fé Railroad was the most powerful locomotive in the world.*

A 'Chapelon Nord'. The Northern Railway of France made greater demands on its locomotives than other French railways, and when Chapelon, of the Paris–Orléans Railway, virtually doubled the power of the Paris–Orléans Pacifics by skilful rebuilding, the Nord did not hesitate to buy a batch, and in the late 1930s ordered more, new, from private industry.

Inset, left: *Dutch-designed and Swiss-built, a 2-6-6-2 Mallet articulated compound tender locomotive of the Indonesian State Railways. The very large low-pressure cylinders are conspicuous.*

Inset, right: *a smaller and older compound Mallet of the same system, a tank engine typical of the earlier years of colonial railways in the then Dutch East Indies.*

The most powerful steam locomotive type ever built: the Union Pacific 'Big Boy'. These locomotives used high-pressure steam in all four cylinders, following the general trend in the USA from the 1920s, when low-pressure cylinders became too big for their valves and for the moving load gauge. The 'Big Boys' could develop 10,000 horsepower, and their firegrates were some 6 m (20 ft) long and 2.6 m (9 ft) wide. Though they were freight-engines, they could easily run at 110 km per hour (70 mph). They were the ultimate expression of the Mallet articulation principle, and several have been preserved.

X4012
4012
UNION PACIFIC
4012
DANGER
KEEP OFF

(called the throttle or the regulator). This is uneconomical, because it causes excessive steam consumption; it is also inadvisable because the large volume of steam passing at speed is usually too great for the passages provided for it, so the engine half chokes itself; and it has a bad effect on the fire, via the blast action, because the blast is too fierce if the regulator is wide open, and too feeble if it is nearly shut.

None of this mattered much in 1829, because speeds were low and fuel economy was not yet being considered seriously. But by the 1840s it was obvious that something better than fixed-travel valve gearing was needed.

It can easily be shown that if the admission percentage is reduced, the reduction in power output is nothing like as great as the reduction in steam consumption, because the steam continues to do work by expanding within the closed cylinder, and the thrust on the piston is maintained, though at a reduced level. A reduction in valve travel reduces the admission period and lengthens the period during which the cylinder is closed, before opening to exhaust. But it also delays the moment of admission, and so cannot be used unless the angular setting of the eccentric is advanced to bring admission back to somewhere near the start of the piston stroke. Hence the need to vary both throw and angular setting of the 'equivalent eccentric'.

STEPHENSON'S LINK MOTION

Expansive valve gears had been applied to stationary engines, but these were not suitable for locomotive use; partly because locomotives ran at much higher rates of revolution and partly because mechanisms which were robust enough for a rigid machine could not stand up to being shaken up on a locomotive. For locomotives, the first solution was almost the best, and continued to be applied throughout the history of the steam locomotive. This was Stephenson's link motion, so called because it was possibly first devised at Stephenson's works. It was first fitted to a locomotive in 1842.

The usual form of gab gear was one in which the engaging pin of the valve drive was not moved by the driver, and the jaws of the gabs, fixed to the eccentric rods, opened out above and below it. The rods were moved vertically in unison, up or down, to engage backward or forward gear. Sometimes the jaws were united in a single oval frame, with the engaging sockets at top and bottom, the frame being pivoted to the eccentric rods, whereas open gabs were rigid with them. Stephenson's link motion replaced the oval frame by a link with a curved slot in it, which remained in engagement with the valve drive in all positions; and so, for the first time, intermediate settings between forward and reverse gear became possible, and these gave 'equivalent eccentrics' as required to advance timing and reduce valve travel.

The proportioning of such a valve gear is a matter of some subtlety, and if gear and valve are properly designed (and that was not always the case) a locomotive may be able to run with the gear in mid-position. It would run thus in either direction, because the cylinders receive puffs of steam just over the dead centres. This will maintain a little power if the engine is running fast, but will not allow it to draw a heavy load.

The only disadvantage of Stephenson's link motion was that it required eccentrics, but so did any other gear arranged between the frames. On the continent it was commonly fitted outside, driving valves above the cylinders; and the eccentrics, being mounted on stub axles outside the crankpins, were smaller and less likely

East African Railways' 59th class Garratt locomotive, with the elongated chimney of a Giesl ejector. The Garratt locomotive was the most important British contribution to steam locomotive development in the 20th century, and the 59th class were among the last and finest of their kind.

to heat up, as well as being better ventilated. A characteristic of the gear, which was often held to be a disadvantage but was in fact an advantage if turned to account, was that the timing was advanced more than was necessary to keep the admission point at about dead centre, when the gear was 'notched up', i.e. made to work more expansively by working near the middle of the link. If the gear was designed with too early an admission in full gear, as was often the case, then full expansive working became difficult because the timing was too advanced. But if full gear admission was late—which was actually helpful when starting a heavy load—then timing could be correct when working with a high degree of expansion. This latter arrangement was a feature of Churchward's excellent locomotives on the Great Western Railway in Britain.

There were two variations of Stephenson's link motion which found quite widespread use. The first of these was Gooch gear, which appeared on the Great Western broad gauge in 1843. In this the link was not raised or lowered to alter the setting but, instead, the rod connecting the link with the valve spindle was raised and lowered so that the die block which engaged with the link slot took the drive from the various positions on the link as required. Whereas in Stephenson's gear the link was curved with a centre approximately in the position of the eccentrics (when it was vertical), in Gooch gear the link curved the other way, with a radius equal to the length of the movable rod. The main advantage of this arose when the gear was placed close beneath the boiler barrel, because the link, being stationary in a vertical direction, required less clearance. On the Great Western at this period very large driving wheels and very large boilers for the time were beginning to be used (made possible by the broad gauge of 2.13 m (7 ft)) with the result that the driving axle was coming very close to the underside of the boiler barrel.

Gooch gear was not much used in Britain outside the Great Western system, but it was quite common in continental Europe, where it was usually arranged outside, in conjunction with outside cylinders. It could be designed to give absolutely constant 'lead' (i.e. steam port opening before dead centre—the usual way in which advance is reckoned) and this is one of the characteristics of the more modern and better known Walschaerts gear. Constant lead regardless of the degree of expansive working has commonly been believed to be beneficial for express locomotives, and so Gooch gear may have been part of the reason for the excellence of Forquenot's 2-4-2 engines for the Paris–Orléans Railway. These engines certainly had a particularly skilfully proportioned form of the gear.

In Allan's straight-link motion the link slot was straight, and adjustment of the gear involved moving the link and the rod joining it to the valve spindle in opposite vertical directions. This too could give something approaching a constant lead, and was at one time much used in Europe. In Britain, it was fitted to numerous designs but probably the most distinguished were the London and North Western 'Precedent' class 2-4-0s, which were always brilliant performers for their size.

These three forms of link motion all required two eccentrics per valve. The link had to be suspended so as to be capable of bodily fore and aft movement, because considerable movement was imparted even to the link centre by the fact that the eccentrics were never opposite to one another (because each was advanced in the direction of its intended working) and also because the eccentric rods crossed and uncrossed as the eccen-

Below: *the Mallet articulated locomotive (top) was conceived as a compound, but the largest examples were built in the USA as simple-expansion locomotives. However, probably the last built, in the 1950s, were of this compound type for the Norfolk and Western Railway. The triplex Mallet banking locomotives of the Erie Railroad (bottom) were six-cylinder compounds. The first of them, in 1913, hauled 16,000 tonnes at 24 km per hour (15 mph).*

A Heisler logging locomotive, with a Vee steam engine driving the trucks via shafts and bevel gears.

trics rotated. The link suspension had to keep it laterally steady. In Gooch and Allan gears the same applied to the rod taking the drive from the link, and the whole assembly could be quite unsteady when worn. This probably explains why, in the long run, only Stephenson gear survived. It was still being built for main line service after the last war.

Eccentrics were dispensed with altogether in Joy's valve gear, which was first used in 1879. Instead, a drive was taken from a point on the main connecting rod. The vertical component of the movement of this point was used to work a die block up an inclined slide, the angle of which was variable. The horizontal component of the movement of this die block, slightly modified, moved the valve. It was an excellent gear for a stationary engine, and was indeed much used in locomotives before 1900, but it had two disadvantages: movement of the locomotive on its springs resulted in movement of the valves on their portfaces, and the connecting rods were apt to break at the point where the drive was taken off. Its great virtue was for inside-cylindered locomotives, because the absence of eccentrics made it possible to bring the cranks closer together and fit longer axle bearings, or, alternatively, to fit a third, central, bearing.

However, though there were very many different types of valve gearing tried with varying success over the years of the steam locomotive, one type dominated the 20th-century history, and that was the one invented by the Belgian engineer Egide Walschaerts in 1844, and brought to its common form four years later (though it was invented a second time, many years later, by Heusinger von Waldegg after whom the gear is named in Germany). This has a single eccentric (or return crank if the gear is outside) which drives a curved slotted link pivoted at its centre on trunnions firmly attached to the locomotive frame. The eccentric is set at 90° to the main crank, and movement is taken from the link by a radius rod which can be set at various positions on the link. This provides an accurate way of varying valve travel and of reversing its movement, but does not in itself provide for the advancement of the timing to keep admission at or a little before dead centre. This is arranged, also with great precision, by taking a drive off the crosshead of the piston rod, the motion of which is combined with that of the radius rod by a vertical combination lever. The valve drive is taken off this vertical lever at a point which gives the correct combination for the valve.

There are four happenings in a steam cylinder: admission of steam; closure of the cylinder; opening to exhaust; and closure prior to reopening to steam. Between the second and third, expansion of the steam takes place; and between the fourth and first, compression of any remaining exhaust steam takes place. Ideally, admission should take place a little after dead centre when running slowly with a heavy load, or at starting. This would be called 'negative lead' and in practice is only possible with Stephenson gear. At speed, admission somewhat before dead centre is desirable; in fact, the faster the earlier. Most valve gears give this effect. The constant lead of Walschaerts gear is a matter of the amount of opening at the dead centre: the timing of the admission point, measured as an angle, in fact advances with 'notching up' to more expansive working.

At slow speeds steam has to be admitted for a larger proportion of the piston stroke in order to maintain the pressure and so produce an even turning effect. At high speeds, cut-off of steam has to be early, and the period of closure of the cylinder must be long enough to get plenty of work out of the expansion of the steam, but the release to exhaust should be earlier than at low speeds in order to assist the 'breathing' of the cylinders, which might otherwise become choked with steam. The compression point is the trickiest of all, because its effect depends upon the exhaust pressure. Compression serves two useful purposes: it cushions the piston, and the weights attached to it, at the end of its stroke; and it fills the clearances at the ends of the cylinders with compressed steam, which raises their temperature and also saves the cost of filling these spaces with fresh steam. For the cushioning action, more compression is needed at higher speeds. A well-designed valve gear will advance the timing of all four events in a way most conducive to smooth running, as the driver 'notches up' the gear with increasing speed.

A three-truck Shay. The Shay type was a speciality of the Lima locomotive works, and had a vertical three-cylinder engine alongside the boiler, which was offset. This drove on to the ends of the truck axles via Cardan shafts and bevel gears. A very flexible locomotive type, it proved the most successful and widely used of all logging locomotives.

A locomotive may be running at speed with full steam, when the driver closes the regulator. This need not be the prelude to a stop—it may precede descending a gradient—and it is important that the engine should run freely under these conditions, but a notched-up valve gear setting which is suitable for passing steam through the cylinders is unlikely to be suitable for the coasting condition. It was a common arrangement to provide 'snifting' valves which enabled the cylinders to draw air and pass it up the blast pipe, but air does not behave like steam, so the valve gear had to be readjusted to something nearer full gear. Automatic or manually operated by-pass valves, which put the two ends of the cylinders into direct communication, were perhaps the best way of enabling a locomotive to coast freely, and had the advantage that they did not produce sudden changes of temperature.

From all the preceding it will be apparent that the locomotive has had to develop in very different ways from those of the stationary engine, which is not normally required to start under full load, to vary its output continuously over a wide range, or to 'coast' at high speed. Taken with the space restriction imposed by the moving load gauge of the railway, and the very severe weight limitations (total weight, weight per 300 mm (1 ft) of length—for bridges mainly—and weight per axle, for the track), these factors have made it difficult to produce a machine which is very efficient thermodynamically. All the same, as more work was required of locomotives, more work was required of the fireman who had to keep the furnace stoked, and a prime reason for improving the efficiency was simply to enable a locomotive to do more for a given effort of the fireman. The development of valve gears in the mid-19th century was the best way of doing this in the first place, but in the last quarter of the century compounding was applied to locomotives on a considerable scale, with further improvement in efficiency, and in the early 20th century a practical design of locomotive superheater produced further economies. Finally, the combination of compounding and superheating was to produce by far the most efficient locomotives of all, and the most powerful in relation to their weight.

COMPOUND LOCOMOTIVES

The word 'compound' as applied to steam engines usually means double expansion. It is so in marine engine parlance, and it applies to locomotives as well. In marine practice, triple expansion became quite the normal thing towards the end of the 19th century. However, in locomotive history triple expansion has only been applied to experimental or unrealized designs.

Double expansion involves expanding the steam first in a high pressure cylinder, and then passing it to a low pressure cylinder of larger dimensions to expand further. A principal advantage claimed for it is that the range of temperatures in each cylinder is reduced, as compared with a similar degree of overall expansion achieved within a single cylinder. There is quite a lot in this, because with the usual design of steam engine cylinder the steam passes in and out through the same passageway, so this passageway is cooled by the expanded steam and therefore absorbs heat from the incoming steam. The heat so lost goes into the exhaust steam. Another point sometimes made is that the greater volume of the low pressure cylinder offers a possibility of greater total expansion, but this is clearly not true unless the valve gearing makes a really early cut-off impossible. What is true is that a greater total expansion is *usefully* possible, because there is less heat loss; or, to put it another way, a given high total expansion will produce more power in a compound machine than in one using single expansion.

Very high overall expansion rates are only possible when the engine exhausts into a condenser, as in marine and stationary practice. A steam locomotive, because it exhausts into the atmosphere (and has to do so with sufficient vigour to draw up the fire) cannot work with an absolute terminal pressure below about 124 kN per sq metre (18 lb per sq in (1.2 Atm.)). The temperature range over which the steam operates is narrower than with condensing engines exhausting into a vacuum, so the advantage of multiple expansion, from the economy point of view, is less, and this explains why triple expansion was never applied in normal service. However, the fact remains that compound locomotives were

Above: *no. 701 of the Nord Railway of France was the first four-cylinder compound locomotive, and the starting point of a development which eventually produced the highest power-to-weight ratio and the greatest fuel economy in locomotives. It was designed by Alfred de Glehn, to the requirements of Gaston du Bousquet, and is preserved in Mulhouse.*

generally more economical than their simple expansion counterparts, and sometimes by a wide margin. No simple expansion locomotive ever equalled the best compounds in this respect, or in respect of power for a given weight.

Another virtue of compounds in practice was that they rode better and lasted better, while their maintenance costs were not generally higher, though their initial cost usually was. Their main disadvantages were that they were harder to design (with the result that some compounds were no better than the equivalent simples), required greater skill and intelligence of their drivers, and had large low pressure cylinders which could not always be accommodated comfortably within the cross-sectional dimensions permitted for the locomotive.

The fact is that compound locomotives were always in the minority. They were unquestionably superior as machines, but it is significant that their longest and finest development took place in France, where educational standards as well as coal costs are very high. However, this is a technical and not an economic or social history of the steam engine, so it is worth considering the special qualities of the compound locomotive in a little more detail.

Once again we have to remember the biological nature of the locomotive, because the advantages of the compound lay in a number of seemingly little things which interacted with the whole machine, Firstly, then, the division of the expansion over two, or two sets, of cylinders and valves not only reduced the temperature fluctuations but also the pressure fluctuations. There was less pressure difference between the two sides of the pistons or valves, and therefore less leakage. Moreover, all the leakage across these parts which took place in the high-pressure cylinders was not to exhaust and to waste, but to the low-pressure cylinders where the 'lost' steam would add to the power. Similarly, the steam which was inevitably wasted in filling the clearance volumes of cylinders (because the compression phase of the cylinder working could not fill the clearance spaces under all conditions of working) was not wasted at all in the high-pressure cylinders of a compound, and less was wasted in the low-pressure cylinders because the range of working in those cylinders was not as great as in a simple engine, and the beneficial effect of compression could be more fully provided for.

Another advantage was simply due to the fact that a high overall degree of expansion could be obtained with longer admission periods in the cylinders: the expansion rates in the individual cylinders were multiplied together. This resulted in a more even turning of the wheels and a steadier pull on the train, which in turn produced a smoother ride and reduced the wear on the locomotive. It also reduced the tendency to slipping and gave compound locomotives a higher practical factor of adhesion.

The presence of the large low-pressure cylinders made it possible to accelerate faster with a compound because, once the wheels had begun to turn, the work done in the low-pressure cylinders could be boosted by using a long cut-off in the high-pressure side and slightly notching up the low-pressure side. The sustained acceleration was superior to anything that could be

Right: *SNCF class 141P were new engines which resulted from a redesign by André Chapelon of a type of locomotive used on a former PLM railway. They had twice the power of their predecessors, developing over 4,000 horsepower in the cylinders.*

achieved with an equivalent simple engine. In theory, of course, a simple engine could have cylinders just as large and should therefore be able to accelerate just as well, but in practice, when this was tried, it proved impossible to avoid excessive slipping and all engines on which this experiment was made had to have the cylinder diameter reduced.

The fact that the low-pressure cylinders were using steam at a low pressure, taken with the characteristics of valve gears and the relatively long admission period with which compounds worked, resulted in a softer blast when starting or working hard. This reduced the tendency of a hard-worked engine to expel unburnt gases and small particles of fuel from the chimney and so improved the boiler efficiency as well as reducing boiler maintenance. This is why the savings of compound locomotives as against equivalent simples, which were 10 per cent or so at light rates of working, rose to 30 per cent or more at maximum outputs.

TWO-CYLINDER COMPOUNDS

These advantages apply whether simple or compound locomotives have two, three, or four cylinders, but it must be said that the last point, about the effect on the fire, might be influenced another way if a four-cylinder simple having the unusual arrangement of eight exhaust beats per revolution were to be compared with a four-cylinder compound which could only have four. However, much of the argument against compounds was against the multi-cylinder machine. Any large compound had to have two low pressure cylinders to accommodate the sheer volume of steam exhausted from the high pressure side; it therefore had to have at least three cylinders, and the great majority had four. The proponents of the simple engine all too often sought further simplicity by having only two cylinders, and this not only produced the starting problems and exacerbated the effect on the fire, but also introduced the balancing problems which cannot be avoided in large two-cylinder engines.

The balancing problem may be simply explained as follows. Some parts of a locomotive's mechanism can be considered as revolving masses; these include things which are eccentric, like crank webs and pins, eccentrics and their straps, coupling rods, and a proportion of the connecting rods including the big ends. These revolving masses can all be balanced completely, in theory, so that there is no tendency for the engine to stop more often with the weights in the lowest position nor for the driving axle to attempt to wobble violently at speed. But such perfect rotating balance is impossible with a two-cylinder engine of normal construction, because extra rotating weights have to be applied in order to balance the reciprocating weights. The pistons, crossheads, and parts of the connecting rods and valve gear all represent masses moving to and fro in a more or less horizontal direction. At any appreciable speed the action of these masses is very disturbing to the locomotive, which proceeds in a series of jerks. To counteract this, weights are placed inside the rims of the driving wheels in such a position that their backwards and forwards movement is in contrary motion to that of the reciprocating masses.

These extra balance weights in the wheels are not, and cannot be, balanced as rotating masses, and so produce the tendency to wobbling of the driving axle at speed which the rotating balance is intended to avoid. The result of all this is that some very strange things happen with two cylinder locomotives, and these things are worse if the cylinders are outside, because in that case the disturbances are produced further from the longitudinal centre of the engine and are therefore more able to cause the whole engine to wag from side to side as it runs. The balancing of the reciprocating masses has to be a compromise: with too little balancing the locomotive will pull the train in a series of jerks accompanied by a marked side to side 'hunting' movement. This is bad for the locomotive, the track, and the passengers (goods trains seldom run fast enough for these effects to be serious). But if there is too much reciprocating balance then the loss of rotating balance in the wheels causes them to pound vertically on the rails, thereby imposing an extra load, known as 'hammer blow', which may be unacceptable to the permanent way engineer or indeed to the bridge designer. Wheels have also been known to lift off the track at high speed between blows.

All this only applies to two-cylinder locomotives, and is due to the fact that the cranks have to be phased at 90 degrees. A four-cylinder locomotive can have its outside cranks phased at 90 degrees, but each inside crank can be at 180 degrees to its outside neighbour. The reciprocating masses will thus always be in contrary motion and will balance themselves. This is at its best if the drive from all four cylinders is concentrated on one axle, as in the London and North Western 'Claughton' 4-6-0s, the Dutch 4-6-0s and, among compound locomotives, the Bavarian Pacifics of Maffei design. But even if the drive is divided between two axles the motion of the locomotive as a whole, relative to train and track, can be very smooth. With three cylinders and cranks at 120 degrees the reciprocating balance is also very good, but there is a tendency for the locomotive to swing from side to side, which some designers have counteracted by introducing a small amount of reciprocating balance weight into the wheels. With three cylinders also, concentrated drive, as on Sir Nigel Gresley's London and North Eastern Pacifics, produces a smoother ride than divided drive as on the London, Midland and Scottish Royal Scots or the LNER Sandringhams.

Of course, it would be possible to balance a two-cylinder locomotive by arranging for special weights to be driven backwards and forwards in motion contrary to the pistons, but this extra mechanism and extra weight could only be tolerated if it served some other useful purpose. The only instance of successful and large scale use of such a device occurred on the Northern Railway of France, where 72 very large passenger tank locomotives incorporated contrary motion weights in the valve gear drive.

From all this it follows that more than two cylinders are needed for fast running in European practice. Large British two-cylinder express locomotives have been known to produce a drawbar pull fluctuation of as much as 12 tonnes (12 tons) at high speeds. In North America, things were not quite the same. Locomotives did not normally produce as high powers in relation to their weight as in Europe, and they were very much heavier and had much heavier tenders. Drawbar pull fluctuations were relatively less in consequence, and the greater length of these large locomotives was conducive to a steadier ride. But in Europe the need for multi-cylinder arrangements was amply demonstrated and, that being so, there was much to be said for applying compound expansion as well.

However, in the early years of compounds they all had two cylinders. The first practical locomotives of the kind were very small indeed: 0-4-2 tank engines for the little Bayonne-Biarritz Railway in Southern France. Their designer was Anatole Mallet and they started work in 1877. To ensure easy starting, both cylinders had to operate from the beginning, so Mallet provided a means of admitting a little high-pressure steam to the large low-pressure cylinder during the first few revolutions. The high-pressure cylinder meanwhile exhausted directly to the atmosphere, but once the engine was on the move it supplied its steam to the low-pressure cylinder which no longer received its high-pressure supply.

F. W. Webb

Other designers devised other starting arrangements, most of which were automatic or semi-automatic in operation, while Mallet's arrangement was under the driver's control, but they all involved supplying steam to the low-pressure cylinder at starting. There were many patents covering starting arrangements, but the Mallet compound did all that was needed.

The two-cylinder compound lasted to the end of the supremacy of steam. There were some large examples. Between the two world wars, numerous 4-8-0, 2-8-0, 2-8-2 tender locomotives were built in Britain for South America, and some 4-8-4 tanks as well. These late examples followed a long and successful development of the two-cylinder compound in Latin America. The United States, before the days of superheating, could show some large examples, including some of the 4-6-2 wheel arrangement. But Europe was certainly the place where most were built. The vast Prussian-Hessian State railway used little else until the early 20th century, and well over 6,000 examples were built to the specifications of their greatest designer, August von Borries. These were mainly 4-4-0s, 2-6-0s and 0-8-0s, many of which only disappeared during the last world war. In Britain, where compounds were never really popular, T. W. Worsdell built substantial numbers for the North Eastern Railway. But probably the longest lasting were those of the Austro-Hungarian empire and its derivative states. Most of these were products of the fertile brain of Karl Gölsdorf, one of the greatest of European railway engineers. They were of innumerable different types, because there were severe weight and gradient problems on different parts of the various systems. There were 0-6-0s and small suburban tank engines at the lower end of the scale, fleet-footed 4-4-0s among the medium sized machines, and heavy 2-8-0 and 0-10-0 freight engines among the largest. Like all two cylinder compounds, they produced only two exhaust puffs for each turn of the driving wheels, and a ten-coupled freight engine climbing a steep gradient with a heavy train seemed to progress with a succession of measured explosions, but they did not throw fire from the chimney. Four years after Gölsdorf's death the largest two-cylinder compounds following his style of design appeared in Bulgaria. They were 0-12-0 tanks, which were later rebuilt as superheated simple expansion engines and lasted for more than 50 years.

THREE-CYLINDER COMPOUNDS

Three-cylinder compounds were a comparative rarity. While the two-cylinder variety became common in Europe, a determined and prolonged effort was made in England by F. W. Webb, of the London and North Western Railway, to develop his own three-cylinder type. With what seems like total engineering perversity he chose to fit two high-pressure and one low-pressure cylinder, which meant that the low-pressure cylinder had to have about twice the diameter of the high-pressure ones. Had he fitted one high-pressure and two low-pressure cylinders, they could all have been about the same size.

In fact, there was reason behind Webb's curious arrangement. The platforms of the LNWR were very close to the track, and only the smallest of outside cylinders could be fitted in, so they had to be the high-pressure ones. Later, Webb fitted two inside low-pressure cylinders, but most of his engines just had one very large one, and this certainly allowed the use of a strong crank axle and long bearings. The drive was divided in most of the three-cylinder engines, and they had no coupling rods, there being a prevalent belief in the 1880s that axles more than 2.4 m (8 ft) apart could not be coupled, at least for high speed running. Not everybody subscribed to this belief, but Webb did, and he wanted to fit longer fireboxes than could easily be accommodated within a short coupled wheelbase, so the coupling rods had to go. He may also, like Dugald Drummond, who built some four-cylinder simple 4-4-0s without coupling rods (i.e. 4-2-2-0s) for the London and South Western, have believed that an engine ran more freely without coupling, and if both axles were powered there was no need to couple them. However, he was wrong (as was Drummond). The lack of synchronism led to curious irregularities of motion as the reciprocating unbalances of the two sets of mechanism either reinforced or cancelled each other and, in the case of the compounds, starting difficulties were caused because there was no provision for admitting steam directly to the low-pressure cylinder. The high-pressure cylinders, exerting the maximum power because there was no back pressure from the low-pressure side, had only the adhesion offered by one pair of wheels with which to start the train. As often as not they slipped, flooding the receiver pipe between high- and low-pressure engines with steam. This flood of steam caused the low-pressure side to slip in turn, draining the receiver and so encouraging the high-pressure side to slip again. To make matters worse, many of these engines were fitted with slip eccentric valve gear on the low-pressure side, on the assumption that the engine would be moved by the high-pressure cylinders before the low-pressure one could come into action. If the engine had backed on to its train, this cylinder was in reverse gear, and if the high-pressure slipped on starting, then the low-pressure side thundered into action—in reverse.

In spite of all this, some of Webb's three-cylinder compounds did very fine work, notably the largest of the 2-2-2-0s. But other classes seemed to need piloting by simple expansion 2-4-0s. As Webb was the very eminent chief mechanical engineer of the largest British railway company, the performances of his compounds gave humbler engineers the idea that compounding could not be made to work reliably, and the effect of this on British locomotive engineering was long lasting and unfortunate.

For the Northern Railway of France, E. Sauvage designed a 2-6-0 locomotive which appeared in 1887. It had a single high-pressure cylinder inside and two low-pressure ones outside, the outside cranks being phased at 90 degrees and the inside one bisecting the major angle between them. It produced four regularly spaced exhaust beats per turn. This engine was the first to have the arrangement which became well known in Britain in the Midland compound 4-4-0s (the most numerous class of three-cylinder compound in the world, and the most numerous British 4-4-0, with 240 examples built). The same arrangement was later to appear in SNCF no. 242A1, of which more below, and it was certainly the layout which was to prove capable of the greatest development. However, on the Nord this locomotive was immediately overshadowed by the advent of the first four-cylinder compounds. As a unique machine, unsupported by other members of its class, it had a rather short life by the standards of French compounds built before 1920, and was scrapped after 42 years' service, in 1929.

The Jura Simplon Railway in Switzerland followed Sauvage's example in 1896, with the first of what was to prove the largest class, numerically, of Swiss steam locomotive. Another 2-6-0, this very successful design differed mainly in having the cranks set at 120 degrees. As this was a hill climbing locomotive used mainly on passenger trains, smoothness of pull was very important and no doubt led Weyermann, the designer, to choose this crank setting. Acoustically, it had the effect of causing the four exhaust beats to come out in the rhythm of a tarantella. The Swiss Federal Railway continued building the design after its formation, and there were eventually 147, some of which survived to

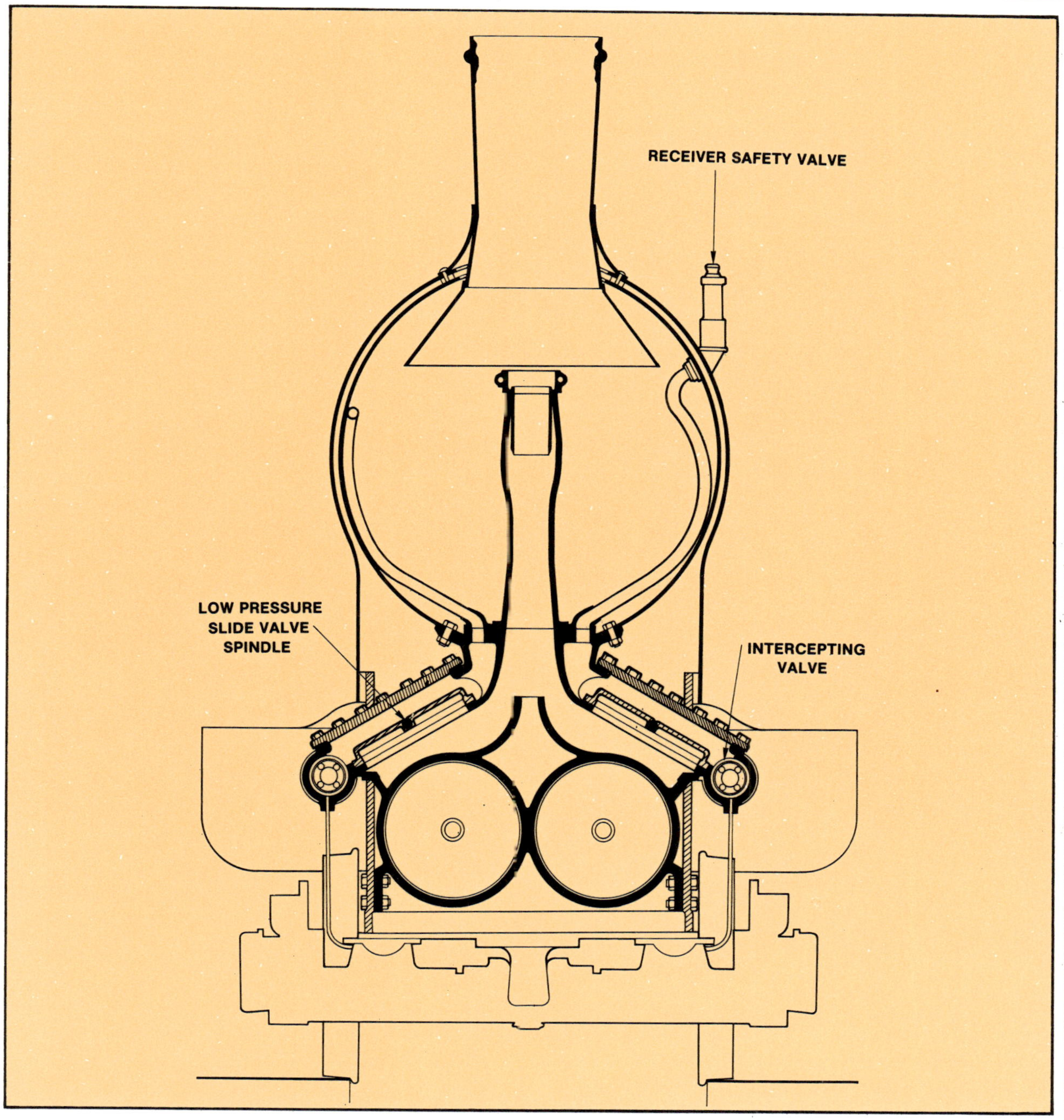

Cross-section through the low-pressure cylinders and slide valves of a du Bousquet compound Atlantic of the Northern Railway of France, 1899. The capacious valve chests and the free passageway for exhaust steam help to explain why the performances of these locomotives were judged so remarkable in their day.

be sold to the Netherlands after the second world war and to remain at work until the mid-1950s.

The Swiss engines had divided drive, but in the Midland compound 4-4-0s the drive was necessarily concentrated. This made possible a very good arrangement of the cylinders in line, with the piston valve of the high-pressure cylinder exhausting very directly into a large receiver which formed the steam chest for the slide valves of the external low-pressure cylinders. All three valve gears were between the frames and the locomotives were of very elegant appearance. Originally, they were described as Smith compounds, and in fact were largely designed by W. M. Smith, chief locomotive draughtsman of the North Eastern Railway, as an enlarged version of a similar 4-4-0 which he had designed for that railway. Smith was a friend of the Midland engineer, S. W. Johnson, the nominal designer of the engines.

The term 'Smith compound' relates to patents covering details of control. This, especially at starting, presented slightly more of a problem with a locomotive having a single high-pressure cylinder, and there was really no alternative to admitting steam directly to the low-pressure side. Unless the high-pressure cylinder could be made to exhaust directly to the atmosphere, which in the Midland compounds it could not, it became ineffective and simply 'floated' with equal pressure on both sides of the piston. In Smith's system the driver had independent controls for the high- and low-pressure valve gears but, after the first few engines of the class, Smith's system was simplified (as the engines came into the hands of drivers who did not belong to a small élite) and independent valve gear control was abandoned. The admission of steam directly to the receiver was no longer via a separate reducing valve, but by a special fitment to the main regulator, which only operated with a very small regulator opening. These engines, first built in 1901, were constructed in nearly 200 further examples after the railway grouping in 1923, by the LMS railway.

Though few in number compared with two- or four-cylinder compounds, the three-cylinder variety included one locomotive which many consider to have been the finest ever built, and which represented the final development of the French steam locomotive, in design if not strictly in date. This was SNCF no. 242A1. Originally built in the 1930s as a single experimental 4-8-2 of the Etat Railway, with three cylinders and simple expansion, this engine was originally an abysmal failure. Long laid aside, it was chosen by André Chapelon as the basis of a rebuild into a 4-8-4 three-cylinder compound, which was meant as the preliminary test locomotive for the projected French postwar steam locomotive range, all of which were to be three-cylinder compounds. In the event, the locomotives of this range were designed but never built, because the policy of the SNCF became one of electrification, and they had also received a very large number of American locomotives which had been ordered during the war by the 'Free French' government which was based in England. These locomotives were capable mixed traffic machines, and the older French locomotives were fully able to handle the express services during the transition to electric traction. So 242A1 was the only Chapelon three-cylinder compound. It was by far the most powerful European steam locomotive, and few locomotives even in the United States ever equalled its 5,500 horsepower; those that did weighed twice as much. This maximum power developed in the cylinders was very evenly divided between the three, and the single inside cylinder made possible a very strong crank axle and large bearing surfaces. It was a very robust and unusually handsome machine, but its life was short, because there really was no use for it once the traction policy had been changed to electric power.

FOUR-CYLINDER COMPOUNDS

In 1886 the first de Glehn four-cylinder compound was constructed for the Northern Railway of France at the Alsatian Company's works at Belfort. This was a 2-2-2-0, incorporating some of the standard parts of the Nord's 2-4-0 express engines. It owed a little to Webb, perhaps, for it had divided drive and no coupling rods. The outside cylinders drove the hindmost wheels. The inside, high-pressure, cylinders drove the leading driving wheels. The de Glehn arrangements provided for separate control of the two sets of valve gear, and also allowed boiler steam to pass directly to the receiver from a second regulator. The high-pressure cylinders could exhaust directly to atmosphere if required, and all this was entirely under the driver's control. The engine could be operated as a four-cylinder compound, a four-cylinder simple, or a two-cylinder simple using either pair of cylinders. It could also work as a reinforced compound, i.e. with a whiff of boiler steam boosting the receiver pressure during compound working. This prototype gave good service for about 40 years and is preserved today, with the leading bogie which it acquired in 1892 in place of its leading wheels.

The engineer of the Nord, Gaston du Bousquet, who was certainly one of the greatest of all French locomotive engineers, had some hand in the design, but revised it considerably when he came to build two further prototypes in 1891. These had the high-pressure cylinders outside, set well back and driving the rear axle, while the low-pressure cylinders were inside and drove the leading driving axle. One of the engines had coupling rods, making it a 4-4-0, and the other did not. The subsequent engines of this numerous class were all coupled. These locomotives were the first to have the most characteristic French mechanical layout, one which was applied to progressively larger locomotives until it reached its maximum size in the 4-8-2s of the Eastern Railway, and its maximum power (4,400 indicated horsepower) in the 4-8-0s of the SNCF, rebuilt from Pacifics of the Paris-Orléans Railway by Chapelon. Those engines had been, in 1907, the first Pacifics in Europe, but their power had been only half of that which the Chapelon alterations were to confer upon them.

The work that Chapelon did, in rebuilding engines already far from new and producing incredible improvements in their performance, justifies a claim that he was the greatest locomotive engineer since Robert Stephenson. If this is true, it is because he alone seemed able to grasp the full significance of those biological interactions between the various organs of the machine, and actually to treat them scientifically in such a way as to be able to predict the extraordinary results he achieved. His locomotives achieved the highest power-to-weight ratio and the highest thermal efficiency ever achieved with steam locomotives, and they were as troublefree in service as those operating them could have desired.

He achieved his results by increasing the steam temperature, greatly increasing the size of the steam passages and the volume of the valve chests, and improving the draughting efficiency of the blastpipe and chimney arrangement dramatically, so that the fire burned better but the back pressure against the pistons was actually much reduced. There were other minor changes, but the cost of this transformation was a mere fraction of the cost of a new locomotive. The Pacifics which he first rebuilt astonished the whole engineering world and eventually every locomotive man followed his ideas to some extent.

Chapelon himself was always quick to point out that some of his predecessors had shown him the way. William Adams (of the North London, Great Eastern and finally London and South Western Railways) had made important improvements in the blastpipe and had provided large steam chests well back in the 19th century, while G. J. Churchward of the Great Western Railway had done much to improve the function of piston valves early in the 20th. But perhaps it was du Bousquet of the Nord who had come nearest to Chapelon's ideas. He had instigated a scientific study of steam flow in his compound 4-4-0s, which had led to improved steam conditions in his celebrated Nord Atlantics, one of the classics of European locomotive design. And he had later produced an extremely potent small-wheeled 4-6-0 in which a still further enlarged steam circuit had permitted daily express running at the maximum speed allowed in France. These last engines, one of which is preserved in France and one in England, were eventually capable of indicating 2,000 horsepower continually, for a weight of no more than 70 tons, and they were the basis of the Nord Super Pacific design.

The Nord pioneer four-cylinder compound was quickly followed by four-cylinder compounds on the Paris, Lyon, Mediterranée Railway. Six prototypes appeared in 1887: two express locomotives, two freight locomotives, and two for the heaviest gradient work. They were all of about the same size and carried the record boiler pressure of 15 atmospheres (about 1,460 kN per sq metre (212 lb per sq in)). All had eight wheels, the express engines being of the favourite French wheel arrangement already largely represented on the PLM Railway: 2-4-2. The others were 0-8-0s. All had the low-pressure cylinders outside and the high-pressure ones inside. To assist at starting, boiler steam could be admitted to the capacious receiver though there was no provision for diverting the high-pressure exhaust to atmosphere, so the high-pressure cylinders contributed little to the starting effort. The valve events in high- and low-pressure cylinders were not independently controlled.

These six engines all showed the sort of fuel economy that the Nord prototype had shown, between 15 and 20 per cent in general service. The advantage of perfect balancing enabled the express engines to run faster and more steadily than the two cylinder 2-4-2s, and the smoother drawbar pull improved the adhesion of the freight and steep gradient engines. From this time onwards the PLM built nothing but four-cylinder compounds, with the exception of some unsuccessful four-cylinder simples which were all converted to compound operation eventually.

The PLM was the largest French railway, and its locomotives numbered many thousands. 4-4-0s, 4-4-2s, 4-6-0s, 4-6-2s, 2-8-0s, 4-8-0s, 2-8-2s, and 4-8-2s figured

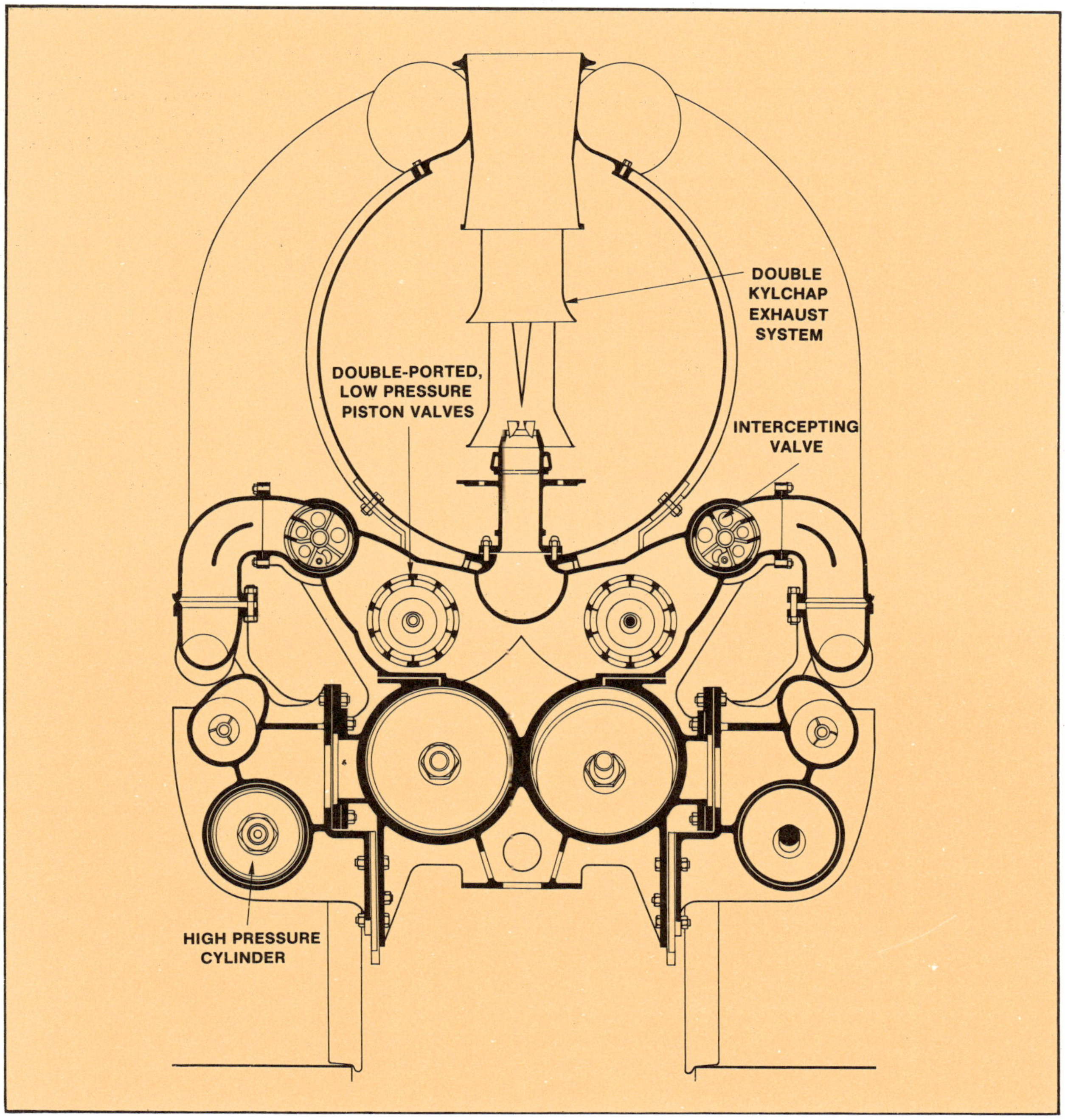

Cross-section through cylinders and valves of SNCF 2-8-2 class 141P, of 1942. These locomotives were a thorough redesign, by André Chapelon and Georges Chan, of earlier locomotives of the PLM railway. The new engines were more than twice as powerful, though they used the same basic boiler design. This was due to large steam passages (evident in this drawing), high superheat, and the double Kylchap exhaust system, with its divided jets and three levels of entrainment of the flue gases. This arrangement reduced back pressure on the pistons, and greatly stimulated combustion.

among the tender engines, and 4-6-4s and 4-8-4s among the tanks. A point of interest in the story of the PLM Pacifics (of which there were, coincidentally, 462) is that 90 of them were built as four-cylinder simples following the introduction of superheating. The first superheated simple had shown greater economy than the first unsuperheated compound, but further comparisons with superheated compounds had shown their superiority, so the simples were converted, but, interestingly, without raising their boiler pressure as usual for compound expansion. In order to re-use boilers pressed to 12 or 14 Atm. (1,179 or 1,372 kN per sq metre (171 or 199 lb per sq in)) the high-pressure cylinders were made larger than in the engines built as compounds from the start and pressed at 16 Atm. (1,565 kN per sq metre (227 lb per sq in)). The rebuilt engines proved more economical by 10 to 15 per cent and more powerful by a similar amount, which was not as good as the engines built as compounds and working with a higher pressure, but sufficed until new boilers were required. The PLM compound 4-4-0s and early 4-6-0s, like the very first of the Pacifics, had the de Glehn-du Bousquet cylinder arrangement, but the rest of the Pacifics had the cylinders in line across the engine, with the high-pressure ones outside. The 2-8-2s and 4-8-2s had the high-pressure cylinders inside, and set well back between the leading driving wheels, to drive the third coupled axle. It must be said that the larger PLM engines were not the best of compounds in France, and only became the excellent locomotives that they were in recent years after they had been modified on Chapelon principles. The last PLM 4-8-2 was of a different type to its predecessors, and this design, again modified in the light of Chapelon's work, was the basis of the SNCF 241P class, the last French express class and a very fine and beautiful looking machine, though not quite the equal in power or robustness of the Chapelon 240P class produced a few years earlier by rebuilding the PO Pacifics of 1907.

Though it originated in France, the four-cylinder compound was built and used almost everywhere to some extent. In Germany it was preferred by those railways which had considerable gradients to contend with, whereas the relatively flat (but enormous) Prussian system generally abandoned compounding with the introduction of superheating. Bavaria, the Palatinate, Württemberg, Saxony and Baden all used compounds for their hardest work. Even Prussia built four-cylinder compound 4-6-0s for the heaviest and fastest express service, and these superheated engines

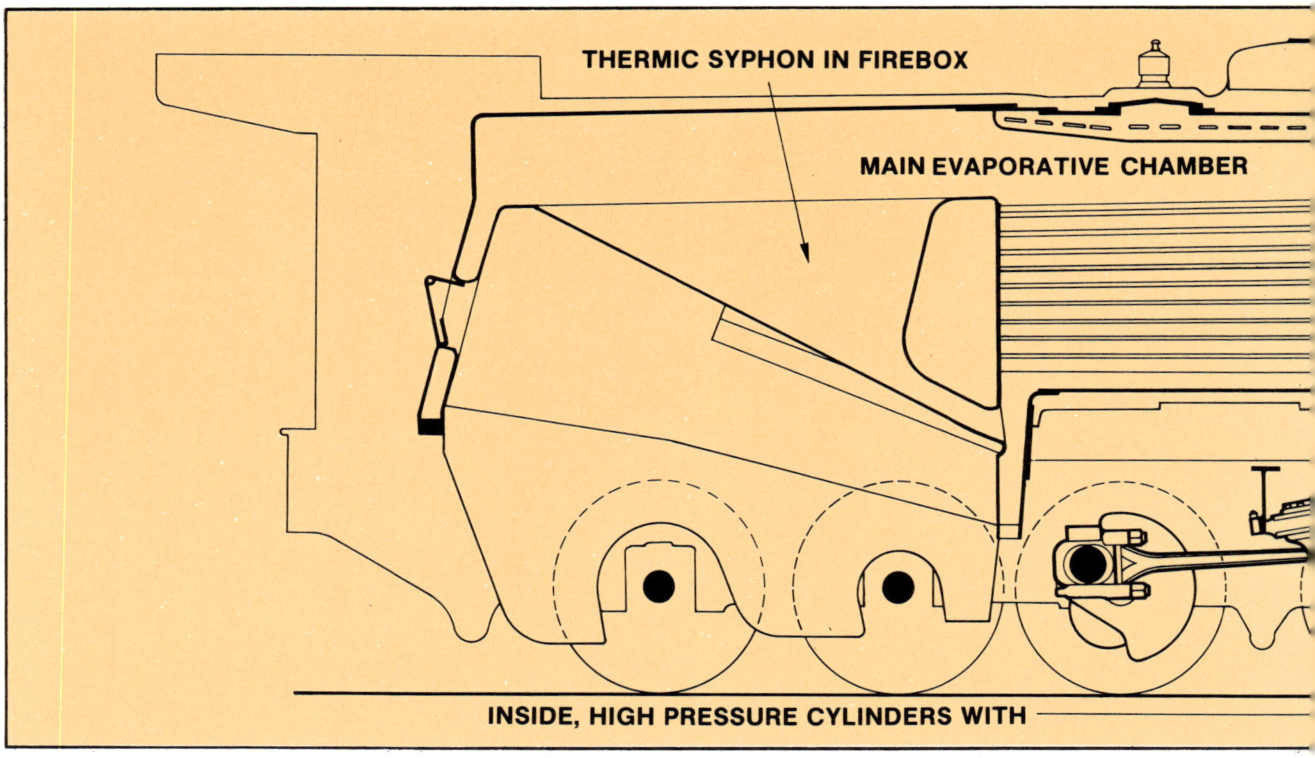

were rated 25 per cent more powerful than the three- and four-cylinder superheated simples of similar weight, built at the same time. But undoubtedly the finest German locomotives built before the railway amalgamation of 1925 were the Bavarian compounds built by the firm of J. A. Maffei at Munich, to the designs of Maffei's director, Anton Hammel. The Maffei style was unique to the firm, and was to be found in their products far outside Germany. These engines had bar frames in the American style, and the four cylinders usually all drove on to the same axle. The Bavarian Pacifics were the best known and perhaps the finest of the breed. They first appeared in 1908 and so effective were they that the Deutsche Reichsbahn found itself obliged to order more as late as 1930. The high-pressure cylinders were inside and steeply inclined, with conspicuous tail rod covers at the front of the engine. Their valves were outside the cylinders, more or less on the centre of the frames and alongside the valves of the low-pressure cylinders, which were outside and horizontal. Bar framing was essential to this design. The very individual appearance of the Bavarian Pacifics was accentuated by pointed smokebox doors and wedge-fronted cabs. Their performance was such that they were never eclipsed by the later German standard Pacifics.

Another celebrated German four-cylinder compound was the 2-12-0 class of the Württemberg state railway, designed by Eugen Kittel and built by the Esslingen works. Forty-four of these were built from 1917 onwards, and they ended their days some 40 years later on the Semmering line in Austria, where the solitary Gölsdorf compound 2-12-0 had long been a visitor. Among Gölsdorf's other four-cylinder compounds there were 4-4-2s, 2-6-2s and 2-6-4s, the last long associated with the working of the 'Orient Express'. From Norway in the north to Italy, Spain and Portugal in the south, from Belgium in the west to the Balkans in the east, four-cylinder compounds were to be found all over Europe in the days of steam, and the last are only now disappearing.

Almost as a footnote to the preceding paragraphs, a brief mention must be made of four-cylinder compounds with only two cranks. There were two main varieties of these, one European in origin and the other American. The European one was the tandem compound, with high- and low-pressure cylinders one behind the other, usually on the same piston rod. Mallet and de Glehn were both involved in their early development, and du Bousquet produced a class of 4-6-0 tank for the Ceinture railway of Paris which was certainly the most long-lasting class of such engines; but it was above all in the United States that the tandem compound became popular, for heavy freight service. The large reciprocating masses of these engines made them worse from the balancing point of view than two-cylinder simples, but for low speed heavy haulage this did not matter. Their vogue in the United States ended with the introduction of compound Mallet articulated engines in the first decade of the 20th century.

Tandem compounds might have receivers of substantial volume, and the European-built ones usually did, but the fact that the high- and low-pressure cylinders were in phase with one another made it possible to dispense with a receiver, the high-pressure exhaust passing directly into the expanding volume of the low-pressure cylinder as the low-pressure piston receded. This could in fact be done with a four-crank four-cylinder compound, where the high- and low-pressure cylinders adjacent on one side of the locomotive were phased at 180 degrees, but the practice was not common except in Norway.

The all-American two-crank four-cylinder compound was due to Samuel Vauclain, chief of the great Baldwin Locomotive Works. Vauclain put one cylinder on top of the other, their two piston rods driving a common crosshead. Usually the low-pressure cylinder was beneath, but if the driving wheels were small and the clearances limited, they might be the other way up. This, too, was a Woolf type (without a receiver) compound design, and the reciprocating masses could not be less than with the tandem arrangement. All the same, it did very well and was much employed in fast service, on 4-4-2 Atlantic and 4-6-2 Pacific locomotives until the increasing power requirements of American railways demanded bigger low-pressure cylinders than could be accommodated.

It was normal practice with compound locomotives to run with later cut-off on the low-pressure side. If this was not arranged for by giving the driver separate controls, it might be built into the valve gear design, so that a single control gave different events in high- and low-pressure engines. The low-pressure engine might

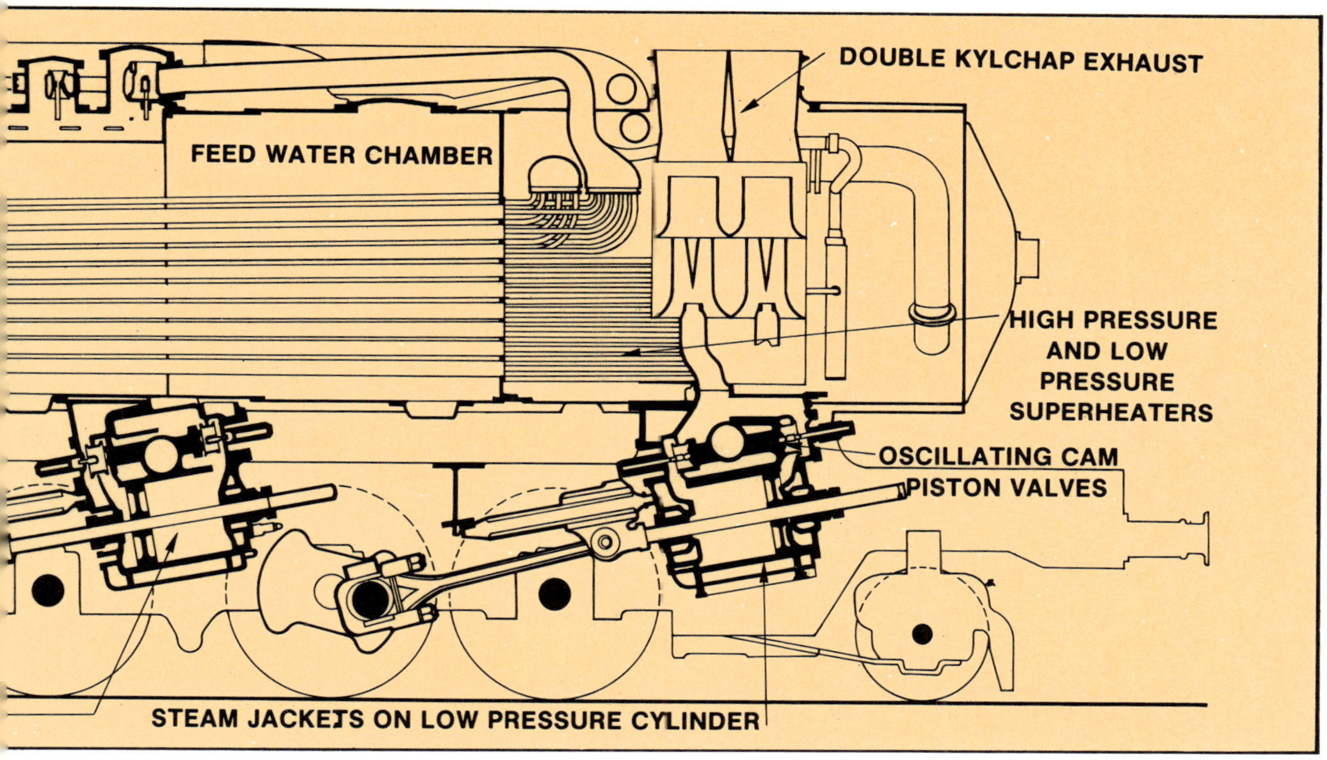

Longitudinal section of Chapelon's experimental six-cylinder compound freight locomotive. This engine was designed to develop high horsepower at low speed, while maintaining high thermal efficiency and fuel economy. Completed during the German occupation, this locomotive made one long journey with a heavy train, and was then stored in the South until after the war. It fully realized its designers' intentions, but electrification made quantity production unnecessary.

not have complete valve gear. The PLM, like Webb of the LNWR, at one time favoured a fixed LP cut-off—around 65 per cent. Later it adopted the arrangement devised by von Borries, in which both h.p. and l.p. valves are driven by the links and radius rods of the same Walschaerts gear on each side of the engine, but have individual combination levers so that the l.p. events can be a little later.

Chapelon's Pacifics and 4-8-0s had poppet valves (i.e. valves lifting off seatings and shutting down onto them again). There were separate valves for admission and exhaust. These were operated by normal Walschaerts gear, via oscillating cams. Rotary cam gears, resembling those fitted to internal combustion engines, have been tried on compound locomotives, but have never been successful. This is because they have not been able to provide variable timing of all four valve events, in the way already described as beneficial for steam locomotives.

EARLY ARTICULATION EXPERIMENTS

The problem of getting a powerful locomotive round sharp curves appeared very early in railway history. The constrictions of tunnels, bridges and so on mean that any increase in size has to be mainly along the length of the locomotive, and there comes a power and size which require that the locomotive should be flexible. The oldest solution to the problem is to work locomotives in multiple, and this is current practice with diesel and electric locomotives. However, these are so designed that they can be connected and worked by a single driver or crew, and that stage was never reached in the design of steam locomotives.

Coupling steam locomotives together with separate crews was seldom satisfactory, though it was common enough. There were problems of communication between crews, and indeed of personality, but the worst thing was the fact that the engines, being mostly two-cylinder simples with outside cylinders (as were the vast majority of freight locomotives all over the world) rode roughly at the best of times, and extremely badly when coupled together. This was partly because the couplings were not designed to link locomotives together, but to link locomotives to trains.

An early and quite good solution was to build back-to-back pairs of locomotives, joined at their cab ends with a specially rigid coupling and with the cabs intercommunicating. The engines had to have side tanks, and the coal capacity was limited, but one driver could control both, and one or two firemen could work alongside him. The Stephensons built such pairs in 1853 for the Giovi incline in Italy, with a gradient of 1 in 29, and they were also built for India (the Ghat inclines) and elsewhere. For a relatively short incline the limited fuel and water capacity did not matter.

The Semmering route in Austria combined 1 in 40 gradients with curves of 190 metre radius, and it was for working this line that a number of articulated locomotive types were devised and tried out in 1851. There were four at the trials: the 'Bavaria' built by Maffei, which resembled a 4-4-0, except that the bogie wheels were coupled together and driven off the driving axle by a chain, and the tender wheels were similarly coupled and driven; the 'Seraing' built by Cockerill of Liege in Belgium, which had two sets of four wheels each driven by two cylinders and forming a swivelling bogie under a double-ended boiler (in fact indistinguishable from what was later known as a Fairlie); the 'Wiener Neustadt', which was similar, except that it had a longer, single-ended boiler and was not completely symmetrical; and the 'Vindobona', which was a fairly straightforward eight-coupled engine.

None of these machines was satisfactory, but just as the 'Seraing' was the forerunner of the Fairlie, so the 'Wiener Neustadt' pointed to the later Meyer articulated locomotives. The 'Vindobona' in a sense won the day, because Haswell, the British-born engineer in charge, adopted a long boiler 0-8-0 type in 1855 and this type was long associated with the Semmering. However, the 'Bavaria' actually gave the best performance, which was not wholly surprising as it had fourteen driven wheels. The complications of its mechanism were too great, but on the other hand the cylinders did not have to swivel with the bogies, so flexible connections subject to boiler steam pressure were not needed, and this was an important advantage.

The 'Bavaria' soon lost its chains, but the idea of using the weight of the tender to increase adhesion persisted. It had actually been applied in 1842 by J. C. Verpilleux in France, who put a complete engine under

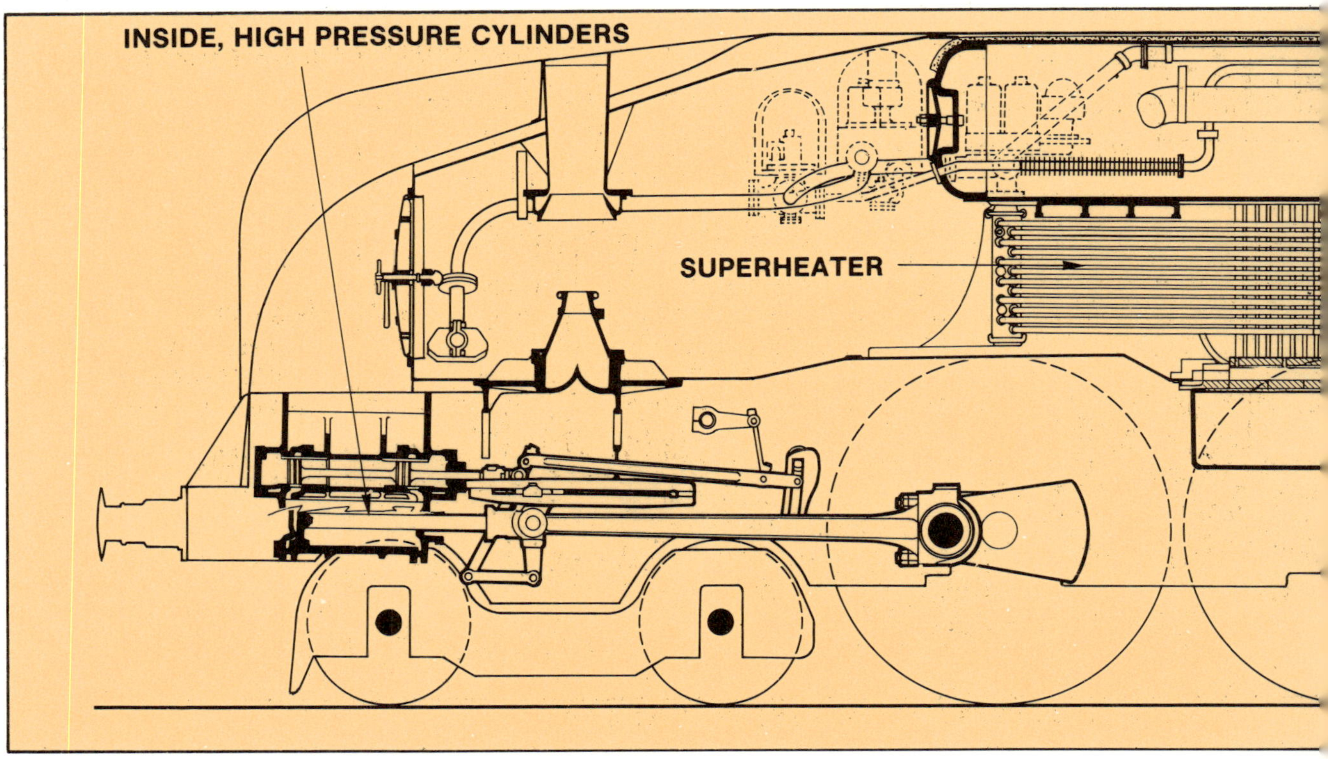

the tender, and this idea was successfully revived and applied to a number of freight locomotives by A. Sturrock on the Great Northern railway in England, in the 1860s. Sturrock 'steam tenders' were tried elsewhere as well. But the most important derivative of this idea was the Engerth locomotive, which was to become common on the Semmering and also in Switzerland and parts of France. Engerth locomotives were articulated to their tenders in such a way that the tender took the weight of the rear end of the boiler, which could thus have a large firebox. The tender wheels were geared to the driving wheels. However, in later Engerth-type engines this geared connection disappeared, and the engine part became eight-coupled. It thus became a form of long boiler locomotive, with the tender supporting the large overhang of a big firebox.

MALLET ARTICULATED LOCOMOTIVES

The most important type of articulated locomotive, built in the greatest numbers by a wide margin, was the Mallet, conceived in 1885 and first built in 1887 for the 60 cm (2 ft) gauge portable industrial systems developed by the Decauville concern. This first machine was a 12 tonne (12 ton) tank engine with eight wheels, but the largest locomotives of the type were eventually to approach a weight of 400 tonnes (400 tons) or over 500 with tender.

The Mallet was conceived as a compound, and that was fundamental to its rapid success. The locomotive frame supported the rear end of the boiler rigidly, and was provided with a set of driving wheels worked by the high-pressure cylinders. A second set of driving wheels, in a subframe pivoted to the main frames just ahead of the h.p. cylinders, was driven by the l.p. cylinders and carried the weight of the front of the boiler by means of a rubbing plate. This front section could, of course, swing laterally on curves, but did not take the boiler with it. The flexible steam connections were not at full boiler pressure, as they supplied, the low-pressure cylinders and also conveyed the exhaust to the base of the smokebox; this was the secret of its success.

In Europe the Mallet was mainly a tank locomotive used for mountainous routes with severe curvature, and in this rôle it was supreme. Most of them were built for narrow gauges and their weights generally fell within a range of 25 to 66 tonnes (25 to 65 tons). They were double four-coupled, or double six-coupled engines, and quite a few still operate on preserved railways in Europe, and on colonial lines not yet fully dieselized. Standard gauge examples were rare and usually heavier. Among the first was a famous double six-coupled tank for the Swiss Gotthard Railway with a weight of 86 tonnes (85 tons). It was built by Maffei in 1890. Still heavier was a unique double six-coupled tank weighing no less than 111 tonnes (110 tons), built in 1897 for working the severe incline at Liège in Belgium; and a series of superheated double eight-coupled tanks for Bavaria reached 134 tonnes (132 tons). There were also a few European tender Mallets, but the great development of the type came in the United States.

This began with a 0-6-6-0 tender locomotive built by the American Locomotive Company for the Baltimore and Ohio Railway in 1904. It was used as a banker and replaced two 2-8-0 locomotives of normal type. Its use reduced repair charges by some 40 per cent and its success was no mean tribute to the American builders of what was to them a completely new type of machine. After this, development was very rapid: larger and larger specimens were turned out, some of them for express service. These engines were so long that they were no longer suitable for sharp curves, on account of the 'throw over' of the boiler. Special provision had to be made to improve the riding qualities, and this involved lateral control springs and damping. These were not branch line locomotives any more; they operated enormous trains at good speeds on main lines.

Among goods engines special mention should be made of the triplex Mallet locomotives on the Erie and Virginian railroads. These had additional eight-coupled wheel sets under the tenders, provided with low-pressure cylinders, so they had two high-pressure and four low-pressure cylinders. The tender set exhausted directly to atmosphere via a vertical pipe at the rear of the whole ensemble, and did not assist in drawing up the fire. The first was built for the Erie in 1913, and weighed 393 tonnes (387 tons). On test it hauled over 16,000 tonnes (tons) over gently undulating tracks at a speed of 24 km (15 miles) per hour. The train was $2\frac{1}{2}$ km ($1\frac{1}{2}$ miles) long. Three more such engines, and a similar machine for the Virginian Railroad, were built in 1916, and they all

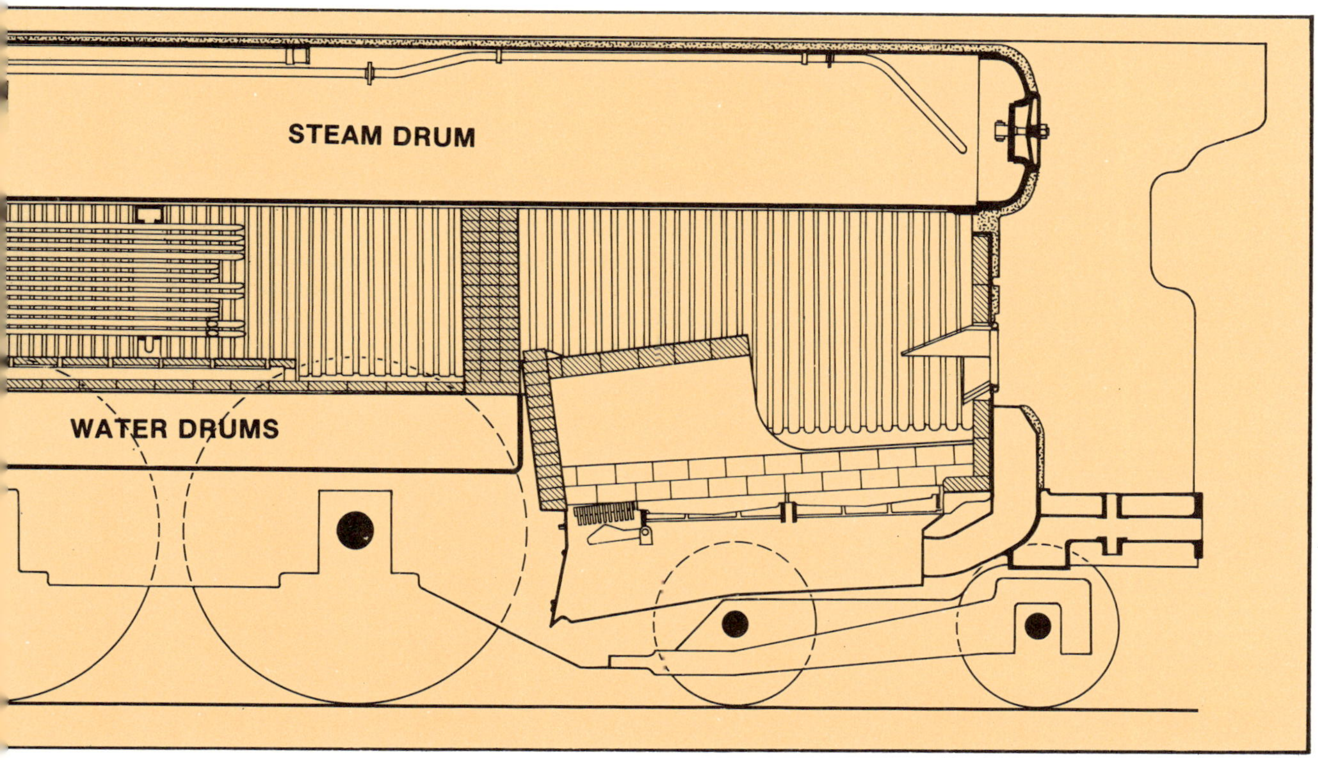

Sir Nigel Gresley, of the London and North Eastern Railway, was the most original thinker of British railways between the wars. The drawing is of his No. 10,000, an experimental four-cylinder compound express locomotive using a high-pressure marine-type water-tube boiler. The economic slump prevented its full development, and it was reboilered and rebuilt as a three-cylinder simple, in which form it lasted until near the end of steam on British railways, and was possibly the most powerful locomotive in the country.

spent the rest of their working lives banking freight trains on the steepest gradients of the two systems.

The six cylinders on the triplex locomotives were all the same size, giving a ratio of 2:1 between l.p. and h.p. swept cylinder volumes. This was rather low for the boiler pressure which was 15 Atm. or 1,448 kN per sq metre (210 lb per sq in), but it kept the size of the l.p. cylinders within reasonable bounds. With a four-cylinder Mallet, however, the limit of size of l.p. cylinders was soon to be reached, and this happened in 1918 on the Virginian Railroad, with the ten double ten-coupled Mallets of class AE. Though the pressure was the same as that of the triplex engines, the ratio between l.p. and h.p. cylinder volumes was 1:2.56, a more normal figure. The low-pressure cylinders required an internal diameter of 1.2 m (4 ft) and this clearly marked the limit of development for the four-cylinder compound Mallet. In fact, double eight-coupled compound Mallets were built up to 1952 (Norfolk and Western class Y6b) but these engines were used for relatively fast running on routes without gradients of exceptional severity. They handled coal trains of up to 7,000 tons at speeds up to 80 km (50 miles) per hour. For the highest powers, simple expansion had to be resorted to, and in the event simple expansion Mallets, with their smaller leading cylinders, proved freer running than the large compounds, simply because of the difficulty of getting the exhaust steam out of enormous cylinders without fitting equally enormous valves and exhaust passages. Moreover, this was a period when there was no longer any difficulty about high-pressure flexible steam connections, but when really efficient blast arrangements had not yet made their appearance. Chapelon would have eased the problem by fitting a triple Kylchap exhaust system, as he did on the SNCF 242A1 (which developed about the same maximum power as the Virginian locomotives). But in the absence of a Chapelon, simple expansion was clearly attractive for large Mallets, and soon came to be applied to somewhat smaller ones as well. So, roughly speaking, the compound Mallet began to give way to the simple one, in the United States, from about 1920, though the last large Mallets built were the compound Y6bs of 1952. These engines, at 270 tonnes (266 tons) without tender, were not in the most powerful class, however: the biggest simple Mallets, the Union Pacific 'Big Boys', weighed 357 tonnes (351 tons).

GARRATT ARTICULATED LOCOMOTIVES

It is difficult to estimate the total number of Mallet articulated locomotives built, but 8,000 would be a fairly intelligent guess. Of the next most important articulated type, the Garratt, probably less than 2,000 were built, because this type was almost exclusively associated with the countries of the former British Empire, and was almost only built in England, and by a single firm: Beyer, Peacock and Co. of Manchester. This firm was an active exporter of locomotives to Europe before 1914, and to remoter markets always, but became best known for the Beyer-Garratt locomotive, as it came to be called.

All Garratt locomotives are tank locomotives. They are in three parts: two engine units, which may have carrying as well as driving wheels and which support the fuel and water supplies; and a central boiler unit which is pivoted on the engine units and transfers its weight to them. There are no wheels under the boiler or the grate and ashpan, and this is a major virtue of the design, because there is no great restriction on the diameter of the boiler barrel (and hence on the cross section of the fire tubes) and nothing to prevent the fitting of a deep, wide firebox with a capacious ashpan. The same sort of freedom applies to the engine units. They can be provided with such carrying wheels as are needed to fit them for fast running in either direction and can have as many as four cylinders each if required. However, the vast majority of Garratts have two cylinders at each end.

The first engines of the type were built for Tasmania, and were compounds, one engine unit having the high-pressure and one the low-pressure cylinders. One of these little engines (they only weighed 33.5 tonnes (33 tons)) is now preserved in the National Railway Museum at York in England. It was built in 1909. The most influential of Garratt designs was the solitary class GA built for service on the South African 1.06 m (3 ft 6 in) gauge. Built in 1920, this 136 tonne (134 ton) 2-6-0 + 0-6-2 engine was subjected to comparative tests against a British-built Mallet type weighing 183 tonnes (180 tons) with tender, which had hitherto been the preferred heavy haulage locomotive of the system. The Garratt took the same loads, but ran faster and burned less fuel. It also rode much better. This test proved decisive: the large and important South African system went over to the Garratt type and its example was copied extensively.

Very many Garratts were small branch line machines, doing the sort of work for which the Mallet tanks had also proved so suitable, but there were some notable large machines. Perhaps the most impressive were the thirty 4-6-2+2-6-4 engines built for general express service on the Algerian railways. They were approximately 30 m (100 ft) long and had driving wheels of 1.8 m (5 ft 11 in) diameter which made them suitable for speeds up to 136 km (85 miles) per hour.

The largest Garratt of all was built for Russia and weighed 266 tonnes (262 tons), but this locomotive was almost equalled by one of the last designs, the New South Wales Railways AD60 class 4-8-4+4-8-4, which weighed 264 tonnes (260 tons). Of 42 ordered, some were kept in pieces to provide spares, because by 1952 the use of diesels in Australia was being strongly advocated. Now, all the Australian Garratts are out of service, but happily many remain in Africa, and though these operate on 1 m or 1.06 m (3 ft 3 in or 3 ft 6 in) gauge, they are not much smaller than most on larger gauges.

SUPERHEATING

Superheating, although it was applied successfully and in its modern form to locomotives in the first decade of the present century, was in fact one of the last major changes brought to the steam locomotive. A practical system, able to withstand the vibrations and sharp temperature changes inseparable from its application to railways, was devised by Dr Wilhelm Schmidt at the turn of the century. Details of its construction, and of other less successful types of locomotive superheater, will be found in the section of this work which deals with boilers.

Superheating means raising the temperature of the steam after it has left that part of the boiler where it is in contact with water. This changes its characteristics from those described in the first paragraph of this book. It has more energy, and cannot condense until it has lost all its superheat. When expanded within a cylinder its pressure falls less rapidly and so it gives more power than ordinary or 'saturated' steam, and it is able to sustain some heat losses (to the walls of cylinders, to steam passages, and to pistons) without significant pressure drop. In fact, locomotives without superheaters often worked with slight but useful superheat, because reduction of pressure between boiler and cylinders by, for instance, a partly opened regulator, will slightly superheat the steam, by expanding it without its doing work.

The Schmidt superheater was first used on the Prussian State Railways, where one of the senior area engineers, Robert Garbe, became very much the disciple and the apostle of Schmidt. The railway was mainly run by saturated steam compounds with two cylinders, plus a few four-cylinder 4-4-2s, all these being designed by von Borries. Garbe produced a class of simple, superheated 4-4-0 which was the equal of the 4-4-2, though its axle loading was higher. This type, the S6, was the second version of the superheated 4-4-0 and was a great success. A superheated 0-8-0, the G8, was also a success, and from then on the Prussian system went wholly over to two-cylinder simple superheated engines except for the heaviest express service, for which four-cylinder compound 4-4-2s (saturated) and 4-6-0s (superheated) continued to prove superior. Among these Prussian engines were two outstanding machines, the mixed traffic 4-6-0 of class P8 and the heavy goods locomotives of class G8[1]. Both of these types were conceived by Garbe, but brought to their well-known reliability and good performance by another Prussian railway officer, Hinrich Lübken.

These Prussian engines were extremely influential, not least because they, with others of the same general type such as 0-10-0s, and 4-6-4, 0-10-0 and 2-8-2 tanks, were spread wide over Europe as reparations after the First World War.

Superheating, then, ranks as the main German contribution to locomotive engineering. It was swiftly taken up everywhere else. In Britain, Schmidt superheating appeared on a Lancashire and Yorkshire goods engine in 1907, but it made its greatest impact when it was applied to a 4-4-2 express tank engine of the London, Brighton and South Coast Railway in 1908. This locomotive was run in comparison with an excellent LNWR saturated 4-4-0, and not only performed the work more easily but astonished everybody by covering some 144 km (90 miles) with a 254 tonne (250 ton) train, without taking water, in spite of a tank capacity of only 9 cu. metres (2,000 gals). It also managed round trips of 425 km (264 miles on a single bunkerful of coal—about 3¼ tonnes (tons).

After this, superheating was adopted on almost all railways for express work, and for the heaviest freight locomotives. Its use spread more slowly to other types, and many older engines were never converted, but eventually only small shunting locomotives were not built with superheaters. This was the pattern in almost all countries.

At first, the belief was common that superheating should not be regarded as an additional technique to be brought to existing designs to increase their efficiency and power output. It was seen rather as a way of doing what was already being done more cheaply. It is hard to see why, in the boom years for railways before the First World War, the emphasis on economy should have been so great, but it was. The result was that many designers saw superheating as a way of reducing boiler maintenance and even first cost, by reducing the working pressure, enlarging the cylinders and expecting the superheat to fill the enlarged cylinders just as effectively. Some locomotives actually had their performances reduced by this procedure. The higher temperature also gave rise to lubrication troubles, which in the case of Garbe's original P8 4-6-0s, and some others, caused seizure and even breakage of pistons. Slide valves, unless of the balanced type (which had much of the pressure on their backs relieved) were generally replaced by piston valves, but the design of piston ring advocated by Schmidt was very prone to leakage after a little use, and the unavailability of suitable oils also caused a great deal of carbonization of steam passages, and sticking of piston and valve rings. All the same, when these troubles were overcome, and boiler pressures were restored, superheating transformed the performance of very many locomotive types.

The view was taken in many countries that superheating made compounding unnecessary, because it produced the needed economy anyway. This might have been true if a specific value could be attached to the needed economy. It was proved over and over again that superheating supplemented the economy of compounding and that the best of superheated compounds was far more economical than the best of superheated simples, but for many railways the simple was good enough, and compounding became far less widespread after the advent of superheating.

OTHER DEVELOPMENTS

Locomotive types were transformed, between the wars and after, by the application of improved blast pipe systems, notably the Kylchap (single or double), the Giesl oblong ejector, and the Lemaître multiple jet exhaust. These devices are described in the section relating to the draughting of boilers, and here it will suffice to indicate that the reduction in the back pressure on the pistons produced notable increases in power and speed. The fitting of a double Kylchap to Sir Nigel Gresley's streamlined Pacific 'Mallard' was undoubtedly the reason for this locomotive achieving the world speed record for steam locomotives—202 km (126 miles) per hour in 1938. The record for an identical machine with an ordinary single chimney stood at 180 km (113 miles) per hour.

In all the development so far recorded, the fundamentals of the 'Rocket' have been preserved: fire tube boiler and direct drive from reciprocating pistons to the driving wheels. But a little space must be accorded to the many attempts to get away from one or other—or even both—of these features. There were indeed numerous attempts, but success only attended a very few, and those were concerned with locomotives having turbine drive associated with normal locomotive boilers, and without condensing apparatus.

In 1930 the Grangesberg-Oxelösund Railway in Sweden received the first of its turbine-driven 2-8-0s, and following its clear success over a number of years, it obtained two more in 1936. Conditions on the railway favoured turbines, in that nearly constant output was required from the locomotives of freight trains over a distance of some 300 km (186 miles). Intrinsically, the turbine offered better efficiency and greater expansion rates than the three-cylinder simple expansion locomotives of the railway, and was also able to support higher superheat. It was placed at the front of the engines, and was geared to a cross shaft carrying flycranks which drove the wheels via coupling rods. The engines weighed 84 tonnes (83 tons) without tender, had a maximum speed of 60 km (38 miles) per hour and developed up to 1,200 horsepower at the drawbar. They remained in service until electrification in 1954, showing an economy of about 10 per cent overall as compared with the simple expansion reciprocating engines, and achieving a total distance of 3,200,000 km (2,000,000 miles) for the three engines. These little-known machines were a triumph for their builders, Nydqvist and Holm, and their record was far superior to that of other non-Stephenson locomotives. Apart from the very large superheaters, their boilers were of normal locomotive type and worked at 13 Atm. or 1,275 kN per sq metre (185 lb per sq in) pressure.

The other success—or at least, non-failure—was the 4-6-2 express engine of the London, Midland and Scottish Railway, no. 6202, built in 1935. This was the third LMS Pacific and was in many respects the same as the rest of the 'Princess' class. However, in place of four cylinders and simple expansion, no. 6202 had a turbine set (there was a small impulse turbine for reverse working without a train) with internal gearing to the leading coupled axle. The main turbine was rated at 2,000 horsepower and had some 30 rows of blading; it exhausted to atmosphere through a double blast pipe and chimney. Because of the complete absence of hammer-blow, an axle loading of 24 tonnes (tons) was allowed—$1\frac{1}{2}$ tonnes (tons) more than the reciprocating locomotives. Weight without tender was 110 tonnes (109 tons) and the boiler pressure was 17.6 Atm. or 1,723 kN per sq metre (250 lb per sq in), as on the rest of the LMS Pacifics.

This locomotive underwent a number of detail modifications, mainly concerning the drive to the wheels, but was in service rather intermittently until 1952, when the main gearing needed replacement. It was then rebuilt as a reciprocating engine. Its performance was perfectly satisfactory, but its coal consumption was always slightly higher than that of its orthodox classmates. This was no doubt because it was required to work at the normally fluctuating power outputs required of express locomotives on routes with considerable gradients.

It is worth noting that these turbine engines had no difficulty in steaming, in spite of the lack of pulsation in the exhaust.

There were about a dozen other locomotives built with direct turbine drive, in France, Germany, Italy, Switzerland, Britain and elsewhere, but none was a success. Mostly they were provided with condensers, but railway practice has shown that condensing is not worthwhile as a means of increasing power or efficiency, because of the weight and size penalties. It is only justified by the need to conserve water in really difficult territory, and has been successfully applied to reciprocating locomotives, notably in South Africa, for that reason only. Condensing tenders were developed by the German firm of Henschel at Kassel, and their type was used in South Africa.

Four large turbo-electric locomotives were built in the United States between 1938 and 1952. These were in effect travelling power stations with electric locomotives beneath, weighing from 254 tonnes (250 tons) to as much as 376 tonnes (370 tons) in one case. They worked well enough, but their existence did nothing to prevent the wholesale adoption of diesel-electric power in the last 30 years. They recall the efforts of J. J. Heilmann in the 1890s to produce a satisfactory locomotive having a reciprocating steam plant generating electricity for traction motors on the axles. His first locomotive was quite celebrated and bore the name 'Fusée', which is French for 'Rocket'—a bold comparison indeed. He subsequently had two of his engines on the Western Railway of France, where they showed some useful qualities, owing to the adhesion provided by their two eight-wheeled bogies carrying a total of 125 tonnes (124 tons). Their drawbar horsepower reached about 1,000, and their maximum speed about 120 km (75 miles) per hour, but no more were built. Their power-to-weight ratio was low and their cost high, but they represented a considerable feat of electrical engineering for their time.

Some attempts have been made at using small high-speed steam engines geared down to the axles. The real successes here have been the American logging locomotives carried on bogies to which a mechanical drive has been brought from a high-speed engine fixed on the main frame. The most numerous type, and the one built in the largest sizes, has been that devised by Ephraim Shay, in which there is a three-cylinder vertical engine set beside a boiler somewhat offset from the locomotive centre line. For a very slow haul on irregular tracks, with high tractive effort, this type proved virtually unbeatable, though the ubiquitous diesel has supplanted it now. Variations of the theme were the Climax and the Heisler, one with a two-cylinder engine arranged across the frames under the boiler, the crankshaft being parallel with the axles, and the other having a vee twin in a similar position, with the crank axle running fore and aft.

Also successful were the Sentinel shunting engines, mostly with only four wheels, which had a special type of boiler and operated at higher pressures than normal in locomotives. This firm made four 2-4-2 passenger locomotives for the Egyptian State Railways, with individual axle drive to the two driving axles. They had normal locomotive boilers and weighed 57 tons without their tenders. As they were assembled from well-tried components, they gave little trouble, but no further examples were built. Attempts at larger locomotives, such as the Paget engine in England, the 'Leader' of the Southern Railway, or the French and German experimental machines, have all been unrewarding.

From the foregoing it will appear that fundamental departure from the Stephenson layout has never yielded much in the way of dividends. Even the Grangesberg-Oxelösund engines produced no greater economy—if as much—than could have been obtained from a super-

H. W. Garratt

A railway breakdown crane, by Cravens Ltd. This type of crane was built in various sizes, of up to 75 tonnes lifting capacity. The diagram shows the various gear trains which enabled the steam engine to raise and lower the hook, alter the elevation of the jib, slew the crane round, and move it along the rails.

heated compound locomotive on the same service. Attempts at using extremely high boiler pressures have also foundered because the Stephenson boiler has been abandoned or altered too greatly. In this class we must cite LNER no. 10,000, the 'Hush Hush' engine, a four-cylinder compound with a water-tube boiler carrying a pressure of 32 Atm. or 3,102 kN per sq metre (450 lb per sq in). It nearly succeeded, but the boiler casing could not be kept airtight as the locomotive sped along the well-maintained tracks of the British east coast main line. Even attempts at water-tube fireboxes, though apparently successful, have never led to a general adoption of them. And ultra high pressures, such as the 60 atmospheres found in the closed circuit part of Schmidt–Henschel compound locomotives, have all too often led to accidents, such as befell the LMS example, the 4-6-0 'Fury'.

Further and fuller accounts of turbines and water tube boilers will be found in the marine engineering and boiler sections of this book.

RAIL-MOUNTED CRANES

By way of a footnote to the section of this book devoted to railways, some account should be given of movable steam cranes. Virtually all steam cranes moved on rails: it was the obvious way of making them mobile in

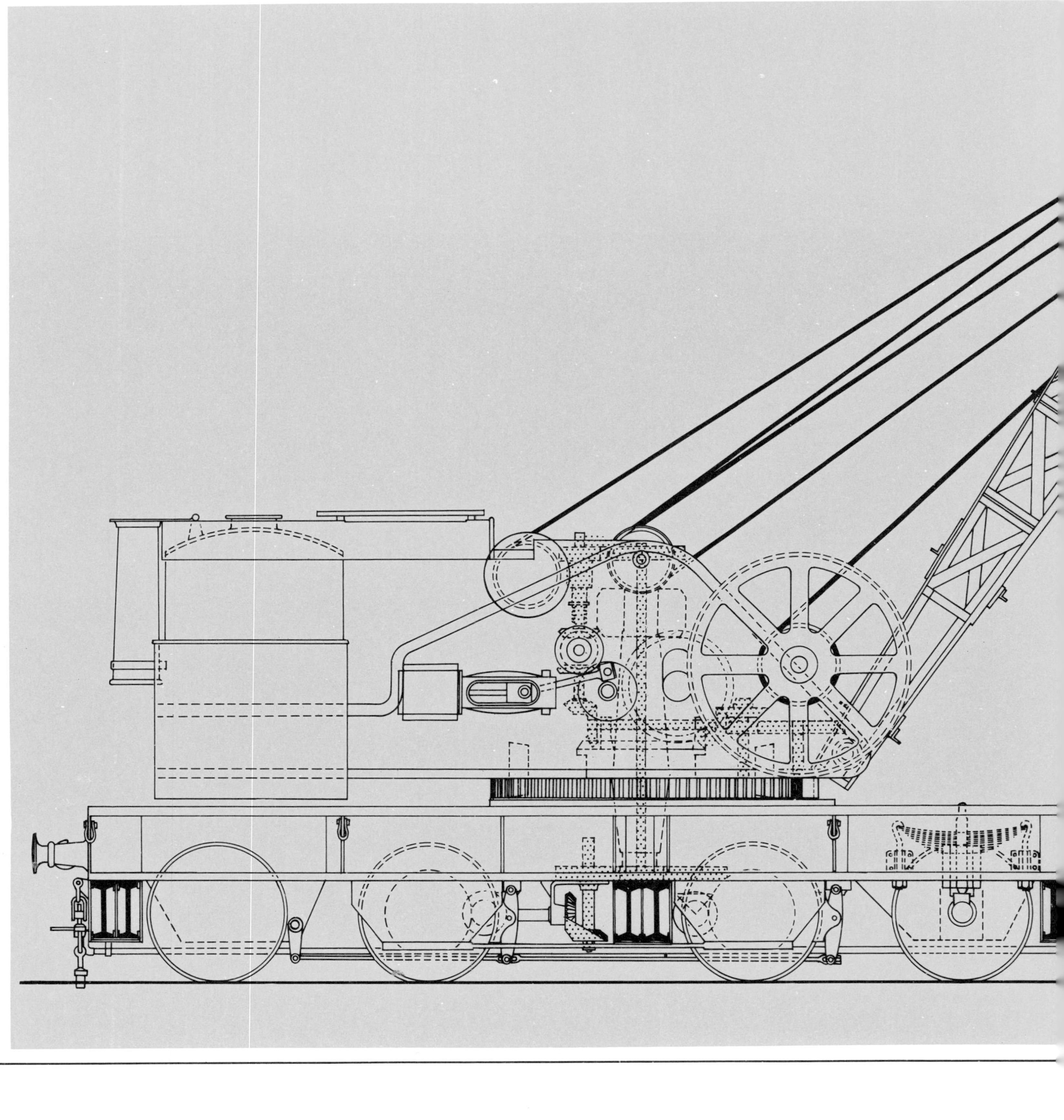

factories or on docksides where their movements were predictable. The railways themselves made, and still make, use of steam cranes, the last survivors being the largest: the breakdown cranes, some of which were capable of lifting up to 100 tonnes (tons). In factories, travelling cranes are often provided overhead, in the form of beams running on rails set against the side walls of the building, the actual hoist being able to run across the beam to any required position. Nowadays these are all electrically operated, but they were once worked by steam.

Practically all steam cranes were of the same pattern, though their size varied greatly. They consisted of a platform able to swivel on its base, with a vertical boiler on one end, roughly counterbalancing the jib and its load, which were attached at the other end. In between was the steam engine with its chain or cable drums, gearing, brakes and clutches. It was usual to provide only one engine, with two cylinders and little if any lap on the valves, so that it could start with full torque at any position of the cranks. By manipulating the clutches and brakes, the crane driver could raise or lower the jib, or the load, and he could also swivel the crane on its base, which incorporated a large toothed ring, and move it along its tracks. An overhead factory crane had no jib and did not swivel, but it still carried its boiler about with it.

In its simplest form, this machine was known as a locomotive crane (quite a different thing from a crane locomotive, referred to later). It was mounted on a low, four-wheeled platform without springs, and had a rudimentary enclosure, often of corrugated iron, for the boiler, or just a roof over it and the driver. Its lifting capacity ranged from about 3 to perhaps 15 tonnes (tons), the limit being set by the resistance to overturning, and the machine built to be safe within that limit. The larger sizes usually had substantial pivot posts (through which the drive to the wheels had to be taken) but the smaller ones simply sat on their bearing rings. For dockside work, much larger and more powerful versions were made, often set high on portal frames straddling railway tracks. They were literally more powerful, not just capable of lifting more but also of doing it faster, and therefore more economically.

The larger cranes used by the railways themselves were of two main kinds: track cranes and breakdown cranes. Both were used for all sorts of civil engineering work as well as their nominal jobs of lifting or laying lengths of track, and lifting derailed locomotives and rolling stock. The track crane was much the lighter, and was more or less the basic steam crane mounted on two four-wheeled bogies. It could handle about 15 tonnes (tons). The breakdown crane was altogether more massive, and might be mounted on something like a steam locomotive's tender chassis, with rigid frames, and often with the addition of a bogie at one end. A very large one might have two six- or eight-wheeled bogies built like tender underframes. These cranes travelled to the site of operations in special trains, with the end of the jib resting on a 'match truck'. When working, they were usually stabilized by outriggers slid out sideways from housings in the frame, and jacked or packed at their ends to prevent overturning. Their engineering was of a high standard throughout, and their driving arrangements complex and well provided with safety devices. Some of these magnificent machines are still in service, and one is preserved in the National Railway Museum at York.

The crane locomotive, of which a few survive in service, was simply a normal tank locomotive with a crane mounted on it. The crane was hand-worked at first, but later it was a steam crane with its own engine, though it took steam from the locomotive boiler. The travelling function was performed by the locomotive, while swivelling, luffing and lifting were performed by the crane engine. This crane might be mounted in the bunker behind the locomotive cab, but it was also commonly mounted in front of the cab, over the firebox, and when travelling to work the jib was lowered to embrace the chimney. An interesting variation had the crane mounted upon the chimney, which served as the pivot post, being secured to the boiler by a substantial casting. These crane locomotives were usually to be seen around railway works yards, but a number were used on industrial railways, the engine doubling as a shunter when the crane was not needed. They were among the more amusing oddities of the steam age, and were usually favourites with their operators, who gave them unofficial names, which were painted on their sides.

The Savannah *(1819), PS* Comet *(1812), and steam tug* Charlotte Dundas *(1801).*

Steam at Sea

THE FIRST STEAMBOATS

All the first successful steamboats were paddle steamers of one sort or another, though reaction propulsion was experimented with and shown to offer possibilities. It is therefore worth recording that paddle wheels were first used with manpower, probably in China in the 8th century. Both side and stern paddles appear to have been fairly common in Chinese warships in the 12th century, because the paddles could be protected as oars could not. Paddle wheels also seem to have been used on inland waterways, which abound in China, to protect canal banks; and, in conjunction with tilt-hammers at the bows, to clear ice from rivers. These were operated by treadmills, although smaller craft may have had handles.

A sort of proto-paddle craft existed in the west a thousand years or more before the harnessing of steam power. This was the floating barge moored in a river with a paddle wheel on each side (or sometimes two on the same side, or only one). The wheels were turned by the current and the vessel was actually a floating watermill. In tidal water it might rise and fall with the tide, and turn either way with the ebb and flow. This device could easily inspire the idea of a paddle-wheel-propelled craft, and there is an account of a Roman vessel operated by paddle wheels turned by oxen, but this was first recorded as hearsay several hundred years after the supposed event.

One idea which arose from the tethered floating watermills was that of propelling a vessel upstream by using millwheels driven by the current to haul on ropes tethered well up the river. This idea appeared at least as early as the 15th century, but was reinvented in the 18th. The use of treadmills to work paddle wheels was tried by the English navy in the 17th century, to power small vessels used for towing other craft within a harbour: tugboats, in fact. In this case the treadmill was actually driven by horses; when Savery, the steam pioneer, revived the idea yet again shortly afterwards, the treadmill took the form of a capstan with manpower.

Watt side-lever engine. Cams for the expansion gear can be identified on the main shaft.

Denis Papin seems to have made a similar experiment, but he also proposed a perfectly feasible design for a small steamboat. He had devised a piston-in-cylinder steam engine in which the water was actually boiled beneath the piston, to make it rise, and the steam was then condensed by cooling, to make the piston fall with the force of atmospheric pressure. On a small scale, as a scientific demonstration, this worked admirably and really operated on the principle of the Newcomen engine which followed some 20 years later. To this little machine—in a somewhat enlarged form—Papin proposed the addition of rods and ratchets which would serve to drive paddle wheels. This is very nearly what Symington successfully did almost a century later.

The natural growth of ideas led directly to the paddle steamer. Had it not been for the long familiarity of the waterwheel, however, it might well have been reaction (or jet) propulsion that dominated the first phase of the practical history of the steamboat. In 1661 there was a patent for ship propulsion by forcing water out backwards with the aid of bellows. In 1729, one John Allen obtained a patent for the same idea, but this time the water was to be pumped by two Newcomen engines (the obvious way to do it, as the Newcomen engine was invented for pumping water). The idea was never tried out, though, perhaps because Allen could not see a way of making Newcomen engines without heavy stone or brick walls, which would have sunk the ship. The scientist Daniel Bernoulli also proposed reaction propulsion, but contrived a design of screw propeller as well.

However, these early proposals serve only to illustrate the climate of opinion, which might be summed up as intrigued by the idea of using steam engines to drive vessels, or indeed to do anything hitherto performed by wind, water or muscle power. Nevertheless, there was no real demand for marine steam propulsion, as there had been for steam pumping machinery to drain mines. Neither was there the prospect which made the steam locomotive attractive at the beginning of the 19th century: that of dispensing with large teams of horses, or short-lived and cumbersome haulage cables. On the contrary, the wind, if capricious, was free, while the space required by machinery could only be gained at the expense of cargo. So, whereas the Newcomen engine was immediately welcomed and copied, the steamboat required dogged persistence to gain acceptance, and the pioneers tended to die in poverty, and even bitterness.

Like the locomotive, the steamboat was brought to maturity mainly in Britain, France, and the United States. France was first: in 1773 a boat with a Newcomen engine was actually built on the Seine in Paris. It never operated, because it was found sunk at its moorings before it could be tried. Whether this was a result of Luddite-type sabotage or the sheer weight of the brick engine and boiler housing is not known. However, the vessel had impressed Claude, Marquis de Jouffroy d'Abbans, who devised a steamboat, the propulsion system of which was based on the paddling of a duck. This was tried on the river Doubs in 1778 but was not a success. Persistence, however, was clearly a characteristic of Jouffroy, and he achieved success in 1783 on the river Saône with a vessel called the *Pyroscaphe.* This steamed upstream and thereby, for the first time in history, demonstrated practical steam marine propulsion effectively.

The *Pyroscaphe* had paddle wheels operated by a single, double-acting cylinder and a ratchet mechanism. The double action is of interest, as at this date proper double action had not appeared in a steam engine, though the principle was being developed by Watt. It seems likely that Jouffroy's inspiration was the artillery tractor of Cugnot (designed in 1770 and tested in Paris) which had two single-acting cylinders maintained in contrary motion by a cross lever and operating the front wheel by ratchets. In a vessel, a single horizontal cylinder with double action would keep the weight lower in the hull, and would save weight as well. Jouffroy's vessel was quite large: 42.5 m (140 ft) long, 4.5 m (15 ft) wide in the beam and with a draught of about 1 m (3 ft).

This promising start was frustrated by the French Revolution. Whatever plans Jouffroy may have had (and though his feat had been noticed and support seemed likely, it was still very much up to him personally to pursue his goal) the political state of the country was not such as to favour experiments, and before long Jouffroy, as an aristocrat, had to make himself inconspicuous and he may indeed have left France. The initiative in steamboat development passed to North America, where experiments were scarcely behind those in France, and Jouffroy made no further efforts until his old age, in 1816. These were not successful, however, and he then retired from the field finally. He died of cholera in 1832, when he was 81 years old, but left behind a model of his *Pyroscaphe.*

JAMES RUMSEY AND JOHN FITCH

Across the Atlantic the two pioneers, and bitter rivals, were James Rumsey and John Fitch. Rumsey made a model of a proposed vessel, and this was seen by George Washington and recorded in his diary for September 6th 1784. At this stage Rumsey seems to have planned some sort of mechanical punt, with pole drive, but three years later he was able to demonstrate a boat with jet propulsion. The machinery consisted of a boiler and a reciprocating pump which forced the water jet out at the stern. There was no mention of a condenser. Cugnot had not used one, and neither had Jouffroy; and though these primitive engines must have used a boiler pressure not exceeding 1 atmosphere, they clearly anticipated Trevithick in an important respect: they exhausted directly to the atmosphere.

Rumsey certainly made his mark, and a Rumseian Society was formed to further his ideas. Benjamin Franklin was one of its supporters. Late in 1788 Rumsey took out British patents and in 1792 he operated a steamboat on the Thames. Then all came to a sudden end with his unexpected death late that year.

John Fitch was born in the same year as Rumsey (1743) but he lived six years longer. He achieved no long-term success either, but at one stage it looked as if he and the steamboat were set for a prosperous future in the United States, and possibly in France also. On the strength of a model, a description and drawings of his ideas he obtained the exclusive right to operate steamboats in New Jersey in 1786, and shortly afterwards obtained the same rights in Delaware, New York, Virginia and Pennsylvania. He built several vessels of increasing size, the first of which had 12 oars arranged upright and driven from a steam engine by an elaborate system of gears and chains. In 1790 he actually operated a regular service for several months on the Delaware river, though with an improved form of vessel. His ideas progressed as far as screw propulsion, with which he was experimenting unsuccessfully in the middle 1790s. He obtained a French patent late in 1791, but the timing could hardly have been worse, because not only was the signature of the King soon to become worthless, but also, as with Jouffroy, people were simply not interested in such matters in the early stages of the French Revolution. Fitch went to France in 1793 with some expectations, but had to work his passage home. He died, a disappointed man, in 1798.

Though neither of these men achieved commercial success, both provided effective demonstrations. This is all that Trevithick did for the steam locomotive, but it entitled him to be named as its inventor. Trevithick did much more, before and after his involvement with

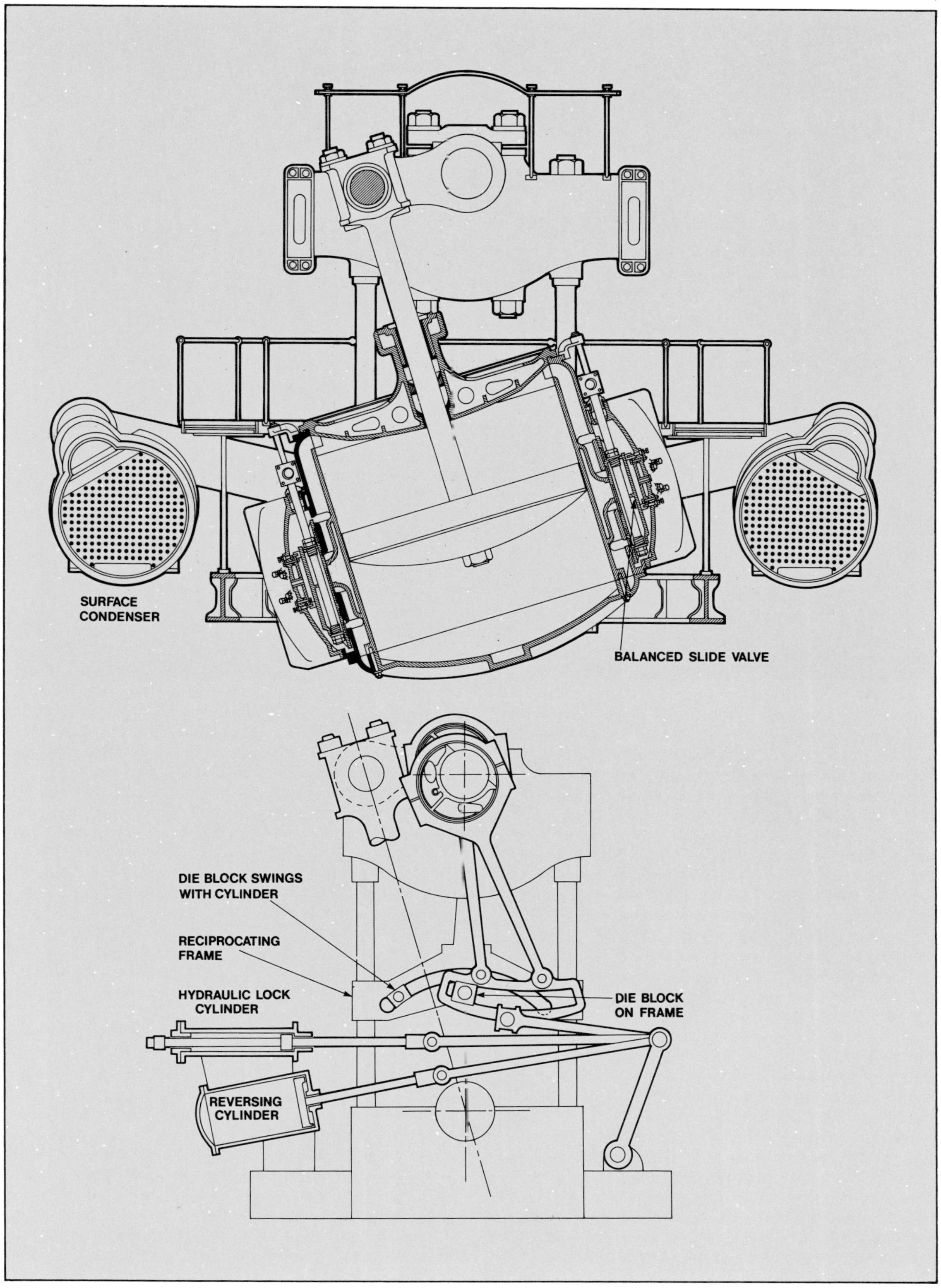

An oscillating paddle engine, from a cross-channel steamer of the 1880s. The cylinder has two slide valves, and is flanked by two condensers. The lower diagram shows how the Stephenson link motion is arranged to work the valves in spite of the swinging of the cylinder. The die block in the link is fixed to a frame which moves up and down. A curved slot in this frame holds a second die block which is linked to the valves swinging with the cylinder.

transport on rails, and that makes his position secure, but Rumsey and Fitch were only associated with the steamboat, which may account for their undeserved obscurity. They, with Jouffroy, deserve to be ranked as the practical inventors of steam propulsion on water. All three showed the qualities of ingenuity and tenacity, and possessed the imagination required to conceive of water travel as a regular means of transport, not an irregular one at the mercy of wind, current and tide. They were unlucky, and they were also in the wrong places. The Americans, especially, were attempting something quite beyond the capability of the engineering industry as it was in the United States at that time, and though France was in a rather better position in that respect, the revolution disabled that country.

The contemporary British contribution to this early history did not suffer from such disadvantages. William Symington (1763–1831) fitted an engine into one of Patrick Miller's double-hulled boats in 1788, and it is said to have propelled it at a speed of 8 km (5 miles) per

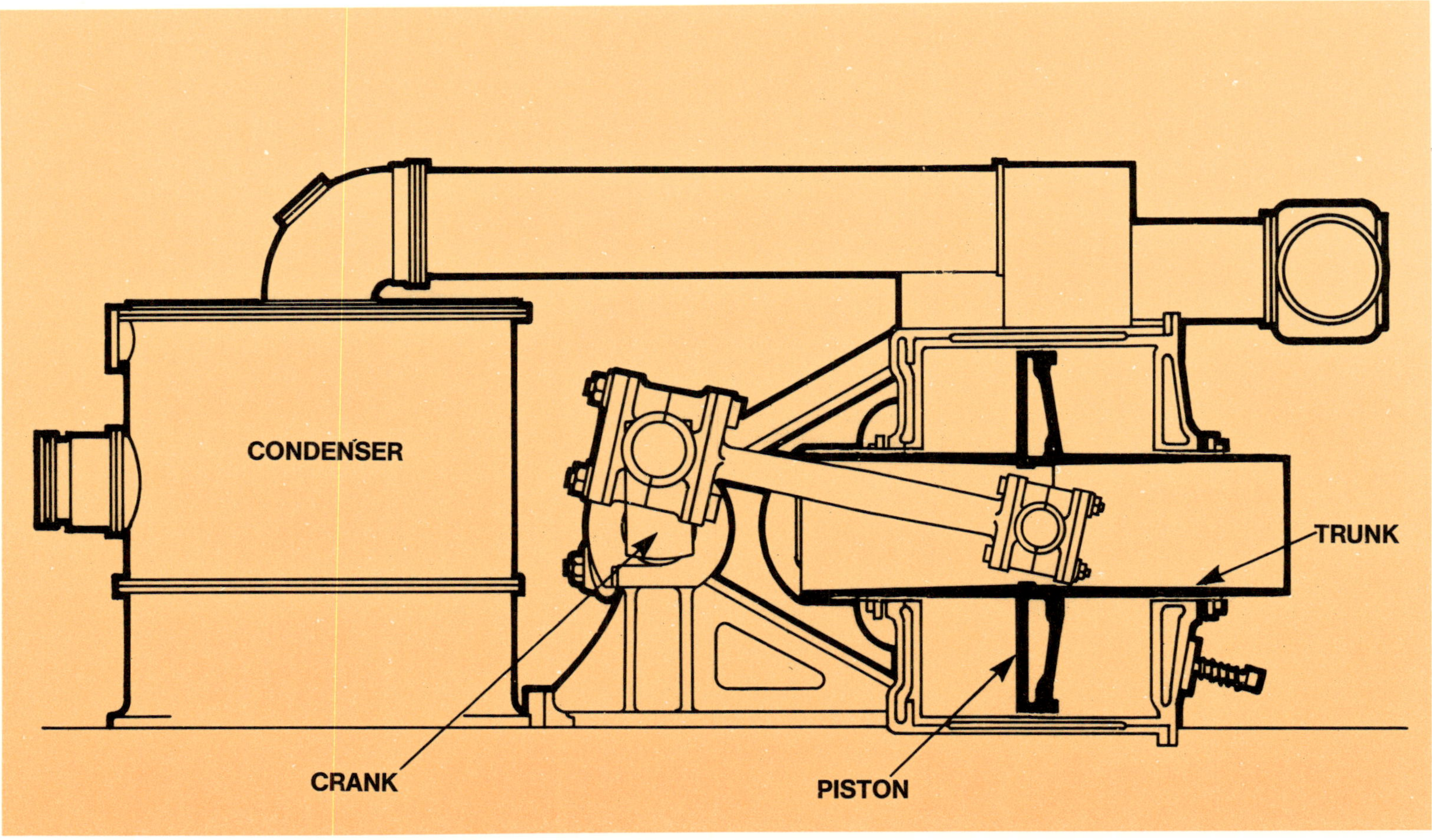

Layout of Penn's trunk engine, showing the small end of the connecting rod within the trunk, or tubular piston rod.

hour. Patrick Miller was an Edinburgh banker who believed that ships should have two or even three hulls, side by side. He was thinking of sailing ships, but he also tried man-operated capstans driving paddle wheels, as Savery had advocated nearly a century before. The first steamboat that Symington engined for him had paddle wheels between the hulls. The boiler was in one hull, and the engine in the other. This engine was almost what Papin had proposed: it had two open-topped cylinders working on the Newcomen principle, by the creation of a vacuum beneath the pistons. Their movement was opposed to provide the equivalent of double action, and they operated two paddle wheels via ratchets and chains. Unlike the Newcomen engines, this machine had a separate condenser, and this infringed Watt's patent; moreover, Messrs. Boulton and Watt declined to interest themselves in steam navigation, adopting their usual stance of opposing innovation. The prospects were therefore not bright.

This engine remained in its original state until 1853, long after its brief use in a boat had ceased. It was eventually dismantled, but in view of its historic importance it was rescued from destruction by Bennet Woodcroft, one of the inventors of the screw propeller, who was also one of the first enthusiasts for the preservation of engineering relics and was long associated with the Patent Office and its collection of Patent models. Woodcroft had the engine reassembled by the famous engine builders, Penn's of Greenwich, who put it into working order. It is now in the Science Museum in London, where it might easily be taken for a model, because the cylinders are only of 10 cm (4 in) bore and about 46 cm (18 in) stroke; and the whole thing could easily stand on a large table.

After Symington had built a larger engine for Miller, the latter decided that Symington's type of engine was unsuitable, and Symington himself later earned a surer place in the history of the steamboat by devising something entirely different for the tugboat *Charlotte Dundas*. This was built in 1801 and was a stern wheeler, 17 m (56 ft) long, 5.5 m (18 ft) wide and 2.4 m (8 ft) deep. The stern paddle wheel was carried on a cross shaft with a crank on one end, and this was operated by one double acting cylinder, 56 cm (22 in) bore, 122 cm (48 in) stroke, via a crosshead and a long connecting rod. The air pump and condenser were below the deck; the boiler was arranged on the other (starboard) side of the vessel; and there were twin rudders. Intended for the Forth and Clyde canal, this steam tug successfully hauled two loaded vessels, each of 71 tonnes (70 tons) load, over a distance of 31 km (19½ miles). It was perfectly satisfactory, but the canal owners were worried by the possible effect of the wash of the stern paddle on the canal banks, and decided, for the time being, to continue to use horses on the towpath. Though soon out of use, this historic craft was not actually broken up till 1861.

The *Charlotte Dundas* really ranks as the first wholly practical steam vessel and its design was marked by that simplicity which always suggests that the experiments are over and the phase of enlarging and perfecting is about to begin. It is tempting to draw analogies with the *Rocket* locomotive, but the development of steam vessels was not at all like that of the locomotive; no single type ever became universal, though there was a long period when triple expansion engines and Scotch boilers provided by far the commonest way of propelling all but the smallest craft. So, though the *Rocket* heralded triumphant careers for the two Stephensons, the *Charlotte Dundas* did nothing for Symington who eventually died in straitened circumstances.

The first steamboat engined by Symington for Miller was about 10 m (33 ft) long, and not at all on the useful scale of the *Charlotte Dundas*. Even smaller was the *Little Juliana*, made in 1804 by John Stevens, which marked the re-entry of the United States into the field. This was only 7.3 m (24 ft) long, but it had a remarkable propulsion system, with a high pressure multitubular boiler, a single-cylinder non-condensing engine and twin screws. Tried in New York harbour, it proved no faster than Miller's steam catamaran (as we must now

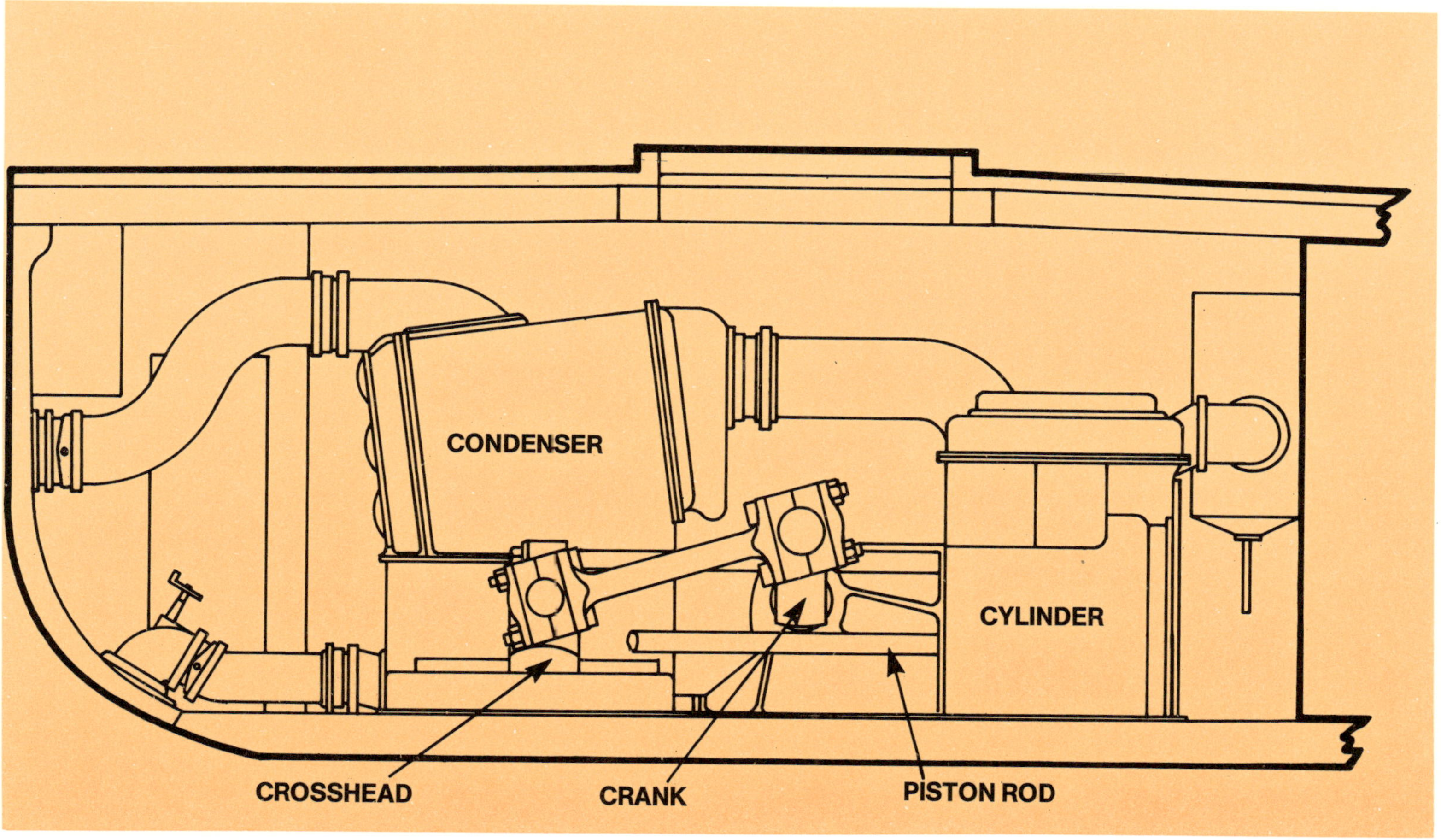

Layout of horizontal screw engine with return connecting rods. The cylinders are to the right, the crankshaft in the middle, and the cross-heads to the left, beneath the condenser.

describe it) of 1788, but its machinery occupied less space. Four years later, Col. Stevens was responsible for the paddle steamer *Phoenix*, which was the first steamer to make a voyage at sea: New York to Philadelphia in 13 days. There were several breakdowns en route. Though Stevens is probably best remembered as the founder of a dynasty of steamboat and railroad operators, his contemporary, Robert Fulton, was the first to operate a steamer on a regular commercial basis, over an extended period. He had spent 20 years in Europe, at first studying portrait painting under Benjamin West but later becoming more and more involved in engineering and especially maritime matters. He had built an experimental submarine in England and had experimented with steamboats in France. When he eventually returned to the United States he had ordered a steamboat engine from Boulton and Watt.

THE FIRST COMMERCIAL STEAMBOAT

Fulton's *Clermont* was built in 1807 on the East River, New York, and this steam engine was installed in it. It was a flat-sided, blunt-ended vessel of 101 tonnes (100 tons) displacement, with large capacity and shaped for relatively slow speeds—more like a barge than a sailing vessel. It was 40 m (133 ft) long but only 4 m (13 ft) wide—a narrow boat, in fact—and its unladen draught was only 60 cm (2 ft). The engine had a single-cylinder, 61 cm (24 in) bore and 122 cm (48 in) stroke, and drove two 4.5 m (15 ft) paddle wheels via a mechanism of bell cranks, flywheel and spur gears which Fulton had devised himself. A plain copper boiler, much like the wagon boiler associated with Watt but perhaps unusual in being made of copper, supplied steam at scarcely above atmospheric pressure. This boiler was locally made, which was no doubt why copper was used, as, though more expensive than iron by this time, it was easier to work and a familiar material in North America (used for brewing equipment and the like).

The first trip was made on 17th August, from New York to Albany, and the ship attained nearly 8 km (5 miles) per hour. It then ran for the rest of the season in regular service on the Hudson River, and after a refit continued in this service for many years. Fulton built other craft, with similar engines but built in America, and his example was followed rapidly so that in a few years steamboats were operating on all the great rivers of the United States.

Five years after the *Clermont*'s first trip, almost to the day, the first commercial steamer in Europe began operating. This was Henry Bell's *Comet*, built by J. Wood & Co. at Port Glasgow, with engine by John Robertson and boiler by David Napier. It could load about 25 tonnes (25 tons), was around 14 m (46 ft) long in the hull and just over 3.3 m (11 ft) wide in the beam, and drew about 90 cm (3 ft) of water. It operated between Glasgow and Greenock on the Clyde, offering two classes of accommodation, and proved so popular that within a year three further paddle steamers were being built to replace it. It was then transferred to various services elsewhere, and worked in the Forth and Clyde estuaries till 1820, when it was wrecked.

Originally it had four sets of radial paddles and its very tall funnel served as a mast if sail had to be set. The paddles were not satisfactory, so two paddle wheels replaced them. The engine, placed on the port side, had a single vertical cylinder 32 cm ($12\frac{1}{2}$ in) bore and 40.5 cm (16 in) stroke, driving the crankshaft, which was beneath it, by means of double connecting rods which transferred the motion from the end of the piston rod (above the cylinder) down to two side levers. A connecting rod worked upward from these to the crank. There was a box base incorporating the condenser, and the air pump was driven off the side levers. Reversing was by slip eccentric (as employed in the *Rocket* and described earlier in this book). This engine was fortunately not lost when the vessel was wrecked, and may now be seen in the Science Museum in London. The boiler was a plain low-pressure type, set in brickwork.

The *Clermont* and the *Comet* demonstrated in their respective continents the possibilities of steam navigation, and both were speedily followed by other and improved craft. In Europe, British engineers played a leading part in developments, as they were later to do with railways, and several of the first steamers of various European countries were built by British consultants or fortune seekers, the hulls being made in the country of proposed use, with engines from Britain.

PADDLE STEAMERS AND THEIR ENGINES

Within a few years of the experiments recounted in the last chapter, steamers were operating on great rivers in much of the world. North America had the most extensive river services: by 1819 there were about 100 steamers there, and 20 years later about 1,000, totalling some 200,000 tons. There were virtually no sea-going vessels in this total. Britain had half the tonnage at this time, but it was largely in sea-going vessels.

Though the term 'paddle steamer' suggests a vessel with paddle wheels at the sides, it could with equal reason be applied to one with a single wheel at the stern, like the *Charlotte Dundas*. There were in fact many American vessels of this type, and elsewhere the stern wheeler found favour for use on narrow rivers with shallow water. Its manoeuvring characteristics were different from those of a centrally propelled craft, being more like those of a screw steamer; and it had the advantage that the stern wheel was unlikely to be damaged by running into shallow water at a river's edge. The engines of these vessels were much like that of the *Charlotte Dundas*, with the difference that two cylinders were provided, one on each side, with the cranks phased at 90°, like a locomotive. The American river and lake steamers remained generally true to Watt's principle of low pressure and condensing, but later examples of the stern wheeler, built for use outside Europe and America in most cases, were sometimes designed to be as light as possible, and used higher pressure without condensing. Locomotive type boilers might also be fitted, because they offered a high evaporative capacity in relation to their weight.

However, side paddles were the more usual provision, and they were almost the only method of propulsion used at sea until about 1840, after which they long remained predominant. In the eastern states of America, river steamers commonly had their paddle shafts operated by engines with overhead beams. These were essentially Watt rotative engines, and though they first appeared in the 1820s, their framing was usually of timber, and remained so to the end of the century. The boiler pressure was steadily increased over the years, from very low at the beginning to 480 to 550 kN per sq m (70 or 80 lb per sq in) in the end; the associated boiler development is dealt with in the section of this book devoted to boilers. The cylinder was upright, on the vessel's centre line fore or aft of the paddle shaft, and the piston rod worked upwards. The cylinder being double acting, Watt's parallel motion was required to guide the top of the piston rod where it was connected to the link which transferred its action to the end of the beam. In later examples a crosshead and guides were used at this point.

The beam was usually visible above the upper deck, even of those high-built steamers. Its trunnions were supported on a pair of A frames massively constructed of timber, to which were attached trestles supporting the crankshaft bearings on the side away from the cylinder. The beam itself was braced above and below, after the manner of the beam of *Old Bess*, the Watt pumping engine which once pumped water round a mill wheel at Boulton and Watt's Soho Foundry in Birmingham, England, and which is now preserved in the Science Museum in London. The only difference is that the beam of *Old Bess*, being a single-acting engine, was braced on the upper side only. Eventually, the beams of American river steamers were made of cast iron but, in order to keep their weight down, exactly the same design of bracing was applied to them. As the vessels moved, their overhead beams nodded majestically up and down. This, the wash of the paddles, and the streaming smoke from their tall stacks combined to present a picture of romantic elegance which has entered American folklore, and, fortunately for us, has been captured in a great many paintings and coloured prints.

As size increased, so did the quest for lightness, and some examples were of very sophisticated design, with slim connecting rods having thin bracing rods to prevent their centres from whipping; and eccentric rods of light lattice construction. The A frame, ahead of the paddle shaft as a rule, remained substantial because it had to convey the thrust of the paddles to the keel of the vessel, while the outer bearings of the paddle shaft conveyed it to the sides of the hull. A representative engine of the 1880s had a cylinder of 91 cm (3 ft) bore and 2.73 m (9 ft) stroke and its normal working pressure was about 310 kN per sq m (45 lb per sq in). Such an engine might be installed in a 60 m (200 ft) hull. As these craft often conveyed mails, and sometimes essential commodities in short supply which were being rushed to their markets by traders anxious to make their fortunes, they were occasionally overdriven. There could also be a great deal of competition between ships and between traders: moreover, in a gambling age, wagers were often laid upon the results of competitive sailing. On such occasions the boilers were pressed above their normal limit, and the fires were also forced to the point where flames could be seen issuing from the tall stacks.

The long reign of the beam engine on American rivers was due to the calm waters which did not demand a low centre of gravity. For sea-going vessels something more like the engine of the little *Comet* was required, and Messrs. Boulton, Watt and Co. produced side lever engines, at first with single cylinders. Other makers followed suit and from the beginning of the 1820s through to the 1860s this type of engine was made in steadily increasing size. As in Bell's *Comet*, the cylinder or cylinders were vertical and the piston rod worked upwards to a crosshead from which depended connecting rods on each side; but the end-pivoted levers of the Comet engine were replaced by centrally pivoted beams like those of contemporary stationary engines. The opposite ends of these beams were united by a crossbar from which a connecting rod worked upwards to a crank on the paddle shaft.

The cylinder and beams were low down in the vessel, and firmly secured. The crankshaft bearings were on A frames (or some architectural equivalent) and bolted to the engine bed. The whole engine was usually made of iron, even from the beginning, and the crossheads were usually guided by parallel motion linkages. There was a jet condenser, and an air-pump driven off the levers. Single-cylinder examples were not common, but even these could be made self-starting by arranging the balancing so as to bring the engine at rest in a position other than dead centre. The usual arrangement was to have two cylinders with the cranks set at right angles. Although the crankshaft and the paddle shaft were all in line across the ship, they were not in one piece. The centre portion or portions, with the cranks, were supported in bearings on the engine frames, while the portions carrying the paddle wheels were supported on the sides of the hull, thus providing the necessary flexibility. Coupling was commonly done by links between fairly closely spaced crankpins, the links being in tension when steaming ahead.

Side-lever engines were especially associated with the firm of Robert Napier, on the Clyde at Glasgow. His

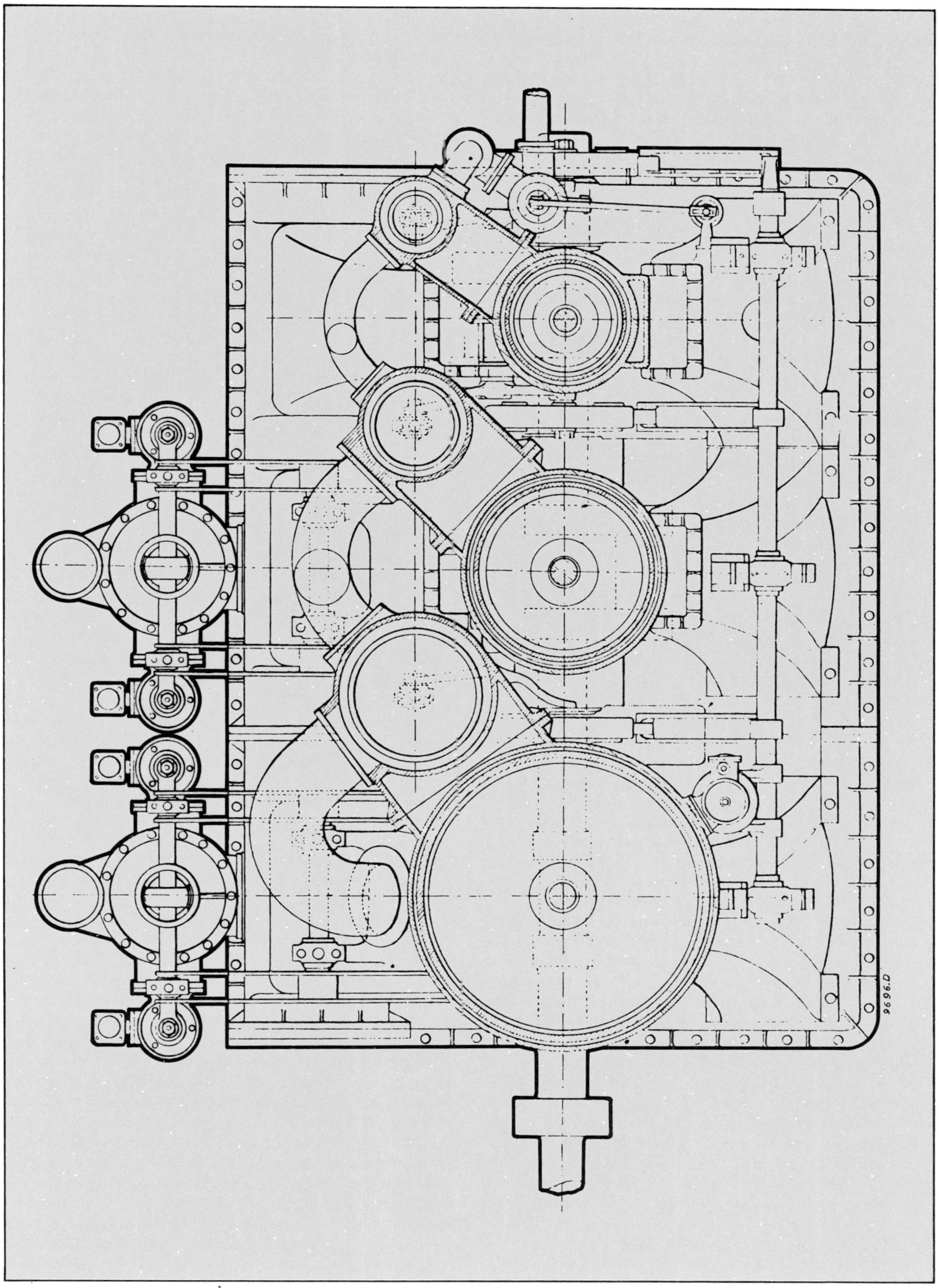

Plan of an Italian-built marine triple-expansion engine of the 1890s. This engine is unusual in having piston valves for all three cylinders, and in having the valves arranged to the side of the engine. This reduces the length of the engine, and hence, ultimately, increases the space available for cargo in the vessel.

first was built in 1824, for the PS *Leven*. It had a single-cylinder 80 cm (31½ in) bore and 91.5 cm (36 in) stroke, and the boiler pressure was probably about 34 kN per sq m (5 lb per sq in) above atmospheric, which, if the condenser was working satisfactorily, would have given an effective pressure on the piston of about 117 kN per sq m (17 lb per sq in). This pinpoints the fact that, though condensing was to become a way to provide clean feed-water for the boilers, it was essential in early marine engines just to enable them to develop the necessary power. In fact, sea-water was used in early marine boilers, and at the low temperatures associated with the low pressure of early days, and with boilers made of iron and not steel, the effect was not disastrous. But the condenser was then and always a particularly vital and often troublesome part of the machinery.

By about 1840, side-lever engines were being made with all the dimensions, quoted above, doubled, except the effective pressure on the piston. The boiler pressure above atmospheric might be doubled, but this only added a third to the effective pressure. An engine of this size would develop about 400 horsepower at 20 rpm, and with 6 m (20 ft) paddle wheels might propel a 404 tonne (300 ton) vessel at 20 km per hour (12 knots). By 1861, when the last Cunard paddle steamer *Scotia* was built, her side-lever engines had two cylinders 254 cm (100 in) in diameter and 366 cm (144 in) stroke. The paddle wheels were 12 m (40 ft) in diameter and nearly 366 cm (12 ft) wide, and the speed was about 27 km per hour (16½ knots), at which the paddles would have been revolving about 14 or 15 times per minute. With a maximum indicated horsepower of 4,900, the *Scotia*'s engines were the ultimate development of the side-lever engine.

NEW ENGINE DESIGNS

As engineering skills, materials and manufacturing machinery improved, so attempts were made to dispense with levers in the drive from cylinders to paddle shaft. Vertical direct-acting engines appeared in the 1830s, but were not long in use and not made in large sizes. Their great virtues were that they weighed less and took up less of the length inside the hull. In the neatest construction, the two cylinders were vertically under the centre line of the paddle shaft and joined together in a unit with pumps and condensers between, the whole being simply secured to the bottom timbers of the vessel, while the crankshaft bearings were on a framework supported on the tops of the cylinders.

The disadvantage of the vertical, direct-acting engine was that the connecting rods were very short, and gave rise to side thrusts at the crosshead far greater than with the side-lever arrangement. The turning action would also be less even, but, on the other hand, the reciprocating masses were less, and this could produce less disturbance of the vessel by the machinery, even at the low rotative speeds of paddle engines. A particularly neat solution of the connecting rod problem, though only really suitable for low-pressure engines, was to make the cylinders open-topped, with the connecting rod attached close to the piston and no crosshead at all. A three-cylinder engine so constructed had a very even turning moment and was self-starting in all positions, even if it were effectively single-acting. But in practice these engines were not really single-acting because, although the greater power would be developed on the downward stroke of the piston (under atmospheric pressure, just like a Newcomen engine), the upward stroke would have the boiler pressure behind it, and if this was 1½ Atmospheres (absolute) there would be an effective pressure difference across the piston of ½ Atmosphere, and the cylinder would be developing rather more than half the power developed on the Atmospheric downstroke (the vacuum under the piston not being, of course, anything like perfect).

The first vertical direct-acting engines were fitted to HMS *Gorgon* in 1837. They had two double-acting cylinders 1.62 m (64 in) bore and 1.67 m (66 in) stroke. The three open-topped cylinders fitted in the PS *Sapphire* in 1842 were 188 cm (74 in) bore and only 91 cm (36 in) stroke. The draught of the *Gorgon* was 4 m (13 ft) while that of the *Sapphire* was only 1.37 m (4½ ft), which explains the very short stroke of the latter's engines.

Direct-acting vertical paddle engines saved weight as well as length: those of the *Gorgon* were estimated to have saved 61 tonnes (60 tons) as compared with equivalent side-lever engines. Almost the same advantages in weight and space saving were possible with steeple engines. These derived directly from Henry Maudslay's table engines which were developed in the first quarter of the 19th century. In a table engine, the cylinder was set upright on a table, beneath which passed the crankshaft. The cylinder worked upwards to a crosshead which was guided at its ends by rollers working in machined guides set upright at each side of the engine, and from these twin connecting rods passed down on each side of the cylinder to two cranks on the shaft. There were slots in the table top to allow this. A variation found later in small-sized engines was to have a single crank, in line with the cylinder axis, beneath the table. The connecting rod was a frame which embraced the cylinder and the table top as it worked.

The object of the table engine was to save space in a factory, and the same object could be attained in a paddle steamer. However, to keep the weight low in the hull, the cylinder was brought beneath the crankshaft. As early as 1831 this type of engine was used on the Clyde for shallow draught steamers in which the piston rod had an upward extension in the form of an egg-shaped frame embracing the crankshaft, the connecting rod and the crank as the engine worked; the connecting rod was hung from the top of the frame, which was guided by a roller crosshead in table engine fashion. It was too cumbersome a design for larger vessels.

Maudslay, Sons and Field, of London, were among the greatest builders of marine engines in the 19th century, and naturally played a considerable role in the development of steeple engines, which derived so closely from their founder's ideas. In 1839 Joseph Maudslay and Joshua Field patented a form of steeple engine in which the cylinder was not quite in line beneath the crankshaft, but slightly to one side. Two piston rods, arranged on a diameter of the piston lying parallel to the line of the crankshaft, rose to an overhead crosshead from which the connecting rod depended. The crank passed between the two piston rods during half its rotation. This allowed the crankshaft to be close above the cylinders of the engine, but the misalignment of the drive inevitably produced a less even turning moment, and made the thrusts upon the crosshead asymmetrical. These effects were less as the connecting rods were longer, but the weight and strength penalty, associated with lengthening the piston rods and connecting rods unduly, was even less acceptable. All the same, this system was successful on the vessels to which it was applied and its unpopularity was probably more a matter of theory than of observed ill effects.

The asymmetry was purely the result of the need to get piston rods past the crankshaft, plus the need to arrange the rods on the piston in a balanced way, to avoid any tendency for the piston to tilt. David Napier, in 1842, solved this by fitting four piston rods, which allowed the piston and cylinder to resume their central position under the shaft. These engines were very popular on the Clyde, but the use of four piston rods involved maintaining four stuffing boxes where the rods passed through the top covers of the cylinders. This was not necessary, because all that was needed was to use two piston rods set diagonally, so that one was on each side of the shaft but both were located on a diameter of the piston itself. In fact, one just had to think of the piston as in the asymmetrical arrangement, but rotated through 45° and brought back into line, and this is what Maudslay's did. That it should have required the passage of some years and the building of many engines, before the simplest solution of a simple problem was arrived at is hard to explain. This, surely, could have been done on the drawing board in a matter of days, but there are more recent examples of designers' difficulties in thinking outside the planes, set at right angles, of the conventional engineering drawing. Engineers are still trained in a way likely to give rise to two-dimensional thinking about three-dimensional problems, and this is a good argument for making models of projected machines. Anyway, for whatever reason, the idea of two piston rods on one side of the shaft had to be followed by that of two on each side of the shaft, before anybody thought of taking two away and just leaving two set diagonally.

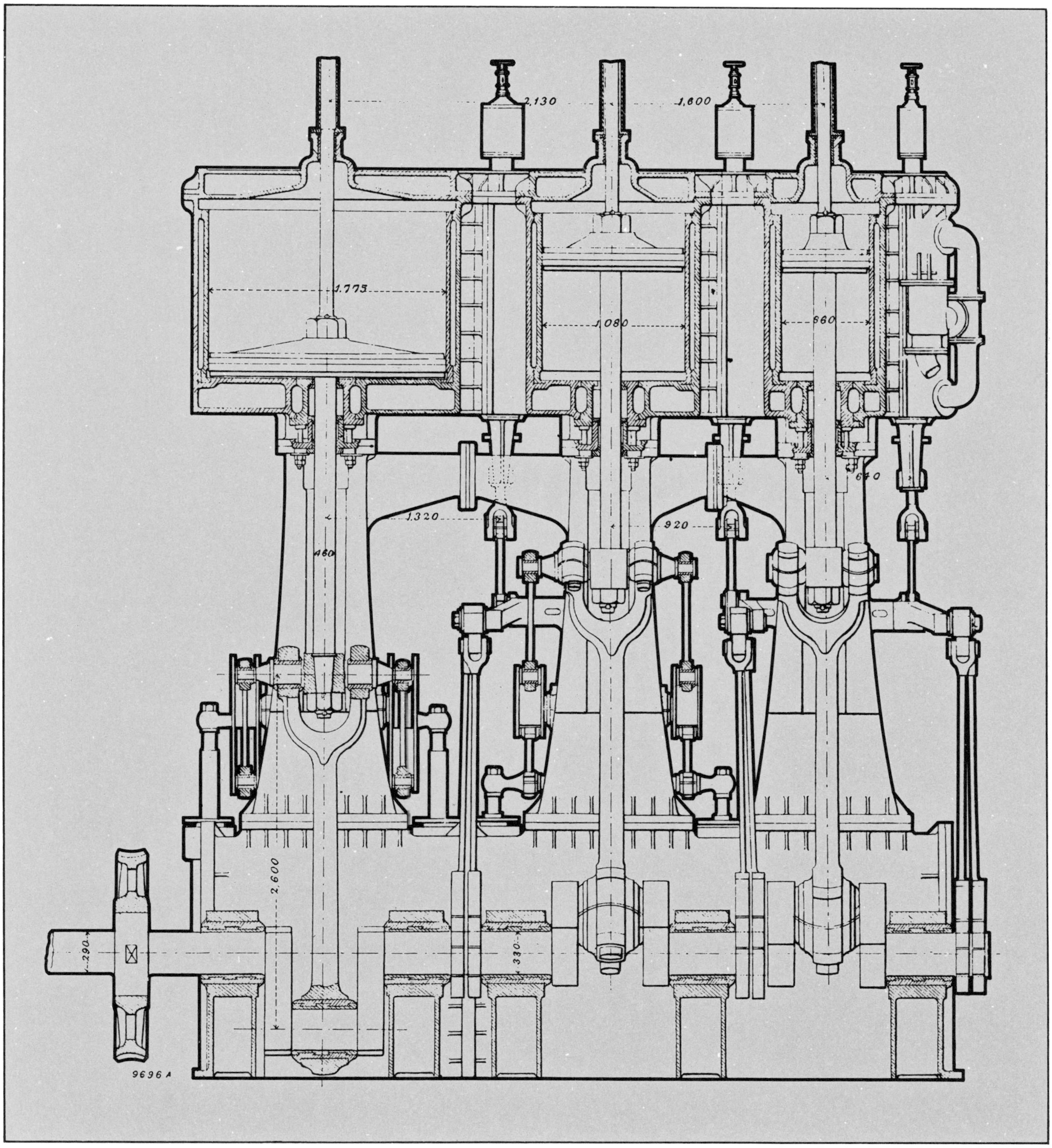

Longitudinal section of the engine shown on page 85. The six eccentrics of the Stephenson link motion can be seen on the crankshaft. Usually, engines with valves at the side used Walschaerts or Joy valve gearing.

Eventually, the design of paddle engines became established as direct drive, either by fixed or by oscillating cylinders, with the cylinders low in the vessel and the line of drive inclined upwards to the crankshaft. But there were still many and various contortions of steam machinery to be tried, especially in the 1840s, before this happy conclusion was reached. Of these, one briefly successful idea was the 'Siamese' engine in its various forms. This had cylinders in pairs, acting in unison and coupled to a single crank. Just as Henry Maudslay's table engine had solved the morphological problem by having one cylinder operating two connecting rods and two cranks, so the Siamese engine solved the same problem by having two cylinders operating one connecting rod and one crank. The PS *Helen McGregor*, built in 1843, had its cylinders placed vertically beneath the shaft, and working downwards to a common crosshead, from which the connecting rod rose between them to a single crank. Everything lay in one plane across the ship in this driving arrangement, and only the details,

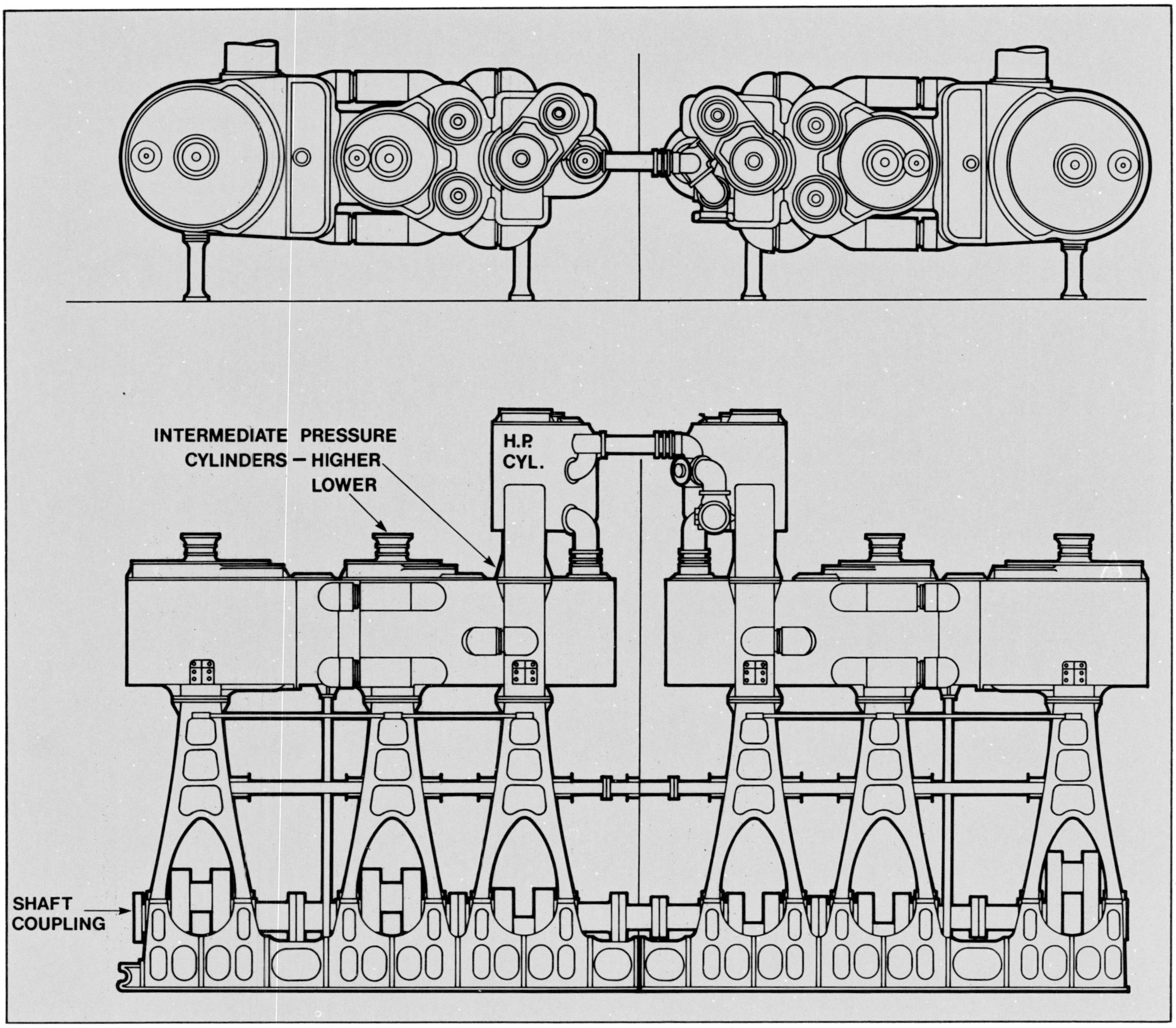

Paired quadruple-expansion engines on a single shaft in the German vessel Kaiser Wilhelm II. *Around the turn of the 19th and 20th centuries Germany made a determined and successful effort to gain the Blue Riband of the Atlantic, and in the process put record powers on to the two propeller shafts of twin-screw vessels. In these engines, the high-pressure cylinders are arranged in tandem with the upper intermediate-pressure ones, thereby reducing the overall length of the engines, which could develop up to 20,000 hp per shaft.*

such as the parallel motions guiding the crosshead, and the various pumps operated by them, encroached at all in a fore-and-aft direction. A single eccentric commanded the valves of both cylinders. This inverted arrangement could only be used in a vessel of some depth, because of the space required below the cylinders for the crosshead to move up and down. The *Helen McGregor* had a moulded depth of 4.5 m (15 ft). The large reciprocating weight could not be entirely balanced on the shaft, but some balance was needed to prevent the engine always stopping at bottom dead centre. Any attempt to balance reciprocating weights with rotating ones, would result in fore-and-aft disturbance when steaming at full speed, for reasons which have been explained in an earlier section dealing with compound locomotives.

Maudslay, Sons and Field patented a Siamese engine in 1839, in which tall cylinders were placed in line, ahead and astern of the shaft, with their piston rods working upwards. Their common crosshead was in the form of a T-shaped box, the downward extension of which fitted between the cylinders and carried the lower end of the connecting rod at its foot. The cylinders were far enough apart to allow the connecting rod to swing as its upper end rotated with the crank. The bottom of the T had to be guided, of course, and the reciprocating masses were as large as in the inverted layout (in fact, generally larger, because of the weight of the T-shaped crosshead). However, Maudslay's had some success with this, especially when arranged in pairs with the cranks at right angles. Between 1840 and 1846, nine vessels were built for the Royal Navy with this type of engine and 40 or 50 other paddle steamers were so equipped by 1850. These were mostly of just under 1,000 horsepower. HMS *Devastation*, a paddle frigate, had four cylinders 1,380 mm (54 in) bore and 1,830 mm (72 in) stroke which could develop some 800 horsepower when the paddle wheels were turning at about 14 revolutions per minute, with a boiler pressure at 50 kN per sq m (7 lb per sq in).

Before considering the types of paddle engine which were to last as long as paddle ships themselves remained in service, it is as well to deal with some of the details which have so far taken second place to the basic morphology in this narrative. Slide valves were universal, and they were operated by eccentrics. With the very

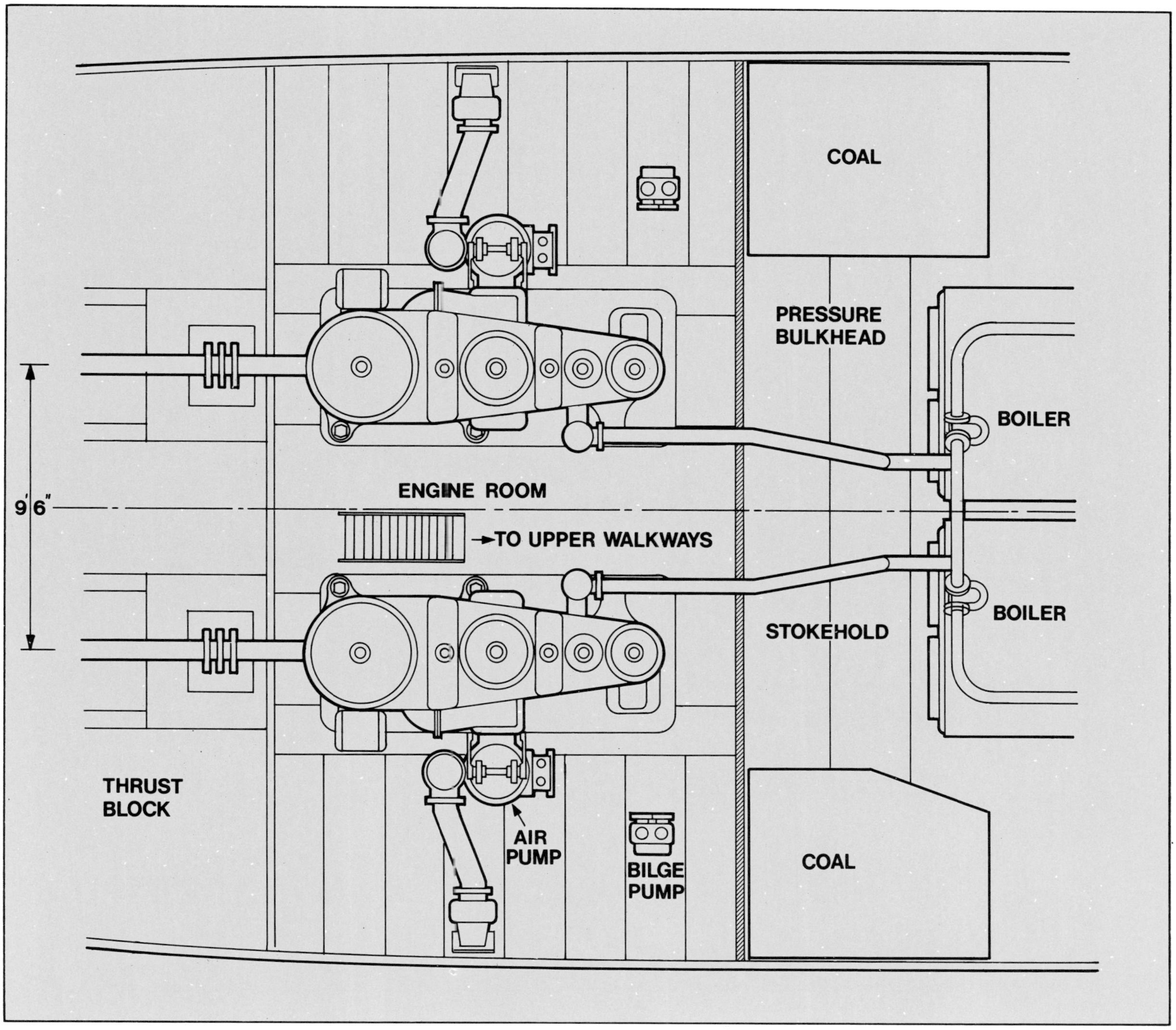

Engine-room layout of a typical twin-screw vessel, c.1900. There are two triple-expansion engines, supplied by two double-ended Scotch boilers. The stoke hold, between boilers and engines, is pressurized to provide furnace draught, and is flanked by coal bunkers. The multi-collared thrust blocks on the propeller shafts are represented diagrammatically.

low steam pressures of the first half of the century, expansive working was severely limited because the valve gears used were either loose eccentrics with manual operation of the valves for reversing (as in Stephenson's *Rocket*), or a form of 'gab' gear, as already described in the section on locomotive valve gears headed 'Efficient use of steam' on page 56. As paddle engines were frequently adjusted and modified by those in charge of them, though, it was possible for some vessels to be far more economical than others, and for some captains and engineers to be expert in coal saving. There was provision for some small degree of expansive working to be introduced by the Watt expansion valve, which cut off the steam supply at its entry to the valve chest, and was operated by a cam, or a choice of cams, on the crankshaft. In slow running engines with small steam chests this device had an effect, but not at all comparable with the effect of an expansive valve gear on an engine working at a higher pressure. Long voyages could not be undertaken without a high regard for fuel economy, and were hardly economic commercially because of the space required for fuel, but a few long voyages were made by naval vessels in early days. It was not until 1840 that the Atlantic crossing began to show signs of ultimate profitability, and the sailing ship was not threatened as an ocean carrier over greater distances until the advent of compound, or double expansion, marine engines.

These early paddle steamers had no auxiliary engines for any purpose, so the main engines operated all the pumps: the bilge pumps, where fitted (other than for manual operation), boiler feed pumps, pumps draining the surplus from the hot well etc. The jet condensers needed large air pumps, because the boilers were fed with salt water containing a great deal of dissolved air, which appeared in the steam but would not condense. Much of this equipment followed stationary engine practice.

Oscillating cylinders are familiar to most people as the commonest type found on toy steam engines. These little cylinders are held on to a portface by a trunnion and spring, and as they oscillate a hole in the cylinder side communicates with steam and exhaust holes in the portface. No other valve gear is necessary. They are usually single-acting, but are sometimes double-acting and have a cylinder cover with stuffing box, through

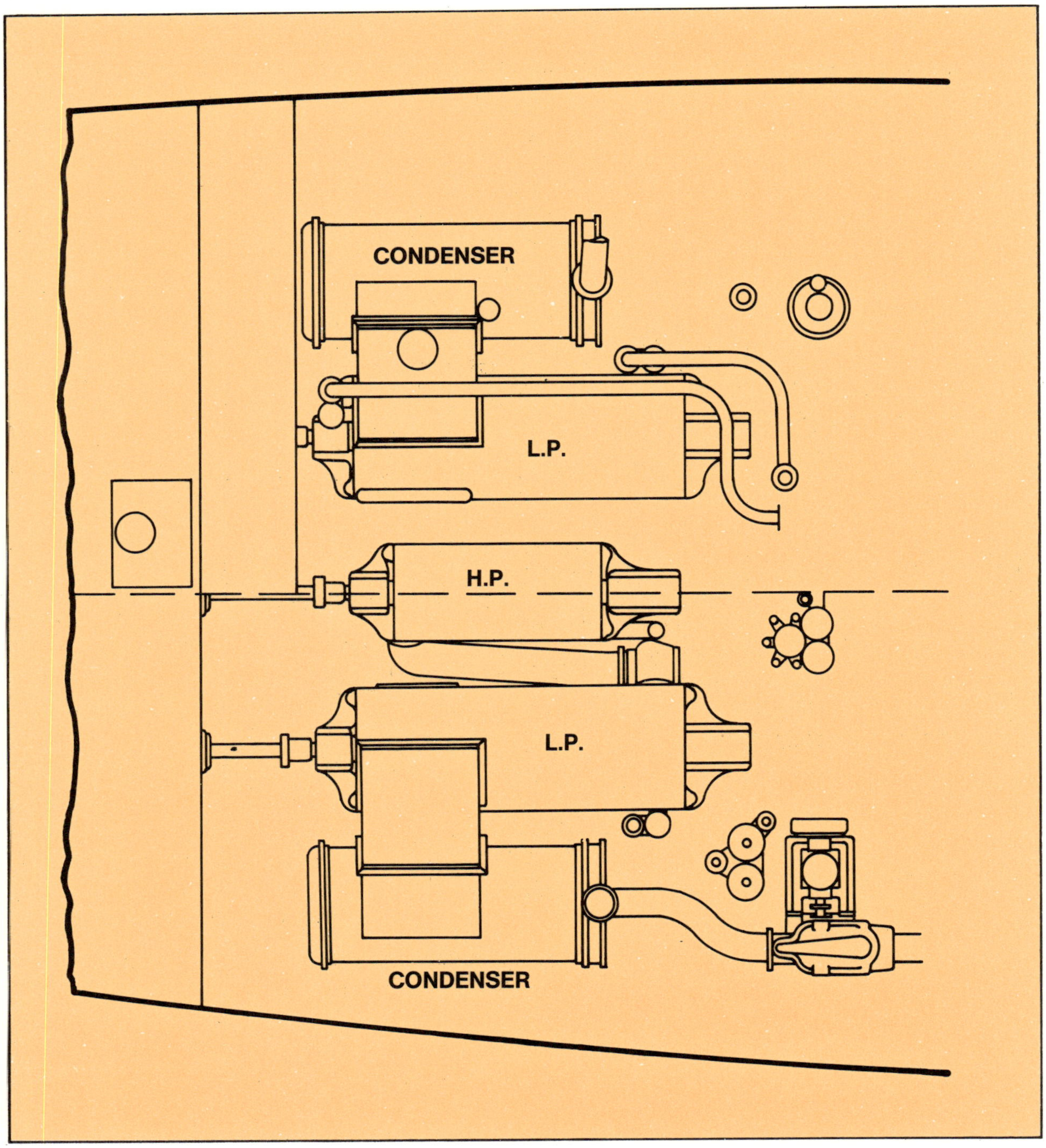

The engine-room plan of the Turbinia, *as fitted with axial-flow turbines.*

which the piston rod works. The distance between the cylinder and the crankshaft can be small, because there is no separate connecting rod and no crosshead.

The idea of oscillating cylinders was proposed by William Murdock in 1785, but it was Joseph Maudslay who first made such engines practical and provided them with proper valves, so that they could operate as efficiently and economically as fixed cylinders. He made several small two-cylinder double-acting engines with the cylinders directly beneath the shaft, for use in Thames steamers, and a number of larger examples followed the success of these first efforts. However, oscillating engines did not really become popular until the type was improved by John Penn of Greenwich in 1838. Maudslay had introduced the valve chest alongside the cylinder at the trunnion, so that long passages led through the trunnion to the cylinder ends. This also widened the cylinder and trunnion assembly and made it occupy more space across the vessel. Penn placed the valve chests on the cylinders themselves and away from the trunnions, devising a linkage for driving the valves from the eccentrics without introducing error due to the oscillation.

From this time on, oscillating cylinder engines remained in favour, and examples of smaller paddle craft of quite late construction and fitted with such engines are possibly still in service today as river ferries. Most oscillating engines were diagonal, i.e. the line of drive was inclined upwards to the shaft, and there were sometimes cylinders both fore and aft of the transverse centre line. But vertical, direct-acting, oscillating engines were still being built near the end of the 19th century, with compound expansion. A representative simple expansion engine of this layout, of the period when boiler pressures were at last rising above the near atmospheric level which was Watt's legacy (but before double expansion was generally applied) was fitted to the PS *Leinster* in 1860. This was an Irish Sea mail boat of 2,032 tonnes (2,000 tons) displacement, 104 m (343 ft) long, with a hull width of 10.6 m (35 ft) at the waterline and a moulded depth of 5.8 m (19 ft) of which 4 m (13 ft) represented the draught. It was quite a large vessel for the time and such a service, being much the same size as the cross-channel ferries of today, and its speed at some 31 km per hour (18 knots) was not greatly inferior to theirs. The two double-acting oscillating cylinders were of 2.5 m (98 in) bore and 2 m (78 in) stroke. With 138 kN per sq m (20 lb per sq in) gauge boiler pressure they could drive paddle wheels of 9.7 m (32 ft) diameter and 3.65 m (12 ft) width at 25.5 revolutions per minute, indicating 4,750 horsepower. These cylinders had two valve chests each, one on each side so as to be balanced with respect to the trunnions. Both valves on each cylinder were worked by a single loose eccentric, which moved a slide up and down. The slide had a slot curved to a radius struck from the trunnion centre, and die blocks in the slot conveyed its movement to the valves regardless of the angle of oscillation.

Three-cylinder paddle engines were not uncommon with fixed cylinders, especially in the practice of the European mainland, but these usually had the cylinders in a transverse line and arranged diagonally. An unusual three-cylinder arrangement with oscillating cylinders was devised by John Scott Russell and patented in 1853. One cylinder was vertically beneath the crankshaft, while the other two were fore and aft of it, and inclined upwards at 60°. As they were, of course, double-acting, the turning effect was the same as if they were evenly spaced around the shaft, as all three worked upon the same crank. Scott Russell was also responsible for the paddle engines of the *Great Eastern*, which had paddles and a screw. This enormous vessel displaced 27,822 tonnes (27,384 tons) and was 207 m (680 ft) long, with a breadth of 25 m (82½ ft) and a moulded depth of 17.6 m (58 ft). Its size was not exceeded until the very end of the century.

Curiously, the paddle engines of the *Great Eastern* were little if at all more powerful than those of the *Leinster*, but they were on a much larger scale. There were four cylinders set low in the vessel and driving up diagonally to the crankshaft. They were arranged in pairs fore and aft, each pair driving one crank and the cranks being at right angles. Their total swept volume was about 2.3 times that of the cylinders of the *Leinster* but the boiler pressure was a little greater and the rotative speed of the 27 m (90 ft) diameter paddles was less than half. In fact these engines originally indicated only 3,400 horsepower, but with maximum pressure they certainly could have produced 5,000. This was only half the installed engine power, as there was a screw engine also. The paddle engines weighed 849 tonnes (836 tons), which was not excessive when taken as a fraction of the displacement of the vessel, but it is quite clear that in most respects this famous ship was too large for the engine technology of its time, and the designers were extrapolating too far beyond the reach of their own experience. The paddle wheels were destroyed in a gale in 1861, and they were replaced by new ones only 15 m (50 ft) in diameter. The total installed power was woefully insufficient for a ship of such a size, but larger engines would have consumed more fuel and made Australian voyages for which the *Great Eastern* was designed impossible. It really needed compound engines.

COMPOUND ENGINES ON THE WATER

For an account of two-stage compounding, or double expansion, the reader is referred to the preceding section dealing with compound locomotives. Though the advantages of compounding for locomotives are more numerous, and in many respects do not apply to marine work, the smoothness of operation was a benefit; and, above all, the increased thermal efficiency (more greatly increased in marine than in locomotive applications) proved crucial in enabling the steamer to take over from the sailing ship. With triple or even quadruple expansion the advantages were still greater.

Compounding of paddle engines occurred mainly with direct-acting diagonal engines, because it came relatively late in the history of the paddle steamer. But there were exceptions. The early steeple engine design, with a frame surrounding the connecting rod and crank, was made as a compound by Napier, there being three cylinders with pistons attached to the bottom of the frame. The central one took steam at 'high' pressure, and the two outer ones received the steam from it. This was a Woolf compound arrangement, receiver compounding not yet having made its appearance and the pistons being inevitably in phase. At a much later stage, compounding was applied to oscillating engines, and as late as 1891 new sets of triple-expansion steeple engines were built for two vessels belonging to the London and North Western Railway. But by the time that compounding was in general use at sea, direct-acting diagonal engines were most commonly being fitted to paddle steamers, while paddle wheels were no longer favoured for general long distance voyaging. The paddle craft remained in favour for short distances and for frequent stops, and was also easier to manoeuvre in awkward situations than a screw steamer, which accounts for its long service on lakes and rivers.

The diagonal compound or triple-expansion engine was not unlike the engine part of a locomotive in layout and detailing. Locomotive valve gears were used and often provided the normal method of control; Swiss lake steamers, for instance, would not only go astern as they drew up to a landing stage, but once stationary would be held with their engines in mid-gear if a quick restart was expected and, once the signal to move was given, the links would be lowered quickly and the ship would accelerate like a train. Stephenson gear was not the only one used, however. Walschaerts was also favoured, and other gears associated with locomotive practice were tried from time to time (for particulars of valve gears, the reader is referred to the section 'Efficient use of Steam' earlier in this book). The condensing arrangements with compound engines were different from those of earlier days, and the multi-tubular surface condenser had replaced the jet type. Surface condensers are described in the section on boilers and accessories. The boilers had long ceased to be fed with salt water, and the condensers provided the feed water as well as adding greatly to the pressure range over which the engines worked. In spite of the much higher boiler pressures—690 kN per sq m (100 lb per sq in) or more—the range of expansion used with compound engines made the power increment due to exhausting into a vacuum very significant indeed.

Typical of the best practice in double-expansion (or compound) direct-acting diagonal engines were those built by W. Denny Brothers of Dumbarton, a firm associated right up to the present day with cross-channel steamers, which it manufactured for foreign owners as well as British. For the Dover–Ostend service Denny built two paddle steamers in 1888 for the Belgian Government, named *Princesse Henriette* and *Princesse Josephine*. They had two-crank compound engines, the high-pressure cylinders being of 1.5 m (59 in) bore and the low-pressure being 2.75 m (104 in). Both had a stroke of 1.83 m (72 in). The volumetric ratio of the cylinders was therefore about 3.35:1, and this was associated with a boiler pressure of 830 kN per sq m (120 lb per sq in (gauge)), i.e. 8½ Atmospheres. This ratio demonstrates the effect of a condenser: a locomotive, exhausting to atmosphere and having such a low pressure would require a ratio of about 1.8:1. With double the pressure, about 2.4:1 would be normal. These ratios allow for equal division of the work between the high- and low-pressure cylinders (if the valve gearing is properly adjusted) and a terminal pressure above that of the atmosphere. The larger low-pressure cylinder of marine practice allows expansion well below atmospheric pressure.

These engines, with cranks set at 90°, were receiver compounds, just like a two-cylinder compound loco-

I. K. Brunel

motive and, like many locomotives, were equipped with Walschaerts' valve gear. The high-pressure cylinders had piston valves and the low-pressure ones slide valves. By this date superheating was incorporated in many marine boilers, and piston valves were found to be more easily lubricated under superheat conditions. Slide valves were usually 'balanced', by having an area of their backs excluded from the steam chest pressure. This was done with spring-loaded strips working against a flat face inside the steam chest cover, the enclosed area being vented. The engines of *Princesse* ships developed some 7,000 horsepower at 50 revolutions per minute, from which it may be deduced that, if the power was equally divided between the two cylinders, the receiver pressure was a little over 138 kN per sq m (20 lb per sq in (gauge)). This sufficed to drive the 91 m (300 ft), 1,016 tonne (1,000 ton) ships at about 37 km per hour (just over 21 knots).

Direct-acting diagonal compound engines also superseded the overhead beam engines in United States lake and river steamers, and were built well into the 20th century for these large craft. Characteristically, vessels of the Fall River Line, like the large steamers on the Great Lakes, were of around 5,080 tonnes (5,000 tons) gross and required 6,000 to 10,000 horsepower to drive them at around 27 km (15 knots) per hour. They were 120–140 m (400–450 ft) long and 15–17 m (50–55 ft) wide, exclusive of paddles, at the waterline. Their engines were mostly three-crank compounds, with two low-pressure cylinders. The Great Lakes steamer *City of Cleveland*, built in 1908, weighed 4,641 tonnes (4,568 tons) gross, had a 1.47 m (54 in) bore high-pressure cylinder and two 2.08 m (82 in) bore low-pressure cylinders. The common stroke was 2.43 m (96 in) and the volumetric ratio of low-pressure to high-pressure side was about 4.6:1, a value which is explained by the high boiler pressure of 1,100 kN per sq m (160 lb per sq in (gauge)). In service, this vessel normally developed a cylinder horsepower of 6,000, at 28 revolutions per minute, and ran at around 32 km per hour (17–18 knots).

Though we are here primarily concerned with the steam engines themselves, it is worth mentioning that paddle steamers from the 1860s on were usually equipped with feathering paddles, i.e., paddles that entered the water and left it in a vertical attitude, rather than remaining in a position radial to their shafts. This effect was obtained by pivoting the paddles in the paddle wheel, and equipping each with a projecting arm. This arm was attached to a radial link pivoted on a centre a little way behind the centre of the paddle shaft (or occasionally ahead of it, depending on the arrangement of the paddle arms). A whole 'spider' of these radial links rotated with the paddle wheel, but about its slightly displaced centre, and feathered the paddles as it did so. The actual pivot was arranged on the outer beam of the paddle box, which meant that there was no conflict with the main shaft.

OCEAN-GOING PADDLE STEAMERS

This is not a history of ships, but of steam engines, but we cannot leave the subject of paddle engines without referring to a few famous craft driven by them, and of these the most celebrated are certainly those used to cross the North Atlantic. The first crossing by a steamboat was that of the *Savannah* in 1819. This was the first of several crossings by steamboats which were under sail for part of the voyage, as they were unable to carry enough fuel for the whole crossing. The *Savannah* was a sailing ship of no more than 324 tonnes (320 tons) gross, which had been fitted, as an afterthought, with a 90 horsepower engine which could drive removable paddles. The boat crossed in May and June, and during a voyage of some 28 days was under steam for about 80 hours. Another vessel reported having seen it on fire in mid-Atlantic, but no such mishap occurred—it was only from its tall chimney stack, somewhat enshrouded by sails, that smoke was issuing. It was new at the time of the voyage, and was to be sold in Europe, but no buyer was found, and it returned, had its engine removed, and was soon wrecked. The United States still takes some pride in this early Atlantic venture, and named their first nuclear-powered ship after it.

The first vessel to complete a transatlantic crossing entirely under steam was the *Sirius*, which left Cork on the 4th April 1838 and arrived in New York on the 23rd, only a matter of hours before the *Great Western*. The *Sirius* was not built for such service; it was a cross-channel boat built, for the St. George's Steam Packet Co., at Leith on the Forth. It was chartered for the voyage to America because the *British Queen*, of the British and American Steam Navigation Co., was not yet ready, and this company was particularly anxious to be the first to achieve an Atlantic crossing entirely under steam. The *Sirius* did what was expected of it, but only just, its bunkers being nearly empty on arrival at New York. It was of 714 gross tonnes (703 gross tons), and was equipped with a two-cylinder side-lever engine of 320 horsepower. The *Great Western* left Bristol four days after the *Sirius* left Cork, and steadily overhauled its rival across the ocean, but the *Sirius* was first by several hours. It made only two trips to the United States, and then returned to cross channel work.

The *Great Western* was a very different type of ship, designed by a genius, I. K. Brunel, of the Great Western Railway. It was much larger than the *Sirius*, with a gross weight of 1,341 tonnes (1,320 tons), and very strongly built to withstand the worst winter conditions. Its side-lever engines, built by Maudslay's, were of 750 horsepower, with two cylinders, and gave the vessel a speed of about 20 km per hour (11 knots). Its arrival in New York was also greeted with enthusiasm, in spite of its being second; in some ways its feat was more impressive because it arrived with coal enough for several days' steaming still in its bunkers, and clearly demonstrated the possibility of regular transatlantic steam navigation. It then fulfilled its promise by doing 74 voyages across the North Atlantic in the following eight years, after which it was relegated to other work.

Two other Atlantic vessels entered service in 1838: the *Royal William* and the *Liverpool*. The first of these was almost a twin of the *Sirius* and entered North Atlantic service in the same sort of circumstances, as a substitute for a larger vessel not yet complete—actually the *Liverpool*. The company involved was a third competitor in the race for the new business, but it did not last long. The *Royal William* was soon returned to cross-channel work and the *Liverpool* was sold to the Peninsular and Oriental Steam Navigation Co. in 1840. The *Liverpool* was much the same size as the *Great Western*, but the *British Queen*, when it was finally completed after difficulties in 1839, was much larger, weighing 1,892 tonnes (1,863 tons). However, the engines were of 500 horsepower only, and it was a slow ship, though still faster than contemporary sailing ships. Its engines were among the first to be fitted with surface condensers, very much a novelty at the time, and, surprisingly, the *Sirius* had them too.

None of the pioneer Atlantic steamship companies lasted long except Cunard, which is today the sole survivor providing a passenger liner service across the North Atlantic, with regular scheduled sailings. Samuel Cunard founded the British and North American Royal Mail Steam Packet Company in 1839, and its first vessel, the *Britannia* set out for the first time from Liverpool on the 4th July 1840, to sail to Boston in 14 days. She was very slightly smaller than the *Great Western* but of roughly equal power. Cunard's guiding principle was safety rather than speed. That, and the Royal Mail con-

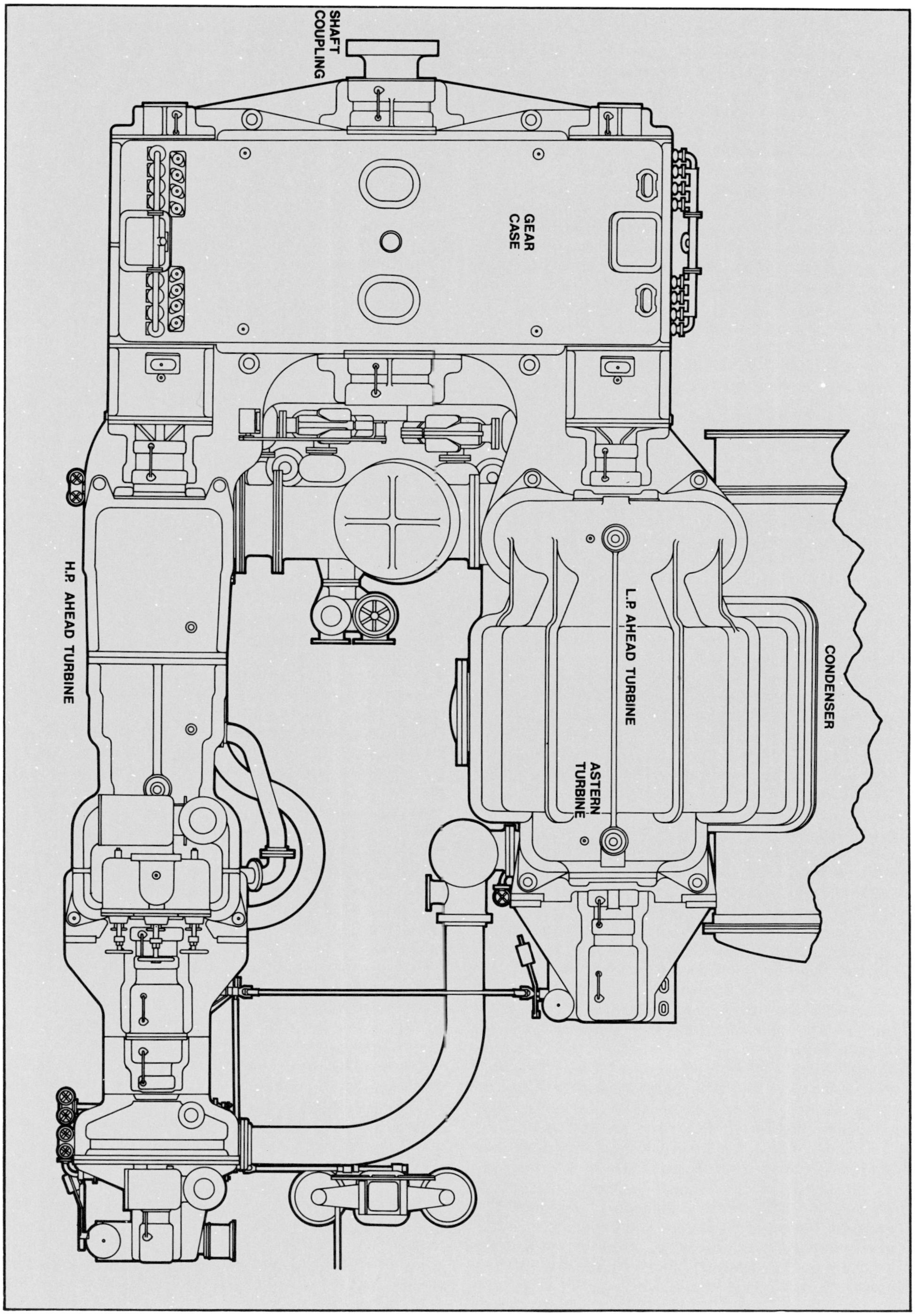

Geared turbines came into use around 1920, and greatly increased the range of application of marine turbines, hitherto used only in fast vessels. This drawing shows a typical single-reduction geared turbine set for a liner. The high- and low-pressure, and astern, turbines, are separated, and the complete set has its own condenser.

tract, secured his success. The *Britannia* and her successors plied steadily back and forth over the years, establishing the tradition of the North Atlantic run. The '-ia' ending was a sort of trade mark, and the *Acadia*, *Columbia*, *Caledonia*, *Persia*, and the *Asia* continued the line of paddle steamers which culminated in the *Scotia* of 1862, of which the engines have already been described. They were elegant ships, somehow more so than the first screw steamers and, as their size increased—from 1,168 to 3,932 gross tonnes (1,150 to 3,870 gross tons)—so did the comfort of their accommodation and the speed of their crossings. The last of them was built in 1861, and thereafter Cunard built only screw steamers for the Royal Mail run.

RECIPROCATING SCREW ENGINES

The invention of the screw propeller has been attributed to hundreds of different people. However, by ignoring published ideas, patent specifications, and sketches which were not translated into reality and tested, one can at least name a few pioneers whose experiments established the practicality of screw propulsion. The first of these pioneers was Edward Shorter, of London, whose propeller was named the 'perpetual sculling machine'. It was tried out in 1802 on a government sailing vessel, the *Doncaster*, in the Mediterranean. This propeller had nothing to do with steam: motive power was provided by eight men walking round a capstan (as it would have been for Savery's paddle craft, a century before, had it ever been built), and a speed of about 3 km per hour (1½ knots) was attained. In the next year, Charles Dallery demonstrated a screw-propelled steamboat on the Seine, but he is not now recognized as the inventor of screw propulsion by steam, even in France.

The two most serious contenders for this title are Colonel John Stevens in America and Josef Ressel in Europe, both of whom achieved convincing results. But practical screw propulsion, which might be able to supplant the paddle wheel, required the solution of more problems than just the design of the screw and the adaptation of an existing engine type to it, difficult enough though those things were. There was the problem of taking the shaft out of the hull under water, without letting the water in. The actual shape of the stern had to be adapted to enable the screw to work in solid, not turbulent or foaming water. The screw had to be balanced, in terms of weight and also in symmetry, to make sure that the water resistance was constant and in line with the shaft, otherwise the wooden hulls would be racked by vibration. And the thrust of the propeller had to be taken by the hull, not by a collar at the stern post, because the friction would be too great, but by something firmly attached to the keel inboard, and possibly borne upon by the end of the shaft itself.

The two men who did most to tackle all these problems and who really established screw propulsion were Francis Smith and John Ericsson, a Swede. Smith's patent was taken out on the 31st May 1836, and Ericsson's six weeks later. Smith had a launch built in London, with a screw fitted in a recess in the keel, near the stern. It was driven by a vertical shaft and bevel gears, from a single-cylinder engine. The screw had two complete turns, and in trials in 1837, one-half of it broke away, with the result, surprising to Smith, that the boat went faster. This led him to a second patent covering a screw with two half turns, i.e. two half turns of a two-start 'thread' and recognizably a two-bladed propeller. Ericsson originally used two contra-rotating propellers, one behind the other, on coaxial shafts. They were multi-bladed and had very large bosses, hollow with internal spokes, to reduce their resistance in the water. In a demonstration on the Thames, also in 1837, his 14 m (45 ft) vessel towed the Admiralty barge from Somerset House to Millwall, a distance of several kilometres (miles), but Sir William Symonds, surveyor to the Navy and true to its traditions of obstinate conservatism, declared the screw a useless device which made it impossible to steer. Dummer, Symonds' predecessor who had baulked Savery, would have been pleased. After a second vessel, the *Robert F. Stockton* of 21 m (70 ft) length, and further convincing demonstrations on the Mersey, Ericsson left for America where he was able to contribute further to the development of screw propulsion and to much else besides.

Smith's patents were bought by the Ship Propeller Company, which ordered a 243 tonne (240 ton) vessel first named the *Propeller* but later *Archimedes*. This ship had two-cylinder engines which drove the propeller through gearing, 26 revolutions per minute of the engines giving 140 revolutions of the propeller. The horsepower was 80. The *Archimedes* cruised from the Thames round the south coast of England to Bristol, calling in at various ports on the way. It sailed to Portugal and had a rather adventurous career, showing off the virtues of screw propulsion, and was even tried against a paddle steamer of the Royal Navy. It made no money for its owners—quite the contrary—but demonstrated most convincingly the effectiveness of the screw propeller, and its influence was great.

The *Great Britain* was its most famous follower. At Bristol the *Archimedes* was seen by I. K. Brunel, who decided at once that he would revise his plans for paddle wheels for the *Great Britain*, and make it a screw steamer instead. This ship, which was floated out from the dry dock where it had been built over a period of four years in July 1843, was the first large iron vessel, with a displacement of 3,676 tonnes (3,618 tons). It was the first screw vessel on the North Atlantic, but was later used on the Australian run (after the fitting of new engines) where it was necessarily under sail for much of the voyage. Its active career was long, and it ended its commercial life as a hulk in the Falkland Islands, but it was recently brought back to Bristol and is currently being restored in the dry dock in which it was built. Its engines are discussed on the following page.

Another important follower was the naval sloop *Rattler*, launched in 1843 and in 1845 pitted against the paddle sloop *Alecto* in a famous stern-to-stern tug of war. The two vessels were not quite equally matched: the *Rattler* was of 1,095 tonnes (1,078 tons) displacement, with 437 horsepower; and the *Alecto* was of 811 tonnes (800 tons) and 200 horsepower. The *Rattler* pulled the *Alecto* backwards at a speed of 4 km (2½ miles) per hour. The result was entirely predictable, and the attempt would probably not have been made if the Admiralty had not been sure of the outcome. But it was a popular demonstration and caught the fancy of the public. If the Admiralty was, for once, being progressive, the reason was fairly obvious. For a warship, a screw had enormous advantages, quite apart from any intrinsic superiority in propulsive efficiency. It was much less vulnerable in itself, and relatively hidden away under the waterline. Its engines could be more safely stowed low down in the ship, and if sail was in use at the same time, the propeller was unaffected by the heeling of the vessel to leeward. The ship could safely be brought right alongside another for boarding in action, and the decks were unencumbered with machinery and could be entirely available for guns and men. The screw was decisively adopted by the Navy, and by 1850 there were about 50 naval vessels (some converted from sail) propelled by screws.

The success of the screw did not indicate any sort of finalization of the design of screw or engines. There was an almost infinite variety of both, but as we are here concerned with the engines above all, the changes in screw design can only be mentioned in passing. But it is worth pointing out that Ericsson's large diameter screws with the blades fixed to a large central drum were not unlike paddle wheels with the blades turned through some 45°. Smith's screws, on the other hand, were definitely more like the Archimedean variety. The most suitable rotative speed was higher for Smith's design, and the torque reaction, tending to make the vessel heel in the opposite direction, must have been less. It was, no doubt, torque reaction that prompted Ericsson to fit contra-rotating twin propellers. But whether the diameter of the screw was the maximum that could be accommodated, as in Ericsson's first ships, or much less, as in Smith's, it required a speed of rotation well above that of paddle engines. The first solution was to drive the propeller shaft through gearing or chains.

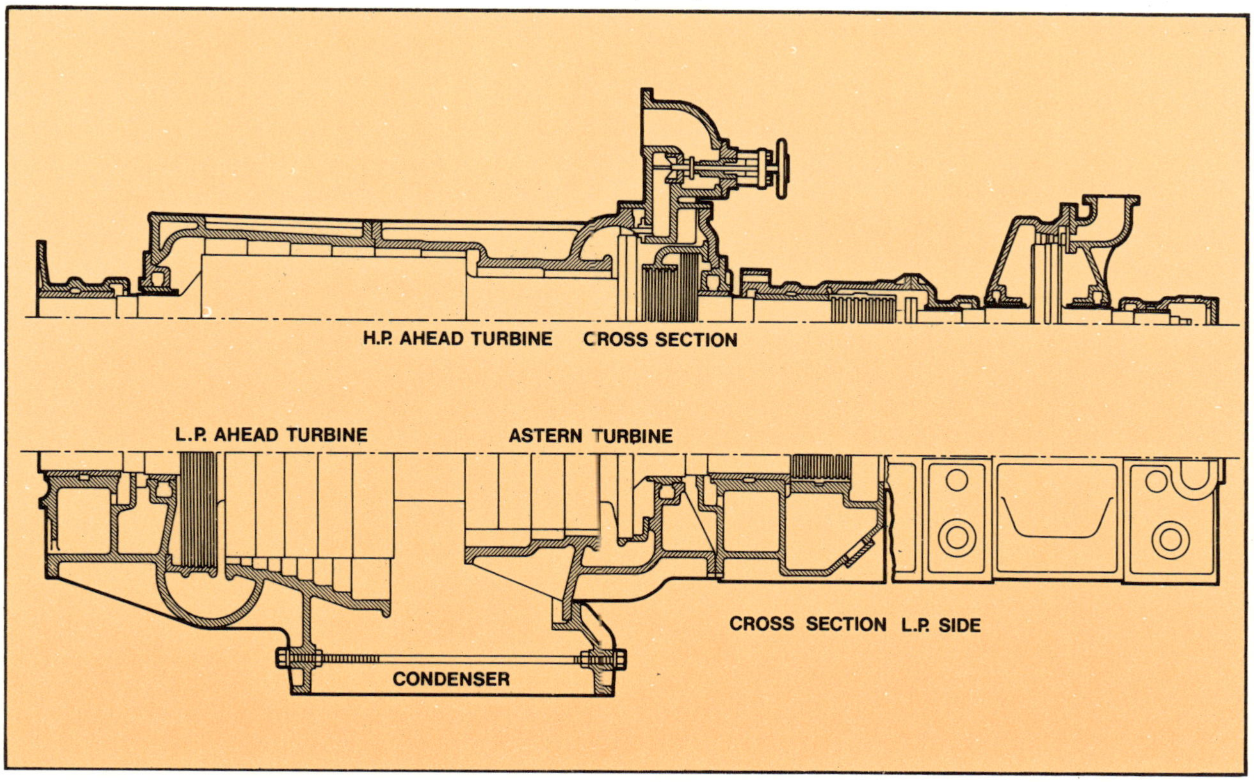

Section through high-pressure side of turbine set.

Section through low-pressure and astern turbines.

The engines of the *Archimedes* had been criticized for the noise from the gearing; the *Rattler* was probably named because of it, for it too had plain spur gearing with a ratio of about 1:3. The larger spur wheel was, of course, above the smaller; mounted on the end of the crankshaft of the engine (which was of Maudslay's Siamese type as described for paddle steamers), with tee-shaped crossheads and two pairs of cylinders 1,070 mm (40½ in) bore and 1,220 mm (48 in) stroke. The thrust of the propeller shaft could be taken directly from its end on to a plate, which was arranged as a sort of dynamometer, as hinged levers attached to the plate enabled the thrust to be measured. This was useful, because the *Rattler* was equipped with screws of different pattern at different trials.

The *Great Britain* was also geared, but in this case the gearing was effected by pitch chains. The thrust of the propeller shaft was again taken upon a plate, which was kept sprayed with sea-water. The engines were designed by Brunel himself, and were unlike any others. The cranks were overhung at the fore and after ends of the engine, and each was driven by two cylinders set low and working upwards, the included angle between their axes being 66°. In the middle of the engine, the crankshaft carried a large drum, 5.55 m (18 ft 3 in) in diameter, carrying four chains which drove the propeller shaft drum beneath, the ratio being roughly 1:3 as in the *Rattler*. The cylinders were 2.23 m (88 in) in diameter and 1.83 m (72 in) stroke, and the valve gearing was by loose eccentrics. With a boiler pressure of 104 kN per sq m (15 lb per sq in) gauge and rotating at 18 revolutions per minute (the screw actually revolving at 53 revolutions), these engines indicated some 1,500 horsepower, which is a low figure for their size and suggests that Brunel was aiming at a high degree of expansive working. The propeller was six-bladed, 4.7 m (15½ ft) in diameter, with 7.6 m (25 ft) pitch, and gave a speed of about 20 km per hour (11 knots). These engines were replaced when the vessel ceased to run on the Atlantic.

Ericsson's most famous vessel was certainly the United States turret ship *Monitor* which, with its very low freeboard and heavy armour, set a completely new style in naval craft and gave its name to a type of warship which is really a floating platform for heavy guns carried in armoured turrets. The *Monitor* had to be very low-built, and Ericsson's engine had two horizontal cylinders placed across the vessel, below the line of the propeller shaft. They were arranged back to back, their outer ends open, so that short connecting rods could work on to a rocking shaft on each side. From longer arms of these shafts, connecting rods worked a single crank on the propeller shaft. The pistons thus moved in unison, and the whole thing was rather like a single, double-acting, cylinder which moved to and fro along its axis while the piston remained stationary in the middle. The stroke was shorter than the crank throw: only 56 cm (22 in), and the diameter of the cylinders was 1 m (40 in). This engine was able to indicate 400 horsepower, which, assuming a rotational speed of 90 revolutions per minute for the 2.7 m (9 ft) propeller, would require an effective pressure of about 240 kN per sq m (35 lb per sq in). This high figure suggests that the engine was not being worked with a great degree of expansion, at this maximum output, but economy was not important for this vessel. The boiler pressure may have been around 207 kN per sq m (30 lb per sq in) gauge, and the engine had a jet condenser.

In one of the most technically significant of all naval engagements, the *Monitor* and its opponent the *Merrimac* were both armoured vessels, but it was some time before armour plating became common on large warships, and there were always smaller naval vessels which were lightly armoured, if indeed they were armoured at all. More than a decade had to pass after 1862, the year of the *Monitor*'s famous battle, before marine engines were allowed to assume the vertical position which seems most suitable for them. In this period, there were three main types of screw engines favoured and, as the main impulse towards screw propulsion was for fighting vessels, the merchant ships followed a little way behind. These types were the return connecting rod engines, like steeple engines laid on their sides and appropriately the speciality of Maudslay, Sons and Field; the trunk engines, patented by John Penn in 1845; and the direct-acting horizontal engines of short stroke, made by most builders.

The return connecting-rod engine needs no further description, and its evolution as the steeple paddle engine has already been described. In its earlier forms it had one or more jet condensers, took steam at around

'Siamese' screw engines for HMS Rattler, *1843.*

138 kN per sq m (20 lb per sq in) gauge and developed 500–1,000 horsepower. The stroke was much reduced as compared with the paddle engines, but the rotative speed was four or five times as high. A representative engine of the kind had a 500 mm (24 in) stroke and 1,140 mm (45 in) bore, with two cylinders, and turned at about 80 revolutions per minute, developing some 700 horsepower. The engines of HMS *Octavia*, built in 1860 by Maudslay's, had two surface condensers and three cylinders of 1,670 mm (66 in) bore and 1,070 mm (42 in) stroke. Steam pressure was still at 138 kN per sq m (20 lb per sq in) and at 70 revolutions these quite large engines produced 2,265 horsepower. Much bigger, two-cylinder engines were fitted by Maudslay's five years later to HMS *Agincourt*, which was a very large vessel for its time, being an armour-plated cruiser of 10,770 tonnes (10,600 tons) displacement. The cylinder bore was no less than 2,570 mm (101 in). There was a reversion to jet condensers in these engines, which produced 6,667 horsepower.

The return connecting-rod engine lasted long enough to come into the age of high-pressure steam and double expansion, and was built by many other builders than Maudslay's. Two unarmoured corvettes, *Boadicea* and *Bacchante*, were fitted with three-cylinder compound engines in 1876, the lack of armour no doubt providing a reason for continuing to employ low-built engines. The high-pressure cylinder was 1,850 mm (73 in) bore, and the two low-pressure ones each 2,340 mm (92 in), the common stroke being 1,220 mm (48 in). The boiler pressure was 483 kN per sq m (70 lb per sq in) on the gauge and the indicated horsepower was 5,300 at 75 revolutions per minute. These engines were modern in their day, with surface condensers and centrifugal cooling water pumps.

Trunk engines did not really lend themselves to compound expansion quite so readily, though there were a few examples. This was because they had to have cylinders of large diameter, and the 'trunk' was more likely to leak than a normal piston rod, making them less suitable for high-pressure steam. The feature which gave these engines their name was a large-diameter tube in place of a normal piston rod. This passed through glands in both cylinder covers and gave enough space to allow the connecting rod to work partly inside it, the small end gudgeon pin at its centre. The crankshaft could be quite close to the cylinder with this arrangement, and trunk engines were used for about 25 years, mainly in naval vessels. Later examples had surface condensers and they were built in sizes up to about 7,000 horsepower.

The screw engines of the *Great Eastern* were direct-acting and horizontal, and were built by James Watt and Co., in 1858. They had four simple-expansion cylinders, arranged in opposed pairs operating the same crank, the two cranks being set at right angles. So that the opposed cylinders were truly in line, one drove the crank with a single connecting rod, while the opposite one drove through two, arranged one on each side of the single rod of the opposite cylinder. Because of the presence of paddle engines in the ship, the screw engine was not as large as might have been expected, and its indicated horsepower was a little under 5,000 maximum. The cylinders were 2,140 mm (84 in) bore and 1,220 mm (48 in) stroke, and the engine drove a four-bladed propeller, 7.3 m (24 ft) in diameter at just under 40 revolutions per minute at full speed.

The most familiar arrangement for a marine engine is the vertical one, with the cylinders at the top and the crankshaft at the bottom, and short connecting rods. This is the layout still used for marine diesel engines today. Although it first appeared in the 1840s in practical form, it only began to make headway in the 1850s, and could not, of course, achieve its major development until the supersession of the paddle wheel. Its application to naval vessels lagged about ten years behind that to merchantmen, because the navy still required the engines of its ships to enjoy the protection of a low position in the hull, and it needed the development of smaller engines and armour-plate to change that.

Left: *the paddle tug* Charlotte Dundas.

Below: *the steam ship* Great Western.

Robert Fulton

Smaller engines (in relation to power developed) depended on two things: higher boiler pressure and higher rotative speed. The former depended on boiler design above all, but also needed compound, triple or even quadruple expansion to make the most of it, and in some 50 years these all came to be provided. Higher speed of rotation was a relative matter: for a given propeller diameter, the speed did rise a little as propeller design improved, but propellers tended to get larger until multiple screws were extensively adopted towards the end of the 19th century.

The technology of the engine itself marched in step with the general progress in metallurgy during the period, no less than with the advance in the science of thermodynamics. Slowly rotating, long-stroke engines of the early 19th century could be made of iron: their connecting rods did not whirl and their crankpins did not have to support the thrust from large diameter and short-stroke pistons. There was plenty of time for steam to get into and out of cylinders and the jet condenser water supply could (except occasionally, when starting or manoeuvring) easily be regulated as necessary by the engineer on watch; he could observe each stroke if he wished. The forces produced by lack of reciprocating or rotating balance were slight at low speeds. But all this was very different with the vertical, direct-acting screw engine, which eventually produced a power approaching 3,700 kW per crank, in some examples.

Regular use of vertical engines began with fairly small vessels. The *Northman*, built in 1846 for trading to the northern islands of Scotland, was so equipped, and coastal craft, notably Tyne colliers, began to have such engines in the 50s. The Science Museum in London has a large (one-quarter size) model of the engines of the steamship *A. Lopez*, built in 1865 by the famous firm of Denny Bros of Dumbarton for service between Cadiz and Havana. There were two double-acting simple-expansion cylinders, each 1.68 m (66 in) in diameter by 1.07 m (42 in) stroke, in the prototype. The short stroke was typical of screw engines generally, as it kept the height of the whole engine down, and allowed a high rate of rotation without excessive piston speed. These terms are, of course, to be taken in the context of the time. The actual rotation on trial was 45.5 revolutions per minute, which was twice that of a paddle engine, perhaps, but only a third that of many later screw engines, while the piston speed was only a third of that attained by railway locomotives of the 1860s. However, for the marine practice of the period these figures represented advanced engineering. The *A. Lopez* had a surface condenser, and the boiler pressure above atmosphere was 124 kN per sq m (18 lb per sq in). An indicated horsepower of 1,427 gave the vessel a speed of just over 23 km per hour (13 knots).

By the date of the *A. Lopez*, double-expansion engines had already made their appearance in screw steamers and, although most of the vertical engines built over the years had the cylinders in line, some of the early compound engines had them in tandem, as did some of the last and largest triple- or quadruple-expansion ones. Two-crank engines like those of the *A. Lopez* obviously lent themselves to the possibility of being fitted with smaller high-pressure cylinders above the existing ones, and this sometimes happened when a ship was reboilered and able to work at a higher pressure. However, the process probably first took place in somebody's imagination and the resulting layout appeared in entirely new engines. The object was primarily to save fuel, with the concomitant of increased range and possibly cheaper actual fuel cost per tonne (ton) consumed, as well as the economy of reduced consumption.

The P. & O. liner *Carnatic* was built in 1863 with such double tandem compound engines. These were built by Humphreys and Tennant, another very distinguished builder of the days when most of the world's steamships were built in Britain. The high-pressure cylinders were 1.09 m (43 in) in diameter, and the low-pressure ones 2.44 m (96 in), the stroke being 0.91 m (36 in). The volumetric ratio of LP to HP cylinders was thus around 5:1. The gauge pressure was 179 kN per sq m (26 lb per sq in), giving an absolute pressure of about 276 kN per sq m (40 lb per sq in) and there were surface condensers built into the upright members of the engine structure. The volumetric ratio of these engines was exceptionally large, and the bore/stroke ratio of the LP cylinders, 2.66:1, must be among the largest recorded. This last feature is not particularly conducive to economy, because the large surface of the piston and of the cylinder cover makes up most of the inner surface of the cylinder, and the piston cannot be steam-jacketed, while the cover, in practice, was not always steam-jacketed. With saturated steam (and though there was a degree of low-level superheat provided, this would probably have disappeared by the time the steam reached the LP cylinders) the temperature falls as the pressure falls, and this means that at the end of a stroke the piston would be relatively cold, with expanded steam on the working side and the condenser vacuum on the other. The cover would also be cold, but without actually being cooled on its outside, and this cylinder would then be put into communication with the condenser for the return stroke, at the end of which, with the piston close to the cover (so that the steam-jacketed walls could have little effect), fresh steam at a higher temperature and pressure would flow in. This steam would have to waste a great deal of heat in warming its new environment before it could do much work. A long stroke and a smaller bore reduces this effect.

In spite of these strictures on their design, these engines were very economical, and the consumption of coal per indicated horsepower-hour was below 0.9 kg (2 lb). The short stroke was imposed by the tandem arrangement. When understanding of thermodynamics was greater, tandem cylinders were only fitted in vertical marine engines when they offered the only feasible way of limiting the length of a very large engine.

IN-LINE COMPOUND ENGINES

The more usual types of two-stage compound marine engines had their cylinders in line only, and had two or three cylinders (and cranks). The two-crank type was much used in small vessels, from steam launches which might require 7 to 15 kW (10 to 20 hp) up to small freighters of two or three thousand tonnes (tons), or passenger ferries of about 3,700 kW (5,000 hp). They were sometimes fitted in pairs in small twin-screw vessels used for fast passenger services over short distances. The three-crank type was built in much larger sizes. The first vessel of the Orient Line, the SS *Orient* of 1879, had an HP cylinder of 1.52 m (60 in) diameter and two LP cylinders of 2.16 m (85 in). The three cranks were set at 120 degrees and the common stroke was 1.52 m (60 in). The volumetric ratio was about 4:1, and the steam pressure 518 kN per sq m (75 lb per sq in) above atmospheric. These engines were rather unusual in having piston valves for all cylinders: balanced slide valves were usual for LP cylinders at least. The indicated horsepower was 5,400, and the weight, or displacement, of the *Orient* was 9,652 tonnes (9,500 tons).

An interesting use of the three-crank compound engine could be made in warships. The *Cristoforo Colombo* was a 2,338 tonne (2,360 ton) naval chaser built in Venice for the Italian navy in 1876. Her three-crank engine was built by John Penn and Sons at Greenwich, and the three cylinders were all the same size: 1.58 m (62 in) in diameter and 1.2 m (40 in) stroke. This engine could be worked as a compound for economical cruising at moderate speed, two of the cylinders serving for the low-pressure steam and the whole engine producing about 1,490 kW (2,000 hp). But when speed was actually required, for manoeuvres or in action, high-pressure steam could be admitted to all three cylinders and the power output doubled. The surface condenser had to be large enough to cope with these conditions as effectively as with normal compound working. Its cooling water circulation pump was driven by an independent engine. Like most vertical marine engines, this one had slide valves worked by link motion (explained elsewhere in this book), but it also had separate expansion valves on the backs of the main valves, independently worked by another link motion.

As late as 1939 a three-crank compound engine was installed in a cross-Channel passenger and cargo vessel, the *Batavier III*. This was a very modern form of the engine, totally enclosed and having the Meier–Mattern valve gear, which uses hydraulic transmission in place of a wholly mechanical linkage to operate the valves. At 120 revolutions per minute, and with a boiler pressure of about 1,720 kN per sq m (250 lb per sq in), this engine developed 2,237 kW (3,000 hp). It had been devised to give the passengers the best possible chance of a good night's rest; the Batavier company had ascertained that diesel engines were unsuitable from this point of view, and even turbines, because of the higher rotative speed of the propellers, were not ideal. Unhappily, this nobly conceived ship was sunk during the war and her interesting engine, built by Werkspoor of Amsterdam, was therefore never subjected to the test of long-term use.

As early as 1862, John Elder secured a patent for triple- and quadruple-expansion engines, but it was not until the 1870s that the former began to be developed. Construction of triple-expansion engines then continued for about 80 years, and thousands were built during the Second World War for use in the famous *Liberty* ships mass-produced in the United States by the Kaiser organization. Perhaps the contributory factors in the choice of steam propulsion for these vessels included the possibility of burning various fuels, but an undoubted advantage was the relative silence of the steam engine: motorships, with diesel engines, could often be heard when still over the horizon. Also, it was possible to get large numbers of these engines built by the American and Canadian locomotive building firms, whereas the large marine diesel was a far more difficult proposition and would have required extensive retooling.

Three-stage expansion produced economies almost as great, when compared with two-stage compounding, as the latter had produced when compared with simple expansion. But it required higher boiler pressures to make it fully worth while. In practice, its introduction was soon accompanied by the adoption of a pressure of 1,035 kN per sq m (150 lb per sq in) and this figure was doubled by the year 1900. For the lower pressures, Scotch boilers were used, but the higher ones, more associated with warships, were attained by the use of water-tube boilers. Associated improvements in general engineering practice included the development of wrought-iron, and later steel, pipes in place of copper ones; metallic instead of soft packing for piston and valve rod glands; better oils and methods of applying them; solid-drawn steel boiler tubes; and the adoption of lighter construction of the reciprocating parts, to assist in more perfect balancing.

Although Britain was still very much in the lead in the building and engineering of sea-going vessels, France was not far behind, especially in the matter of quality and the application of scientific principles, though the volume of the French output was small. And it was in France that the first triple-expansion engine was fitted in a vessel, by Benjamin Normand, a member of a celebrated family of naval constructors at Le Havre. In 1871 a Seine steamer was fitted by Normand with a triple-expansion engine having two cranks, two of the cylinders being arranged in tandem. In the next few years he fitted several other engines of this kind into small ships, and there is no doubt that he deserves to be called the pioneer of the triple-expansion engine. However, the classic triple-expansion engine was a three-crank machine, although the largest had four cranks of which two were connected to two low-pressure cylinders. The three-crank engine was first fitted into a launch, the *Mary Ann*, in 1872. A. C. Kirk, who was also experimenting with new types of high-pressure boiler, fitted such an engine into the *Propontis* in 1874. This vessel had Rowan water-tube boilers pressed to 1,035 kN per sq m (150 lb per sq in) and the engine had cylinders of 585, 1,040 and 1,550 mm (23, 41 and 61 in) diameter with a stroke of 1,070 mm (42 in). The volu-

THE "CLERMONT."

Fulton's Clermont, *built in 1807, on the Hudson River. This vessel marked the beginnings of the commercial operation of steamboats.*

Midnight race on the Mississippi. This Currier and Ives print is somewhat naive, but it does not exaggerate the consequences of forcing the boilers.

metric ratios were therefore 1:3:7 approximately; not particularly well chosen, the intermediate cylinder being a little too large for equal power distribution. The boilers gave trouble, and had to be replaced with others operating at only 620 kN per sq m (90 lb per sq in), after which the *Propontis* was not able to demonstrate any advantage over two-stage compounding.

This setback did not deter Kirk, who was remarkably well qualified by experience and ability to make major advances in marine engineering. He had been apprenticed with Robert Napier, a draughtsman with Maudslay, Sons and Field, manager of John Elders at Fairfield where the *Propontis* was built, and eventually manager of Robert Napier and Sons, where he built the *Aberdeen*. This was the vessel which did most to convince naval architects and engine builders that the triple-expansion engine was likely to give the greatest economy, and likely to provide the greatest power in relation to weight, of any marine engine type so far tried.

The *Aberdeen* was significant in that it was built for a shipping line which had hitherto used a series of sailing clipper ships for trading with Australia. It was a cargo vessel with a total capacity, including fuel, of 4,100 tonnes (4,000 tons), and it sailed for Melbourne in 1881 and arrived after 42 days' steaming, having taken on fuel at the Cape. The *Aberdeen* lasted some 40 years, and kept its original engines to the end. Those engines had cylinders of 760, 1,140 and 1,770 mm (30, 45 and 70 in) diameter and 1,370 mm (54 in) stroke. The volumetric ratios were 1:2.25:5.4, or, to put it another way, 1:2.25 HP to IP, and 1:2.4 IP to LP. This was much better than the *Propontis* and allowed even distribution of power over the three cranks. The boiler pressure was 860 kN per sq m (125 lb per sq in) gauge, and on trial this engine indicated 1,342 kW (1,800 hp) with a fuel consumption of only 0.56 kg ($1\frac{1}{4}$ lb) per 0.7 kW (1 hp) hour. On the maiden voyage to Australia, and including the Cape coal which was probably of lower calorific value than that used for the trials (which would certainly have been the best available) the consumption averaged 0.76 kg (1.7 lb) per indicated horsepower hour, while the output averaged slightly more than at the trials.

After the *Aberdeen*, the shipbuilders in north-east Britain, who specialized in cargo vessels, built many ships with triple-expansion engines and they appeared on the North Atlantic passenger services in the Clyde-built German liners of 1886, *Aller*, *Trave* and *Saale*. These vessels closely followed the three-crank compound-fitted *Umbria* and *Etruria*, which were actually more powerful, with no less than 10,480 kw (14,500 hp) or nearly 3,728 kW (5,000 hp) per crank. In 1889 twin screws appeared on the North Atlantic run, in the two beautiful Inman liners *City of Rome* and *City of Paris*. The latter of these two ships was the scene of one of the most spectacular of all engineering disasters, which was fortunately without loss of life, when one of the propeller shafts broke. Its associated engine immediately raced away and reached a speed at which most of it disintegrated. It was subsequently calculated that the enormous low-pressure piston had elongated its rod to the point where it struck the cylinder cover, and this had initiated the collapse of the whole low-pressure side. However, the engine was replaced and this 35 km per hour (20 knot) ship continued in service.

These engines were very fine three-crank triple-expansion machines of 7,084 kW (9,500 hp) each, of orthodox design, but their 2.86 m (113 in) low-pressure cylinders were certainly about as big as was practicable, and it soon became the practice to provide two low-pressure cylinders for the largest engines, either with four cranks or with three (the high-pressure cylinder then being arranged above one of the others, in tandem). For naval vessels, the need to build low precluded the tandem arrangement and there was often an armoured deck close above the tops of the cylinders. The four-crank triple-expansion engine was much favoured for destroyers, cruisers and even battleships, and these engines were kept low also by having a shorter stroke than their equivalents in the merchant services. To compensate for this, their rotative speed

The Cunarder Persia *exemplifies the elegance of the last large paddle steamers on the North Atlantic route.*

was higher. Destroyer engines were built with attention to the saving of weight, because the whole purpose of the design of these ships was to attain high speed. One of the first British destroyers was the *Daring*, built by Thorneycroft in 1893. She had two four-crank triple-expansion engines, each developing 1,640 kW (2,200 hp): their piston stroke was only 410 mm (16 in) but they revolved at nearly 400 rpm. The total weight of the machinery, including boilers and condensers, was only just over 115 tonnes (113 tons)—for 3,280 kW (4,400 hp). This just equals the best power-to-weight ratio achieved with any class of steam locomotive, but that was nearly 50 years later, with higher boiler pressure of 1,900 as against 1,380 kN per sq m (285 lb per sq in as against 200), but double expansion only. The volumetric ratios of these engines were precisely 1:2 between stages of expansion, and thus 1:4 overall. One must regret that none of these magnificent engines has been preserved.

At the other end of the scale of naval craft, the last British battleship to be fitted with reciprocating engines was HMS *Defence* of 1908, and her engines, though much larger and totalling 20,134 kW (27,000 hp) at a lower rotation of 125 per minute, were also of the four-crank triple-expansion kind.

Quadruple expansion first appeared in 1884. Ten years later, a five-crank engine of this type (though a small one) achieved a coal consumption of no more than 0.51 kg (1.15 lb) per indicated horsepower hour. But the triple-expansion type was never supplanted, and the two types continued to exist alongside, even in identical service. Perhaps the ultimate in triple-expansion engines were those fitted to the triple-screw liners *Olympic* and *Titanic*. There were two of them, driving the outboard propellers and each developing 11,185 kW (15,000 hp), on four cranks. They were compounded with a low-pressure turbine driving the central propeller, and this added 11,930 kW (16,000 hp), giving a total for the ship of 34,300 kW (46,000 hp). The high-pressure cylinders were no less than 1.37 m (54 in) in diameter—probably the largest high-pressure cylinders ever provided—but the low-pressure ones established no records, because there were two of them on each engine.

The possibility of superimposing cylinders in tandem on merchant ships led to some interesting configurations. The Cunarder *Campania* of 1893 had five-cylinder three-crank engines with triple expansion, but the Germans, when at the end of the 19th century they set out to gain, and for some years held, the 'Blue Riband' of the Atlantic, produced the most extended engines of all. The *Kaiser Wilhelm II* of 1903 had twin screws, with 14,900 kW (20,000 hp) on each. To obtain this, each shaft had a double engine, totalling six cranks and eight cylinders, with quadruple expansion. The high-pressure cylinders were above the others.

Most of these great engines had piston valves for the high-pressure and slide valves for the low-pressure cylinders. Stephenson's link motion was the commonest, but Walschaerts' and Joy's gear were also used, and several other types of gear with special expansion valves. Trip gear, drop valves and other features of stationary engines were seldom tried.

TURBINES

The development of enormous reciprocating engines for marine use was finally paralleled by something similar for the generation of electricity. And it was for the latter purpose that, as recounted elsewhere in this book, the steam turbine was invented by Charles Parsons.

The story of Parsons and his struggle to regain control of his patents belongs in another chapter, but it is worth pointing out that the Parsons turbine was and is the ultimate expression of the principle of compound expansion. Turbines are designed as impulse or reaction machines, or as combinations of the two modes, but these are details, however important, of the design of their blading. The essential thing about the Parsons turbine, and all subsequent compound turbines which are derived from it, is that the expansion of the steam takes place in very many stages, corresponding to rows of blading. This division of the pressure drop over many rows of blading lowers the velocity of the steam at each stage to a level at which its energy can be efficiently absorbed by the machine, and also results in more or less constant temperature conditions at each stage. The steam flows axially along the turbine, passing in turn through stationary blading fixed to the casing and moving blading fixed to the rotor. As the steam expands, it requires more space (just as the multiple-expansion reciprocating engine passes the steam through cylinders of increasing size) and this is obtained by increasing the diameter of rotor and casing, and lengthening the blades. In practice, this is not done by tapering the whole thing, but by having steps of increasing size, each with several rows of blading.

The *Turbinia* has already been mentioned in the introduction to this book. It was the first turbine-propelled vessel and its public performance at the Diamond Jubilee Review in Spithead was the most effective propaganda for this new way of driving ships. It was not, on that occasion, in its first form; Parsons, having been obliged to develop the radial-flow turbine as an answer to the loss of his patent rights, at first fitted the *Turbinia* with a radial-flow engine driving a single shaft, directly, at 2,400 revolutions per minute. The power and speed proved insufficient, so he fitted three axial-flow turbines (having recovered his rights for such machines) which, compound in themselves, were also compounded together and drove three shafts directly, high- and intermediate-pressure turbines driving the outer shafts and the low-pressure one driving the central shaft. Each shaft was equipped with three widely spaced propellers, which had to be of small diameter, as the shaft rotated at around 2,000 rpm, being directly coupled to the turbine. The double-ended water-tube boiler worked under forced draught, the fan for which was worked directly off the propelling machinery. The effect of this was that, the faster the vessel went, the faster the coal was burned, and there are many accounts of this small craft charging through the water with flames flaring from its short funnel. Its maximum speed on test was about 61 km per hour ($34\frac{1}{2}$ knots). The presence of this newcomer at the Diamond Jubilee Review seems to have been allowed, but the nature of the demonstration had clearly not been anticipated and a naval picket boat was put out to stop the *Turbinia* in full cry. The result was ludicrous: the picket boat was nearly run down, her ensign staff was cut in two, and her commander prepared to swim to safety but was surprised to find his craft still afloat and relatively safe after a very near miss.

At this point a brief reference should be made to Parsons' work on propellers. With the high rotational speeds he was using, there was difficulty in getting the propellers to 'bite'. Parsons found ways of watching and photographing the action of a propeller under water, and soon observed the nature of the cavitation that impaired efficiency and also eroded the propellers themselves. Cavitation is of two main kinds: that due to air drawn from the surface of the water and that in which the cavities are filled with water vapour. This last is particularly destructive because the bubbles of vapour collapse with a sort of explosive force. Parsons made a very great contribution to propeller design and to the science of hydrodynamics with his researches, and they were vital to the success of the *Turbinia*.

The subsequent success of the marine turbine was extremely rapid, though not without its early misfortunes. By 1914 the reciprocating engine was obsolete as far as the fighting vessels of the Royal Navy were concerned. Two destroyers were ordered in 1898, to be equipped with Parsons turbines. They were the *Viper*, with the same dimensions as the 50 km per hour (30 knot) destroyers with reciprocating engines: 64 m (210 ft) long and displacing 376 tonnes (370 tons), and the *Cobra*, slightly larger. The *Viper* achieved a maximum speed of 64 km per hour ($36\frac{1}{2}$ knots) with an estimated power of 7,457 kW (10,000 hp) on short trial runs, and continuously maintained 59 km per hour ($33\frac{1}{2}$ knots). The larger vessel achieved similar speeds. Unfortunately, both these ships were wrecked in 1901 and there were suggestions that the turbines might be at fault, by producing gyroscopic effects which prevented accurate manoeuvring at speed. In the event, the basic construc-

The paddle steamer Waverley, *which is preserved and used for enthusiasts' trips, photographed at Newhaven.*

tion of the hulls was called into question. There can be no doubt that getting an extra 10 km per hour (6 knots) required an increase in power and propeller thrust which was quite out of proportion to the increase in speed, and it was perhaps not surprising if existing destroyer hull design proved inadequate. The loss of the *Cobra* was the more unlucky, as it cost 67 lives, including that of the manager of Parsons' Marine Steam Turbine Company.

The effects of these disasters would have been greater but for the fact that a turbine passenger steamer was undergoing trials at the same time. This was a famous vessel, the *King Edward*, built by Denny Bros of Dumbarton and engined by the Parsons company, for service on the Clyde. Its engines consisted of a central high-pressure turbine driving a shaft carrying two screws, exhausting into two flanking low-pressure turbines, each driving a single screw. About 2,610 kW (3,500 hp) was developed, and drove the 660 tonne (650 ton) ship at just over 35 km per hour (20 knots). The success of this steamer, the first turbine-propelled merchant vessel, led quickly and predictably to the building of a slightly larger consort, the *Queen Alexandra*, and the entry of turbine vessels into cross-Channel service with the *Queen*, built for the Dover–Calais run in 1903.

The first North Atlantic liners to be turbine-equipped came in 1904: the Allan liners *Virginian* and *Victorian* of 13,208 tonnes (13,000 tons) and 8,948 kW (12,000 hp). But the Cunard line took a greater step with the much larger *Carmania*, of 30,480 tonnes (30,000 tons) and 8,203 kW (21,000 hp), and this vessel was really the first of the larger and larger, and even faster, turbine liners crossing the Atlantic during this century. Their story is only now ending with the *Queen Elizabeth 2*.

The progress of the marine turbine was greatly helped by national rivalry. The Germans had taken the Blue Riband of the Atlantic in a tremendous effort with reciprocating engines, from which the Cunard Company had held aloof, maintaining as always that its business was not speed but safety and regularity (in which it remained unrivalled), while the White Star line continued to emphasize comfort. The Government, however, was less content, and it may be surmised that the Cunard Company was not uninterested in the Government's feelings on the subject. A 'Merchant Cruisers Committee' was appointed to look into the question of building two real record-beating liners, to be subsidized because of their potential usefulness in wartime. The outcome of the deliberations of the committee was a pair of very fast, turbine-driven 30,480 tonne (30,000 ton) ships, the *Lusitania* and *Mauretania*.

These two famous liners appeared in 1907, and the *Mauretania* soon established itself as slightly the faster, gaining the Blue Riband and holding it for 22 years of active service, longer than any other vessel (though the most recent holder of the title, the *United States*, will probably hold it indefinitely, as no attempt to challenge it is now likely to be made). Its speed was given as 44 km per hour (25 knots) but, at the end of its career, and actually after losing the title, this splendid old ship made its fastest passage on the Atlantic, and maintained nearly 51 km per hour (27½ knots) for much of the crossing. Its turbines had a combined output of 50,700 kW (68,000 hp), and were coupled directly to its four propellers on four shafts, two of which were driven by high-pressure turbines and two by low-pressure ones.

When speaking of passenger vessels the quoted tonnages can be confusing. The commonest way of indicating the size of a vessel is to refer to the gross register tonnage, which is not a weight at all, but a measure of cubic capacity of enclosed space. In these terms, the *Lusitania* and *Mauretania* were 30,500 tonne (30,000 ton) ships, and this provides a useful index from the passenger operator's point of view (and is also the basis of many commercial transactions relating to a ship). But these vessels actually weighed some 38,508 tonnes (38,000 tons), which is the figure known as displacement tonnage, and is more informative from the engineering point of view, and more useful when comparing powers, speeds and tonnages. The effect of power on speed, and the enormous cost of high speed, may be illustrated by a few simple comparisons: the *Carmania*, weighing 30,480 tonnes (30,000 tons), required 15,660 kW (21,000 hp) to drive it at 33½ km per hour (18½ knots); the *France* of the Compagnie Générale Transatlantique was slightly lighter at 27,400 tonnes (27,000 tons), but to give it a

The Rhône *on Lake Geneva is a splendid example of a Continental side-wheel paddle steamer. Built in 1929 by Sulzer Frérès of Winterthur, it has a diagonal compound poppet-valve engine and accommodation for 1,000 passengers.*

speed of 41 km per hour (23½ knots) required over 34,300 kW (46,000 hp). The same speed for the much larger *Aquitania*, weighing 53,850 tonnes (53,000 tons), only required 44,700 kW (60,000 hp), which illustrates the economy produced by length, the *Aquitania* being 272 m (901 ft) long, and the *France* 217 m (714 ft) long. The *Aquitania*, incidentally, was the longest-lived of all the large Atlantic liners (its career extended from 1914 to 1950), and it was the last of the great four-funnelled ships. At the end, it seemed very much a survivor of a past age of elegance, both externally and in its Palladian and Elizabethan interior décor.

No less significant than the old *Mauretania* was the battleship HMS *Dreadnought*, the construction of which commenced in 1905. It has been said of this ship that it immediately made every other capital ship obsolete, because of the fire power concentrated so largely in big guns, and the strength of its armour-plating. If subsequent battleships were all called *Dreadnoughts* after it, this was perhaps why, but it also outclassed every other capital ship in the world because it was powered by turbines. These were of 17,150 kW (23,000 hp) and gave to the 18,288 tonne (18,000 ton) warship a speed of over 37 km per hour (21 knots)—5km per hour (3 knots) more than the best of her rivals among capital ships. Moreover, the fitting of turbines was said to have saved 1,016 tonnes (1,000 tons) in weight, and £100,000 in cost. There were four propeller shafts, with two coupled to high-pressure and two coupled to low-pressure turbines. For low-speed cruising, special cruise turbines were fitted to two of the shafts, and all shafts had astern turbines, manoeuvrability being of prime importance. (Turbines cannot be reversed, so separate turbines are fitted on the same shafts for going astern. In liners, these can be very small, because tugs usually help docking, but some other vessels, and especially warships, need more astern power). The fitting of cruise turbines pinpoints the fact that turbines are really meant to provide a constant power output, and are less flexible in this respect than reciprocating engines. They are inefficient at low speeds and reduced outputs.

As already indicated, the Royal Navy went over completely to turbines, in which Britain had a decisive lead, though France and the United States were not far behind. Soon, the installed horsepower of the largest and fastest naval vessels was exceeding that of the Atlantic liners: in 1914 the battle cruiser *Repulse* was provided with nearly twice the power of the *Aquitania*, to give her 26,923 tonnes (26,500 tons) a speed of 54 km per hour (31 knots). Destroyers, too, required high power, and by the same date a 1,016 tonne (1,000 ton) destroyer might have 19,388 kW (25,000 hp) to drive it at 62 km per hour (34 knots). We have already seen how, 20 years before, destroyer machinery with reciprocating engines and water-tube boilers was producing 30 kW (40 hp) per tonne (ton) weight. By 1914 this figure was approaching 52 kW (70 hp) with turbines and perhaps 25 per cent higher steam pressure. Of course, the destroyer's machinery was built to the limit of mechanical stressing, and not intended for a long life. A battleship's engine might be three times the weight for a given power.

Water-tube boilers were the normal thing in warships by this time, but liners and smaller mercantile vessels were mainly fitted with Scotch fire-tube boilers and worked at somewhat lower pressures. But it was not so much pressure as weight and rapid steam raising that caused the naval preference for the water-tube type. In fact the pressures were not very different; 1,380–1,720 kN per sq m (200–250 lb per sq in) were common with Scotch boilers, and water-tube boilers were not often pressed above 2,070 kN per sq m (300 lb per sq in). Oil-firing had replaced coal-firing entirely in fighting ships by 1914, but it was not until about 1930 that almost all passenger ships were oil-fired, and cargo steamers still operate on coal in some parts of the world today.

Though the turbine itself did not alter very significantly during this period, the mode of its application did. As already mentioned, low-pressure turbines were sometimes added to triple-expansion engines, most notably in the *Titanic* and *Olympic*, where the turbine drove the central screw of a three-shaft arrangement. This system was applied to many other liners, and was a speciality of the Belfast firm of Harland and Wolff. Exhaust turbines were fitted to humbler craft too, such as some of the Dutch cargo vessels on the East Indies run, when much money could be saved during long weeks at sea by running the turbine when the main engine was at full power. This also allowed the flexibility of the triple-expansion engine to be available in the European rivers and estuary waters, and among the islands of the East Indies.

The fact was that the turbine's need to spin fast was not always compatible with the propeller's need to get a good grip on the water or to run occasionally at reduced speed. The two requirements could be brought close together in a fast Atlantic liner but, even there, there was room for improvement. Splendid though the direct-drive ships were, propellers and turbines could both be improved by interposing gearing or some other

Facing page, top: *the paddle engines of the* Great Eastern, *1858. This model is in the Science Museum, London.*
Facing page, below: *one of the finest models in the Science Museum is this, which represents the British-built three-crank compound engines of the Italian warship* Cristoforo Colombo.
Above: *the classic marine triple-expansion engine was fitted in vessels large and small from the late 19th to the mid-20th century. This is a relatively modern and small example, from the steam yacht* Glen Strathallan, *of 700 hp. It is seen here in the Science Museum in London.*

Above: *the Cunard liner* Mauretania, *the first large liner to be equipped with turbines, held the Blue Riband of the Atlantic for 22 years, from 1907 to 1929. No other vessel ever held it for so long, except the last, unchallenged contender, the* United States.

transmission which would allow the turbine to spin fast (with a possibility of reduction in diameter) and the propeller to turn more slowly (and be larger in diameter).

Parsons' patent of 1894 allowed for the interposition of gearing in the propeller drive and in 1897 a 7 kW (10 hp) geared turbine set was made for a small steam launch. Helical-toothed gearing with a ratio of 14:1 enabled a single turbine, rotating at 20,000 rpm, to drive twin screws. Twelve years later an old cargo vessel was converted to geared turbine drive and showed a fuel economy of 15 per cent as compared with its previous condition, with a triple-expansion engine (though not, of course, a recent example of the type). In 1911 two cross-Channel steamers of the London and South-Western Railway were fitted with geared turbines, and could be compared, to their advantage, with a direct-drive turbine vessel on the same service. There were other successful applications and during the First World War the Royal Navy took up geared turbines extensively. It became evident that the resulting improvement in the propulsive efficiency of the propellers was about 50 per cent, and this was achieved with the relatively low gear ratios of a single reduction set. Double reduction geared turbines followed soon after, and were first used in 1918.

The other important way of matching the properties of the turbines with those of the propellers is by electric transmission. Quite apart from the possibility of the equivalent of a variable gear ratio, this arrangement obviates the need for separate astern turbines and also reduces the length of the shafts, because the electric motors can be situated well aft, and only power cables have to run to them. The turbines are part of a generating set, not different in principle from one in a power station on land (the boilers, however, are very different and have to be of marine type). This system carries a weight penalty, but not a severe one and, though it has never been extensively adopted, it has been applied to some very fine passenger ships such as the P. & O. liners *Viceroy of India* of 19,322 tonnes (19,000 tons) gross, and the sister ships *Strathaird* and *Strathnaver*, the Furness Withy liners *Queen of Bermuda* and *Monarch of Bermuda*, and to liners of the American Dollar Line and the German Norddeutscher Lloyd. But the finest ship to have turbo-electric drive, and one of the finest ships ever to sail the North Atlantic, in both its design

and speed, was the French Line's *Normandie* This ship held the Blue Riband several times during the 1930s.

The three largest ships on the North Atlantic run, and so the three largest liners in the world, were the Cunard liners *Queen Mary* and *Queen Elizabeth*, and the *Normandie*. They were all just over 81,000 tonnes (80,000 tons) gross, and about 300 m (1,000 ft) long overall. Of the three, only the *Queen Mary* survives, preserved as a tourist attraction in California. Their sole successor and today the only passenger liner operating a scheduled North Atlantic service, is the *Queen Elizabeth 2* which, at 67,056 tonnes (66,000 tons), is only a little smaller and ranks fourth in size among all Atlantic liners. All four ships were designed for a service speed of around 51 km per hour (29 knots), but the first three, when they were competing for the Blue Riband, made several crossings at considerably faster speeds. Some comparisons and details of these four superb ships may serve to illustrate the ultimate development of the steam turbine plant for passenger ships.

The *Queen Mary* had 24 boilers pressed to 750 kN per sq m (400 lb per sq in); the *Normandie* had 29, with the same pressure; the *Queen Elizabeth* had only 12, pressed to 3,100 kN per sq m (450 lb per sq in); and the *Queen Elizabeth 2* has only three, pressed to 6,200 kN per sq m (900 lb per sq in). The last-named ship is the only one to have two propellers only, but they have six blades each. The other three all had four, four-bladed, screws. The *Normandie* obtained 119,000 kW (160,000 hp) from its turbo-electric machinery and its rival, the *Queen Mary*, obtained 134,000 kW (180,000 hp) from its single-reduction geared turbines. This did not ensure its supremacy, however, since the *Normandie* was often just as fast, and only the Second World War put an end to an exhilarating rivalry. In fact, the hull design of the

Facing page, below: *Parson's steam yacht* Turbinia *at speed. This was the first turbine-powered vessel, which convincingly demonstrated its speed at the Golden Jubilee review of the Fleet at Spithead in 1897.*

Above: *the* Normandie, *French holder of the Blue Riband in the 1930s. Its design, technically and aesthetically, was outstandingly modern and no Atlantic liner was ever more admired.*

Below: *a destroyer at sea. This class of warship always demanded powerful machinery for high speed, and today gas turbines are beginning to replace steam.*

Below: Queen Elizabeth 2, *the last Atlantic liner in scheduled service, probably owes its survival to its ability to cater for the luxury cruise market as well.*

Normandie was more efficient, and it wasted less energy in producing waves, which cancelled out the slight power disadvantage of its machinery.

Each shaft of the *Queen Mary* had an independent set of turbines: there was no cross-compounding. Each set comprised four compounded ahead turbines and two astern ones, and the rotation speed was 1,800 rpm, reduced to 180 on the propeller shaft. The sets on the *Queen Elizabeth 2* are more powerful, for there are only two of them and they total 82,000 kW (110,000 hp). Each has a high-pressure turbine and a double low-pressure turbine, and they are among the most powerful and compact units ever built. Perhaps the most remarkable thing about the ship is that so much power has been built into so little space: one has to descend deep into the lower decks to find any space not occupied with cabins, swimming pools, games rooms, bars, lounges or other appointments provided for the comfort of the traveller.

The turbines for this ship were designed by an organization which is known under the acronym 'Pametrada'. The P stands for Parsons. But though he was the originator and great exponent of the compound turbine, and especially the compound reaction turbine, his work stimulated others who were to make contributions scarcely less important to marine engineering, and of

those the two outstanding were the Frenchman Rateau and the American Curtis, both associated with compound impulse turbines. Naturally, their turbines were most favoured in their own countries, but their contributions were influential in the subtler science of turbine design everywhere. And as tribute has been paid to the French vessel *Normandie*, so we may end with one to an American one, *United States*, which so decisively captured the Blue Riband in the 1950s, and proved able to steam at 80 km per hour (45 knots). Details of its machinery have remained secret: it was built with defence funds, and not simply as an economic exercise.

Left: *the* United States. *Its speed across the Atlantic far exceeded that of its predecessors, and is never likely to be surpassed by a liner, now that air travel has virtually put an end to the North Atlantic route.*

ROBIN GREENAWAY & SON
STOKE ROW. HENLEY-ON-THAMES
EU2683
Foden
Sentinel
Steam Waggon
BPD 969

Steam on the Road, on the Land and in the Air

The story of steam on the roads of Britain is one of a long fight—against inadequate roads and bridges, against technical problems, against penal legislation. If we look at the world as a whole, the story is patchy because many countries enjoyed a transition from the horse- and ox-driven vehicles direct to those with internal combustion engines without the intermediate step of using steam. Few if any countries saw steam on road wheels used to the same extent as in Britain after the easing of legal restrictions in 1896; indeed, the wide use of steam power for heavy road haulage was exclusive to Britain, mainly because of its supply of cheap indigenous fuel.

The use of agricultural traction engines and ploughing engines, the antecedents of the farm tractor, were however more widespread. With two primary centres of development, Britain and the USA, a highly successful type of machine evolved which by the year 1890 had become commonplace in rural districts. Though the general principles were identical, engine arrangements and details differed considerably according to the side of the Atlantic from which they originated. British traction and portable engines were exported in large numbers to Australia, New Zealand, South Africa and even Russia, while American engines spread into Canada. The latter is probably the only country today where, thanks to preservation-conscious individuals, engines from both Britain and the USA may be seen.

Compared with stationary steam engines for pumping, winding, factory drives and the like, the traction engine arrived late on the scene. Not until the 1870s did designs begin to be established. The basic arrangement stemmed from the much earlier portable engine, first used by Trevithick around 1800 when he was producing his early high-pressure engines. It was he who combined the engine and boiler into one unit and he who first put the assembly on wheels, either for self-drive or for towing from site to site by horse team. Power was taken by belt from the flywheel or through gearing. Even before Trevithick, the idea of carrying passengers was held to be the obvious application of steam on the road and most early experiments were to this end. Many vehicles were produced, but they were nearly all freaks with a tendency to be designed by 'ideas men' rather than by practical engineers. Without the skill in translating ideas into practical reality (possessed in abundance by Newcomen, Watt and Trevithick, for example), any new concept is almost certainly doomed to failure.

While few of the early steam carriages fall into any properly defined path of advancement, it becomes evident from studying the various designs produced that many had features which were ahead of their time.

Recognition that a more practical application of steam than passenger carrying was heavy goods haulage, in the form of a vehicle which carried its payload on its own chassis, did not occur until late in the 19th century in the surge of activity following the changed legislation. By then, designs of traction engine for heavy haulage and farm work had crystallized into a definitive form. By and large, the most successful steam lorries were those in which British traction engine-builders simply scaled down their existing designs and fixed a body to the rear—a parallel process to the evolution of the steam railcar on the railways.

The early 20th-century steam lorry and bus designs using solid rubber tyres were, for the most part, remarkably popular considering the low prevailing speeds. And when internal-combustion-engined vehicles began to fill the roads in the 1920s and 1930s, steam did not go down without a fight. Both Britain and Germany produced highly developed commercial vehicles which were propelled by steam.

As a prime mover for aircraft, the steam engine literally never got off the ground—the petrol engine was clearly more suitable. Several inventors did, however, consider that a flash boiler and small steam engine could be made to drive an aircraft propeller and some units were built. Sir Hiram Maxim's flying machine would certainly have flown if its designer had let it.

Today, with perhaps 2,000 steam vehicles of all types preserved and regularly operated by enthusiasts, it is possible to study the work of the early pioneers with greater insight. How many expensive setbacks must have occurred because of people's unfamiliarity with the vehicles: the water in a boiler being allowed to get too low, bridges and culverts giving way, the vehicle going out of control or running away down a hill. Today an experienced engineman will help the newcomer.

Members of what may be called the 'traction engine family' demand a degree of accord between Man and Machine that must be rare in the history of technological progress. Let there be no doubt about the great joy and satisfaction in driving a traction engine, for example, from A to B single-handed without mishap. A steam engine and boiler are a demanding combination once the fire is lit; add to that the size and weight, the constant vigil on the steering wheel, the rudimentary braking and above all today's traffic conditions and one becomes oblivious to a blister on the hand or a few oil spots on the shirt. The vast majority of the public, too, still take delight in seeing a well-cared for steam vehicle in use. Our forbears managed to perfect the art of traction engine building, operation and maintenance with roads in unimaginable condition, and before the time when one could build up or join metal by welding, and they deserve more credit than is usually accorded them.

EARLY LIGHT STEAM VEHICLES

The first full-size, mechanically-powered vehicle to run on a road was one constructed by Nicholas Joseph Cugnot, a French military engineer, in Paris in 1769. Much earlier, in about 1680, a missionary in China called Verbiest had propelled a model by steam issuing from jets in a container and which blew upon vanes fixed to the periphery of a wheel. Denis Papin, mentioned in earlier chapters, also apparently made a model in 1698. He used his piston-in-cylinder arrangement with a ratchet drive to the rear wheel.

Facing page: *two later types of steam road vehicle in Britain. In front is a Foden steam lorry of the 1920s with a loco-type boiler and compound overtype engine. More advanced was the Sentinel S4 lorry of the 1930s behind it, which had a high-pressure vertical boiler, enclosed undertype engine and shaft drive.*

A model of French pioneer Nicholas Cugnot's three-wheeled steam carriage of 1769 in the London Science Museum collection. Drive was by a ratchet-and-pawl arrangement.

Cugnot also used a ratchet and pawl to take the drive from his two-cylinder engine to the single front wheel. As the photograph shows, his carriage must have been highly unstable since the large copper boiler overhung the front wheel, and the whole engine unit formed a swivelling front carriage. The boiler was always ready to scald the occupants should the driver accidentally use it as a bumper! However, the vehicle did go, although the boiler could not sustain motion for longer than 15 minutes.

The French government ordered a second carriage which was ready for trial in 1770. It is reported to have attained 5 km (3 miles) per hour but trials came to an abrupt end when its instability caused it to overturn in a Paris street. Both the carriage and poor Cugnot were locked up to prevent further mischief. The vehicle was very advanced for its time with its two 330 mm (13 in) diameter brass cylinders and use of high-pressure steam—that is, steam at higher than atmospheric pressure. It is now preserved in Paris at the *Conservatoire des Arts et Métiers*.

In 1784 a spirit-fired model steam carriage began making trips in Redruth, Cornwall. It was the brainchild of William Murdock, Boulton & Watt's resident assistant in Cornwall, who was in charge of erecting the firm's pumping engines at the mines. This tiny vehicle was a three-wheeler like Cugnot's but the boiler and single-cylinder engine, of 19 mm ($\frac{3}{4}$ in) bore and 51 mm (2 in) stroke, were at the back. Motion was transmitted to the rear axle and two 241 mm ($9\frac{1}{2}$ in) diameter drive wheels by a crank and connecting rod hinged to an overhead beam, pivoted at the front.

With this carriage too, things did not always go as planned. 'One winter's evening after returning from work', Dr Samuel Smiles wrote in 1884, 'Murdock went with his model locomotive to the avenue leading to the church, about a mile from the town. The walk was narrow, straight and level. Having lit the lamp, the water soon boiled and off started the engine with the inventor after it. Shortly after he heard distant shouts of terror. He found that the cries had proceeded from the worthy vicar who, while going along the walk, had met the hissing and fiery little monster which he declared he took to be the Evil One in person!'

In 1786 Murdock was pressured by his employers to discontinue his experiments, though the state of the roads at that time would probably have prevented the success of a bigger version, even had Murdock made one. His model, probably the first steam vehicle intended for use on the roads to be made in England, is preserved at the Birmingham Museum of Science and Industry. (Subsequently, Murdock was the first person to light his house by gas.)

Watt's aversion to Murdock's experiments may have been influenced by the fact that he was constantly being urged to turn his own attention to steam road vehicles. He published a specification in 1784 which provided, among other things, for the use of spur gearing between the drive shaft and main axle with a clutch and means of changing gear for different road speeds. Watt appears not to have built even a model, but he clearly gave thought to the subject.

In 1786, William Symington in Scotland made a model steam coach with two horizontal cylinders working on the atmospheric principle and rack transmission, but he turned his attention to steamboat propulsion before going any further. In America, Nathan Read and Oliver Evans also experimented on steam road vehicles before turning their attention to railway stationary engines and marine engines. Read actually built a model steam carriage, and his vertical multi-tubular boiler of 1788 was one of the earliest of its type.

Evans was America's leading exponent of high-pressure steam. Failing to find a backer for his road vehicle, he mounted one of his engines in a flat-bottomed dredging barge he was building, put it on temporary wheels and then used the engine in it to propel the barge 2.5 km (1.5 miles) from his works to the riverside. Consequently he is credited with being the first man to run a self-propelled vehicle on the road in America.

Other early pioneers were Robert Fourness and James Ashworth who, in 1788, took out a patent for a small four-wheeled steam vehicle and subsequently built a small example at Halifax. It had three single-acting vertical cylinders fed by high-pressure steam which drove a cross shaft three-crank geared to the rear axle. It ran short distances but Fourness' premature death brought its development to a halt.

Richard Trevithick, whose fertile brain contributed so much to the steam engine generally, was also a major contributor to the development of road steam. He built several models at Camborne, leading to his celebrated full-size steam 'carriage' of 1801. This was the first vehicle that looked like a locomotive to be produced. It had a horizontal cast-iron return-flue boiler, which also acted as the frame and was carried on four wheels; a single vertical cylinder set into the top of the boiler to reduce condensation; a crosshead on the piston rod giving motion to the rear wheels via connecting rods and cranks; and the funnel in front, into which the exhaust was led, to create a blast and hence draw the fire. According to Fletcher, the bellows device for creating a blast was discarded after the first trip. This suggests that Trevithick discovered by accident that piece of equipment so essential to steam locomotives and traction engines: the blast pipe.

Other refinements on this pioneer vehicle included a safety valve; heating the feed water by passing the exhaust through it; a feed pump worked from the crosshead to maintain the water level in the boiler; a fusible plug in the crown of the flue to give warning of low water (which could cause an explosion); and a self-acting four-way cock for admitting steam alternately into the top and bottom of the cylinder. According to a local newspaper report, the steam carriage, when coupled to a carriage containing several people, could maintain 6 km (4 miles) per hour up a 'hill of considerable steepness' and 13–14 km (8–9 miles) per hour along the level. Steam pressure was 414 kN per sq m (60 lb per sq in).

Trevithick and his partner in this enterprise, Captain Andrew Vivian, then moved to London where they patented their invention in 1802 and built an improved version in 1803. In this the engine was tucked beneath a carriage body which was independently sprung; the cylinder was horizontal, and the whole engine unit was lighter and more compact. Again the cylinder was inserted into the boiler, which this time was of wrought-iron, and a forked piston rod was used with the ends working in guides so the crank axle could be close to the cylinder. Spur gearing took the drive from the crank axle to the main driving axle. The wooden drive wheels were about 3 m (10 ft) in diameter to make negotiation of the rough roads of the day easier. The stoker rode at the back and the steersman in front.

Trevithick's London carriage, as it became known, was put together at Felton's carriage shop in Leather Lane near Grays Inn Road, and it made many successful runs around the streets of London. Unable to gain financial backing, Trevithick soon turned his unbounded energy to railway locomotion, using the experience gained with his road vehicles. He was more successful in this field and, by 1811, had a locomotive running in London on demonstration. Meanwhile, the engine from his steam carriage went to drive a hoop rolling mill, which task it performed for many years.

An interesting oddity, patented in 1813, was 'Brunton's mechanical traveller' which was basically a Trevithick locomotive with horizontal boiler, which had two legs at the back with feet engaging the ground so it could push itself along as the piston reciprocated. Not surprisingly, it only achieved 3 km (2 miles) per hour and the motion must have been jerky in the extreme.

With the practicality of steam carriages now proved beyond any doubt, the years 1820–40 saw a minor boom. Griffiths is accredited with first fitting change-speed gears to a steam vehicle, in 1821, while differential gear to help negotiate corners was introduced by M. Pecqier in France in 1829. Ackerman steering, a German invention, was first used by Gough in 1830.

Development of the steam road vehicle was helped by outside factors, such as the invention of the laminated spring by Obadiah Elliott in 1805, and improved road construction under civil engineers Telford and McAdam. Artillery wheels with iron tyres, much stronger than conventional wheels, had also been patented in 1805 by Samuel Miller. These developments helped to produce lighter, more comfortable spring-mounted steam passenger vehicles, able to move faster. Though the use of high-pressure steam kept the engines themselves compact, it was more difficult to restrict the size of boilers, and many shapes and forms were tried.

Steam carriages were now able to be larger. In London in 1826, Sir Goldsworthy Gurney built a coach about 6 m (20 ft) long, capable of taking 6 inside and 15 outside passengers. His patent D-shaped boiler supplied steam to a 12-horsepower horizontal engine beneath the carriage, driving the 1,500 mm (5 ft) rear wheels directly. He also provided legs rather like Brunton's, but when

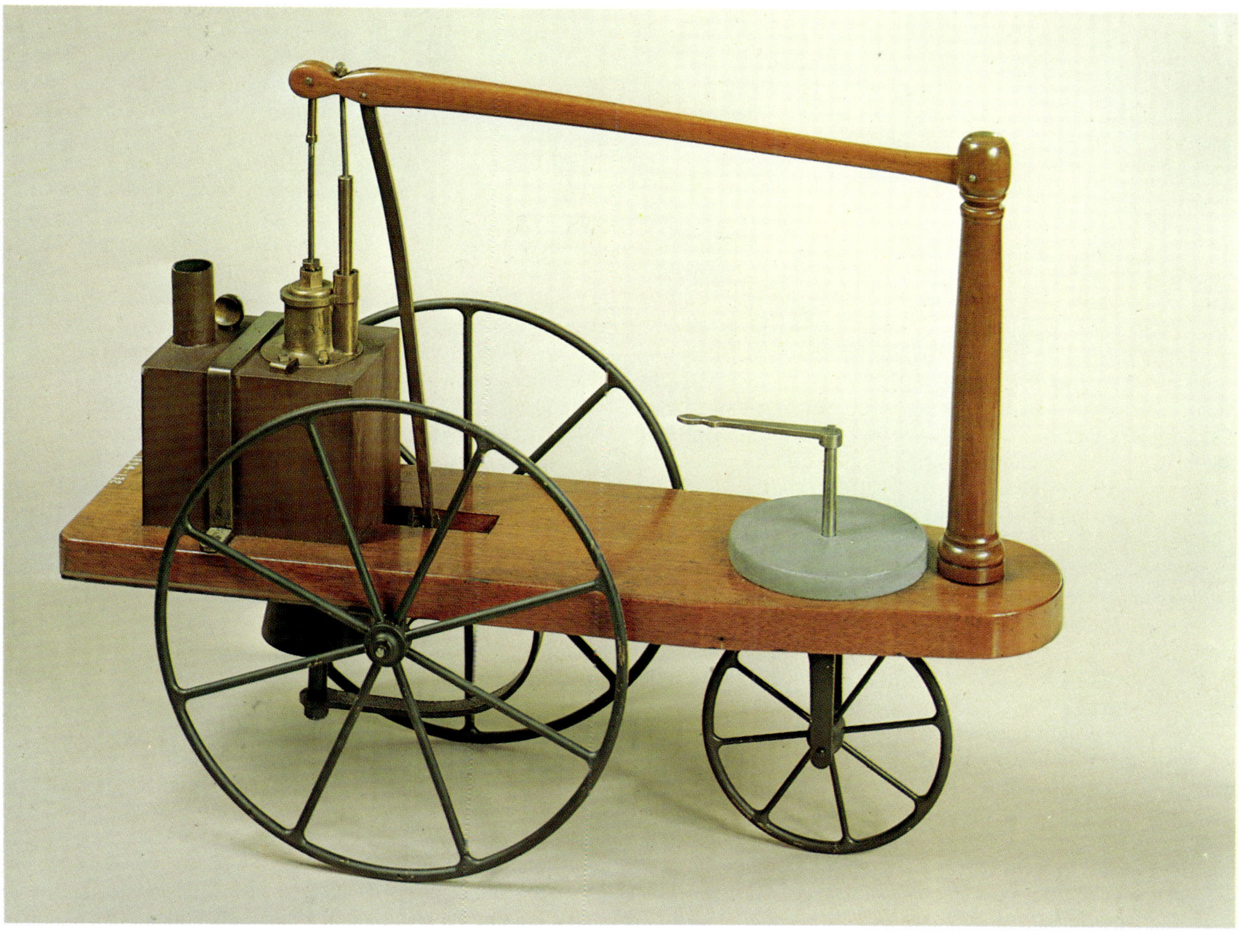

A model of William Murdock's spirit-fired steam carriage of 1784, also in the Science Museum.

One of the most promising early experimental aircraft was Sir Hiram Maxim's steam-driven biplane, which tried to take off from its track when it was tested at Bexleyheath in 1889. Naptha was used as fuel for the water-tube boiler and the 5.2 m diameter propellers were driven by a pair of 350 hp compound engines. Lack of finance caused the project to be abandoned.

his carriage first ascended Highgate Hill without the wheels slipping, they were removed. This carriage is reputed to have travelled from London to Bath and back.

In 1831 Sir Charles Dance started a regular service of Gurney carriages between Gloucester and Cheltenham. The vehicles consisted of a four-wheeled steam 'drag' coupled to an ordinary carriage. They ran for four months, carrying some 3,000 passengers, and frequently covered the 14 km (9 miles) in only 45 minutes. However, intense local opposition forced the service to be stopped—stones were laid across the road in the path of the carriages, and prohibitive turnpike dues were introduced (a foretaste of things to come). In those days, many ordinary citizens opposed the new-fangled 'steamers' on all sorts of grounds: their weight was said to damage roads and bridges, they made smoke, and horses were scared of them.

Probably the most successful of steam carriage schemers was Walter Hancock of Stratford in East London, who devoted 16 years to the subject, building 10 during that period. What set him on the road to success was his boiler, patented in 1827; this consisted of a number of flat chambers 50 mm (2 in) wide, arranged side by side. Into the 20 mm ($\frac{3}{4}$ in) wide spaces between chambers projected bosses touching one another, thus forming abutments and giving extra heating surface; the pressure used was 690 kN per sq m (100 lb per sq in) and, since the boiler was only 0.7 m (2 ft) square and 1 m (3 ft) high, it was highly suitable for its task.

Hancock also tried oscillating cylinders to save space, but reverted to fixed cylinders in 1832. By 1834 he was running his 'Autopsy' for hire in London, achieving speeds of up to 19 km (12 miles) per hour and, soon after, his 'Automaton' took 20 people along the Bow Road at 34 km (21 miles) per hour. His carriages rode well, were quiet and made little smoke or steam: the steam bus had arrived.

Gaining public acceptance, however, was still a problem. Heavy tolls and lack of financial support in Britain and lack of interest in France caused a period of stagnation from 1840 onwards. As we shall see later, steam-engine builders were concentrating on traction engines, developed from the portable engine, for heavy haulage and as such were exempt from tolls.

Some attempts to introduce steam carriages were also made in America. In 1853 J. F. Fisher in New York produced a light steam carriage with a water-tube boiler and two-cylinder engine, driving 1,500 mm (5 ft) wheels directly. But it was no advance on what had been done in Britain and proved too fragile for the appalling roads of the time. Fisher, Joseph Batin and Richard Gudgeon all built more vehicles, culminating in an 18-seater by Gudgeon in 1866, but nothing came of them.

In 1865 the steam road vehicle in England received another blow, in the shape of the infamous 'red flag' act. Aimed at controlling the behaviour of heavy traction engines on the road, it restricted speeds of all locomotives to 3 km (2 miles) per hour in towns and 6 km (4 miles) per hour in the country, and required a man holding a red flag to walk in front to warn other road users of the vehicle's approach. There were size and weight restrictions too, and local authorities were also empowered to control the hours of passage through towns and villages as they saw fit. The only steam-carriage development now possible was by a hardy few using private roads, who hoped that the law would one day be changed.

Twenty-five years later, Fletcher wrote bitterly: 'All the high-speed engines of recent times have been built for service in foreign countries—our foolish and meddlesome laws prohibiting sensible speeds in this country—hence Russia, Greece, Turkey, India, Ceylon, France, New Zealand and Germany are all ahead of Britain in this matter'.

It is remarkable that, in some respects, what development there was came out of the original starting points in the history of road steam, and no real progress was made. The work of Hancock and others was forgotten and large, clumsy boilers and single front-wheel steering reappeared. In the rest of Europe, however, where there was no such penal legislation, steam engine development flourished, particularly in France.

M. Lotz of Nantes built several steam tractors and omnibuses and was running regular services with them in 1864. Four years later, Joseph Ravel of Neuilly, near Paris, introduced a boiler fired on petroleum and fitted it to a small two-seater carriage, bringing the advent of the motorcar as we know it a step closer. But war with Germany stopped further work on it.

Steam buses were once common in London. The biggest fleet was that of the National Steam Car Company, under whose name Thomas Clarkson operated nearly 200 steam buses made in his works at Chelmsford before the First World War.

In 1873 another Frenchman, Amédée Bollée, reintroduced Ackerman steering in carriages he had under construction. Later he re-invented shaft drive from the engine to the driving axle as well as four-wheel drive, both of which he applied.

MOTORCYCLES, CARS AND AIRCRAFT

The first steam motorcycle is accredited to a Frenchman called Perreaux, who fitted a small engine and petroleum-fired boiler to a 'boneshaker' bicycle. Between 1878 and 1885, pedal tricycles enjoyed a spell of popularity and several attempts were made to fit steam drive to them. But, by 1885, with more and more engineers in Europe devoting their energies to mechanical propulsion, and the arrival of Gottlieb Daimler's first coughing single-cylinder petrol engines, the steam car had to make an immediate impact in order to survive even a short time. That it did enjoy a fair innings before being swamped by its rival is due to Leon Serpollet and his development work on the flash steam boiler.

The idea of pumping feed water into red hot tubes and so producing 'instant steam' was not new; it had been postulated by an Englishman, John Payne, in 1736. From 1823 onwards Jacob Perkins conducted experiments by pumping hot water through square iron tubes kept red hot in a furnace, and he managed to produce superheated steam at a pressure of no less than 10,340 kN per sq m (1,500 lb per sq in). However, the short life of the tubes and the difficulty of controlling temperature, feed water flow and pressure prevented his boiler from enjoying any practical application.

Serpollet enjoyed great success with his boiler which he used in his cars and continued to improve painstakingly for a number of years. His cars gained many prizes and in 1902 he held the world land speed record by travelling at 121 km (75.06 miles) per hour. His boiler contained a single long tube fashioned into a series of crossover loops. The flash-steam boiler's great advantage was its rapid start from cold (half a minute being common instead of an hour or more to reach working pressure), and its lightness, compactness and freedom from explosion risk. However, it was more difficult to regulate than a normal boiler containing a constant reserve of water and steam; moreover, the load on a car boiler tends to vary constantly because of changes in gradient, cornering and giving way to other traffic. The secret of success lay in having adequate means for regulating the amount of water pumped to the boiler and the rate at which the fire supplied heat to it.

In 1880 Parkyns managed to regulate automatically the supply of liquid fuel to the fire in accordance with boiler pressure. In about 1900, Serpollet, Pearson, Cox, and others who used flash boilers, controlled the burner by hand and often kept the boiler tube red hot; no attempt was made to govern the temperature of the steam leaving the boiler. Most of the boiler troubles at that time were caused by failure to regulate accurately the steam pressure and temperature.

By the time the restrictions in Britain on mechanically-propelled vehicles were lifted in around 1896, the British had been left out of the steam car race and most such machines used on its roads were imported.

In 1906 another steam car won the world land speed record, this time a US-built Stanley steamer which achieved 204.3 kph (127.66 mph) at Ormond Beach, Florida. Stanley steamers were in production by 1899, the first cars of any type with a light tubular chassis. The engine was a two-cylinder simple (i.e. non-compounded) design using a common valve chest and Stephenson's link motion. Early Stanleys had the engine arranged vertically, with chain drive from a sprocket midway between the cranks to the differential on the rear axle. One now in London's Science Museum has cylinders 62 by 88 mm ($2\frac{1}{2}$ by $3\frac{1}{2}$ in) bore and stroke which, with a steam pressure of 1,241 kN per sq m (180 lb per sq in), could develop about 6 bhp. The racing car of 1906 had a much bigger engine, 113 by 164 mm ($4\frac{1}{2}$ by $6\frac{1}{2}$ in) arranged horizontally, the pattern for the later Stanleys.

The White steam car, also made in the USA, was both the swan-song and the 'Rolls-Royce' of steam cars and appeared in nearly every country of the world. Its main technical features were use of a condenser to save frequent stops for water and provision for burning kerosene as well as petrol. It resembled a normal petrol-driven car in general form, having the engine head boiler under the bonnet in front with shaft and bevel gear drive to the back axle. The White was still being produced in 1910.

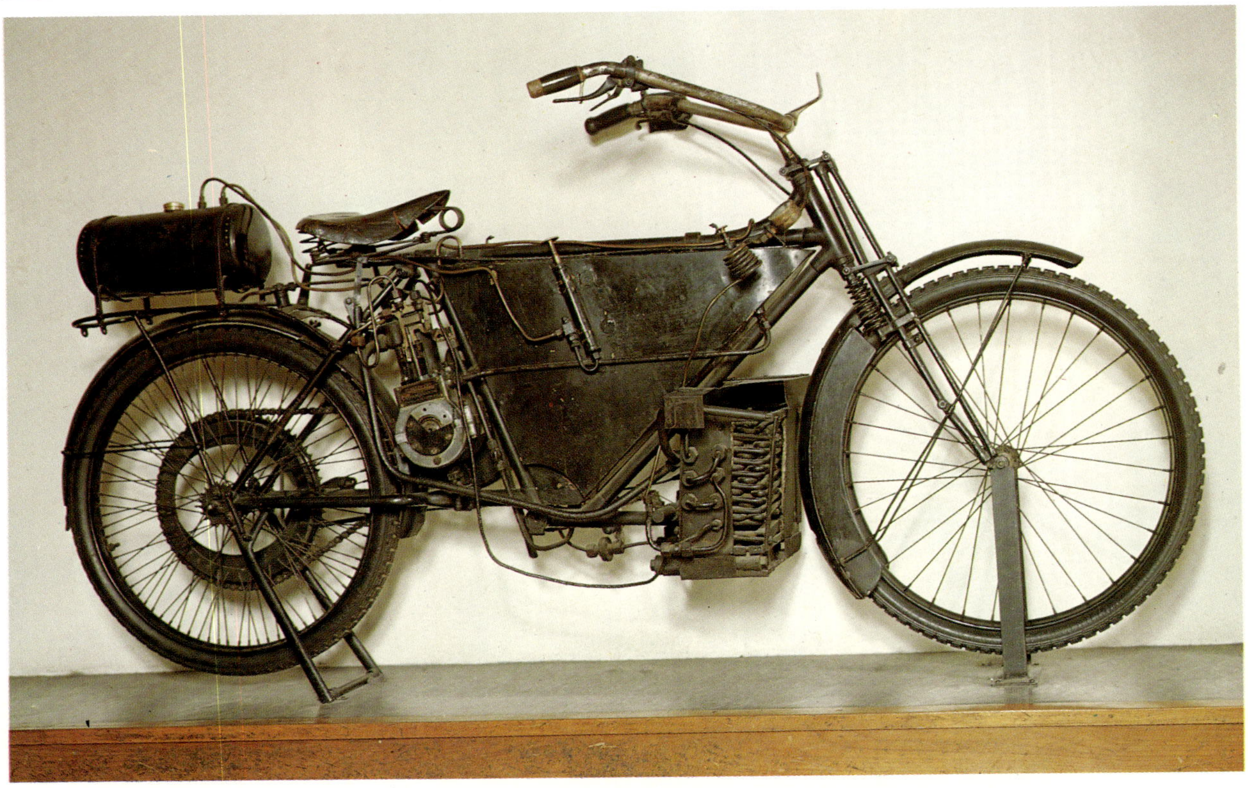

A Pearson & Cox steam-driven motorcycle of about 1913 with flash boiler, now in the Science Museum collection.

A steam wagon built by the Yorkshire Patent Steam Wagon Company in which the boiler was made Tee-shaped and placed crosswise to improve weight distribution and make better use of the limited space. This made the boiler more complicated. The firebox is situated in the upright part of the Tee, with the ashpan just in front of the leading axle.

VANS, LORRIES AND BUSES

Commercial vehicle development using steam propulsion had in the meantime pursued a separate course. Prior to 1895, all the heavier steam-driven vehicles built on the European continent were large public service vehicles for carrying passengers and little attention seems to have been paid towards lighter examples for carrying goods. But a Paris–Bordeaux–Paris race in 1895, in which the ic engine scored over steam, also aroused interest in the potential of small, light vans.

In England, Thornycroft, anticipating relaxation of the law, began development work on a steam van and built a prototype in 1896. This had a vertical boiler behind the driver and a horizontal engine underneath. Other makers of lorries soon followed: Leyland (the beginning of British Leyland today), Clarkson and Capel, and the Liquid Fuel Engineering Company (LIFU). Test trials for lorries, wagons and omnibuses took place in France in 1897; De Dion and Bouton and

A typical Foden 5-tonne tipping wagon of 1912. The tipping gear was worked either by a separate engine at the back of the cab or, as in this example, by attaching a belt to a pulley next to the flywheel.

Scott both had omnibuses in these trials. In 1902 steam buses were tried in London (70 years after Hancock), and the London General Omnibus Company ran a fleet, mainly of Clarksons, until 1909 when new Metropolitan regulations ousted them on weight grounds. By this time a rival company was operating buses with improved Daimler petrol-engines.

Nevertheless, by comparison with the short reign of the steam bus, from 1897 the steam lorry enjoyed some 30 years of popularity in Britain, until a new system of taxation by unladen weight was introduced. Even then, its replacement by petrol- and, later, diesel-driven vehicles, was only gradual. A few coke-fired steam wagons were still being operated by North Thames Gas Board after the Second World War; and a brief construction revival occurred in 1950 when the Sentinel Waggon Works at Shrewsbury fulfilled an order for 108 S4-type lorries for the Argentine.

The British steam wagon, as it was termed, fell into two principal categories: the undertype with a vertical boiler in front of or behind the driver and a horizontal engine slung underneath the chassis; or the overtype which followed traction engine practice with a horizontal loco-type boiler in front of the driver, and cylinders and motion on top. Only Clarkson used a flash boiler. Vertical boilers were mostly of the water-tube type, in contrast to fire-tube loco-type boilers.

The arrangement of cylinders in the undertypes varied. Most had two cylinders, simple or compounded. Drive from the crankshaft was taken through two-speed change gears with the final drive by pitch chain. There were a few departures from the norm, notably the Yorkshire steam wagon which had an undertype engine and the boiler arranged transversely across the front of the vehicle, and the Fowler which had a vertical boiler and cylinders in V formation at the rear of the cab.

Vertical boiler wagons tended to be poor steamers and, by 1906, most manufacturers were following Mann and going over to the horizontal loco-type boiler. The final and most successful form of overtype had the boiler shell forming an extension of the chassis frame, like a mini-traction engine but with cab and seats in place of a tender. It was produced by Aveling & Porter (the firm begun by Thomas Aveling), Burrell, Wallis & Steevens, Allchin, Garrett, Robey and, the biggest overtype producer, Edwin Foden of Sandbach in Cheshire. Every one of these was an established builder of traction engines and so 'knew the ropes' when it came to turning out practical steam vehicles. With the sole exception of the Sentinel Waggon Company of Shrewsbury, whose excellent water-tube boiler and dynamic designs were to outlast all the rest, the vertical boiler firms faded from the scene.

In the 1920s, as the competition from petrol became more intense, the loco-boiler lorry builders also gradually stopped production, leaving Sentinel to carry on alone after 1931. Just as this firm had amazed the world with its two-cylinder undertype Super Sentinel design in 1923, so in 1930 it did so again with a massive 15-tonne (15-ton) eight-wheeler, the DG8, with four steered wheels in front and four driven wheels at the rear, all on pneumatic tyres. This forerunner of today's juggernauts was more than 9 m (30 ft) long and carried 0.76 tonnes (15 cwt) of coal and 1,045 litres (230 gal) of water.

Two years later came the 6-tonne (6-ton) S4 with totally enclosed crankcase, four poppet-valve cylinders and shaft drive. With this fine vehicle, only a thin plume of steam above the small chimney showed that it was a steamer at all! The water-tube boiler was behind the driver who had a comfortable cab (at least in cold weather) and a rotating firegrate gave automatic stoking. There was no gear changing and with absolutely silent running, the looks on the faces of other lorry drivers as an S4 accelerated from traffic lights had to be seen to be believed! Some S4s remained in service until well after the Second World War, and a handful came back into operation during the 1956 Suez crisis. The late order for the Argentine was followed by the firm becoming part of Rolls-Royce, which meant that spares became impossible to obtain, so S4s are now, like traction engines, collector's items. Despite all the efforts made by Sentinel to keep steam a viable force on the roads, there was one thing the firm could do nothing about—it was very dirty work for the driver.

In the 1930s and 1940s, several German firms produced heavy steam commercial vehicles and steam railcars with engines fully in the Sentinel idiom. In 1948 Lenz and Butenuth developed two four-cylinder single-acting engines on the Uniflow principle (described on page 169) which was a conversion of a Henschel ic

A Stanley steam car of 1904 with steam provided by flash boiler and radiator-condenser.

engine. According to statistics published by the Germans, the Butenuth steam lorry gave a fuel cost of 2.2 pfennig per kilometre (0.62 mile) compared with 2.7 for an S4 Sentinel and 17.3 for a petrol lorry. The Butenuth had a corner-type water-tube boiler and could burn lignite, lignite briquettes or bituminous coal, with gravity feed and a sloping grate.

STEAM ON LAND: BRITISH TRACTION ENGINES

Prior to Trevithick, the established form of steam engine was a cumbersome, slow-moving beam engine requiring a condenser and a river of cooling water as well as a plentiful supply of low-pressure steam from a boiler to keep it at work. In 1802, when steam engines of even small power were described as being 'about as portable as a parish church', Matthew Murray of Leeds took out a patent for an engine which was 'transferable without being taken to pieces'. It could be used 'for any process or manufacture requiring circular motion, or for irrigating land, or for the various purposes of agriculture'. But it was Trevithick who actually began building portable engines, derivatives of his stationary high-pressure engine with the exhaust turned up the chimney to create a draught on the fire. By 1812 he was producing small portable engines on wheels for agricultural purposes, weighing only 0.76 tonnes (15 cwt) and costing 60 guineas.

A steam tramway locomotive built by John Fowler & Co. of Leeds is occasionally steamed at the Crich Tramway Museum in Derbyshire, where it is preserved.

A few years later, Trevithick went to South America, leaving much of his work unfinished. Owing to the country's disturbed state at that time, there was little incentive for anyone else to start building portables. But in 1830 Nathan Gough did, and his design consisted of a vertical boiler and cylinder on a wheeled timber frame. A forked connecting rod drove upwards at an angle to an overhung crank on the end of a shaft mounted across the boiler and carrying a flywheel at the other end. A pendulum ball governor was fitted, as with rotative beam engine practice.

The traction engine was, in fact, a logical development of the portable steam engine, conceived by Richard Trevithick as a derivative of his high-pressure engine in which boiler and engine were combined in one unit. Although Trevithick made his engine self-moving with

Driver's-eye view of a Burrell traction engine. The steering wheel is on the left (the reversing lever is out of the picture). Above the steam-pressure gauge can be seen the throttle, or regulator and, on the right, part of the gear-change mechanism.

his locomotive-cum-carriage in 1801, the method of applying motion when the first portables became self-moving in the 1840s bore no relation to his use of connecting rods and cranks between the crosshead and the wheels. Instead, rotary motion of the crankshaft was transmitted to the wheels by pitch chain and sprockets.

Like the diesel-powered farm tractor of today, the steam traction engine intended for agricultural duty spent some of its time on the road, hauling agricultural implements from job to job, as well as on the land itself, where it was used to drive machines of various kinds. (The traction engine's progeny, steamrollers and road locomotives designed specifically for road use, are discussed elsewhere in this chapter.)

With the chain off, the engine could drive farm or other machinery by a belt from its flywheel, as did the portable. But with the chain fitted, it could move itself along instead of requiring a team of horses to pull it, though at first a horse between shafts fixed to the swivelling forecarriage was still needed to steer. (It was said that once a horse became used to being pushed along by an engine, it was never any use for pulling again!) Once it could move itself, of course, the engine acquired various appendages: a tender for carrying fuel and water and a manstand for the driver-cum-stoker who had hitherto operated the engine from the ground. Thus was born the traction engine, a machine which, with surprisingly little refinement, continued in production for almost a century. The humble portable also continued to be built, and by the end of the 19th century, one firm alone had built 33,000 of them. They were produced in every manufacturing nation in the world, using the same basic formula of a horizontal locomotive-type boiler and 'overtype' engine on top, usually with the cylinders arranged over the firebox where the heat was greatest. They were made in various sizes up to 200 horsepower, some of the largest examples being made with removable wheels, which gave rise to the term 'semi-portable'. In some parts of the world, notably in Europe, fitting refinements like feedwater heaters and superheaters became commonplace.

Thousands of portable engines were exported to underdeveloped countries where low cost, simplicity and ability to burn a wide range of available fuels were important factors. Today they are still to be found driving sugar machinery and sawmills, and the boiler is fed on waste so that they cost nothing to run, at least until the boiler gives out.

In the early days, the combined high-pressure steam engine and boiler presented a formidable problem when it came to negotiating the bad roads. A boiler full of fire and scalding hot water with heavy pieces of cast-iron carrying revolving shafts bolted to it is clearly unsuited to being shaken about, and a high proportion of the many patents taken out for traction engines were on various forms of resilient wheel, to ease the shocks. As roads improved, so did the mechanics of the machines which ran on them, and the springing arrangements which ultimately became standard on traction engines and road locomotives are delightful in their simplicity.

A Tuxford-built portable engine as used in the early days of mechanized farming and now preserved in the Musée des Arts et Metiers in Paris. Note the Watt-type governor in the top right of the picture.

Facing page, top: *the earliest traction engines were little more than an adaptation of the portable engine, with cylinder(s) over the firebox and drive to the rear wheels by pitch chain from a pinion on the crankshaft. This is a recently-built replica of a Savage traction engine of about 1870. The steersman rode in front.*

Facing page, below: *a typical farmyard scene of the Edwardian era re-enacted on the rally field with a traction engine driving a threshing machine by belt. When the work was finished the engine would be unbelted, coupled up to the tackle and driven to the next place of work. The engine was built by Marshall of Gainsborough in 1905.*

Even on a tarmac road, an engine on steel strakes vibrates alarmingly when in motion, and it is a constant source of surprise to some of today's engine owners that their charges are still in one piece after going a short distance. The fact is, of course, that everything is made stronger and heavier than it theoretically needs to be, to counteract the effects of metal fatigue (quite the opposite of aircraft practice where, for lightness, every component is tailored right down to the limit).

Portable engines in Britain also acquired their own peculiar system of power rating. This was probably based either on the number of horses required to pull one from place to place on unprepared ground, or on the number of horses that would otherwise be needed to perform its work.

In actual practice, engine builders worked to a rule of thumb that one nominal horsepower was equal to 'ten circular inches of piston area'. Even in the early days, with steam pressures of around 310 kN per sq m (45 lb per sq in) an engine of, say, 5 nhp would doubtless produce more than that in terms of Watt's horsepower unit of 745 W (33,000 ft lb per minute). As steam pressures rose, the divergence between actual horsepower produced and nominal horsepower based on piston area became even more absurd. However, the nominal horsepower rating was established for portables, traction engines and road locomotives until the end of steam on the road, though steam tractors, wagons and rollers were classed on a weight basis. One enterprising builder of showman's road locomotives for fairground work refused to accept the 8 nhp rating the cylinder dimensions deserved and rated it at 65 brake horsepower, which was much more realistic. In North America, where portable and agricultural traction engine development closely paralleled that in Britain, power ratings were based more on the engine's actual capability.

The main stream of development of the traction engine family from the utilitarian portable engine is clear enough, but a host of short avenues were explored (on both sides of the Atlantic) in which engines took many different forms. In this short treatise, only the mainstream development in Britain will be followed, with US practice described at the end. Digressions on the various avenues will be confined to the few which enjoyed some success.

The increase in the use of steam power in the farmyard owes much to the setting up of a firm of agricultural engineers destined to become famous, J. R. and A. Ransome of Ipswich. In 1837, what are now the Orwell works were established and, at the English Agricultural Society's first Royal Show held at Oxford in 1839, the firm won a gold medal. Future Royal Shows were to provide a powerful stimulus to portable and traction engine development by encouraging the competitive spirit between manufacturers.

Ransomes showed their first portable engine at the 1841 show which was held in Liverpool. This had a vertical boiler. At the following year's show, at Bristol, a pitch chain and sprocket drive to the wheels, and a horse-steered forecarriage, had been added: this was probably the first traction engine.

For a time, there were few further developments. Then, in 1848, Clayton & Shuttleworth of Lincoln produced a portable of more or less modern form with a horizontal cylinder, wood-lagged but not jacketed, on top of the firebox of a loco-type boiler, driving forward to a 'bent' crankshaft carried in cast brackets on the boiler top. It was built in three sizes of 4, 5, and 6 nhp and the boiler worked at 310 kN per sq m (45 lb per sq in).

In 1849, at Leeds, Ransomes showed another self-moving engine: this one had been built locally at the Railway Foundry of E. B. Wilson, and resembled a railway locomotive in that it had two high-pressure cylinders arranged under the boiler. The drive, however, was another important milestone because there were two gear trains from the crankshaft to the hind wheels, so two speeds could be selected, with the crankshaft able to run free in the neutral position. The engine could also be steered from the footplate behind the boiler. Also resembling railway practice was the incorporation of a four-wheel tender to carry coal and water. When the engine was required for stationary duty, the tender had to be detached and the rear wheels jacked clear of the ground so that one wheel could be used as the belt pulley.

It is not known who was the first person to apply chain drive to a portable engine of accepted form, but one of the first was a bold innovator, Thomas Aveling of Rochester, whose firm was later to become world famous for its steamrollers. In 1859 he fitted chain drive to a Clayton & Shuttleworth portable, and a year later produced his first engine designed as a traction engine with a single cylinder arranged over the firebox, driving to a forward-mounted crankshaft. In 1862 he moved the cylinder to the front of the boiler with the crankshaft over the firebox. This simplified the drive and became the standard arrangement, later of course in association with gear drive.

Though chain drive was used by a number of manufacturers well into the 1870s, a gear drive overtype engine was made as early as 1854 by Richard Bach of Birmingham, using the parts of one of his 8 nhp portable engines. This was no ordinary engine, for it featured 'Boydell's patent endless railway', an ingenious arrangement of six huge hinged pads fitted to each wheel to prevent the engine from becoming bogged down. In fitting the special wheels, Bach found that the spur gear ring on the rear wheel came close enough to the crankshaft for him to fit a drive pinion on it which could be slid in or out of mesh with the spur ring.

A number of other firms (later to become famous) also made engines with Boydell's system; they included Charles Burrell of Thetford, Richard Garrett of Leiston, Clayton & Shuttleworth and Tuxford, both of Lincolnshire. There was a preponderance of builders in East Anglia because the flat terrain of that part of England was highly suited to mechanical cultivation. Boydell engines had single front wheels with similar pads at first; later, two plain wheels were used in combination with Boydell hind wheels. They must have been noisy contraptions on the road but they were excellent for

SAVAGE

Two ways of utilizing steam for ploughing: **above,** *a pair of Fowler ploughing engines tow the 'anti-balance' plough to and fro across the field between them, while* **below,** *a Mann steam tractor tows the implement.*

moving about or towing implements in muddy fields.

By the 1870s, traction engine driving wheels had adopted the familiar 'straked' tread shown in some of the illustrations, the width of the groove between the strakes and their angle being so arranged that, in theory at least, there would be no vibration when travelling over a smooth surface. In practice there was considerable vibration, as today's enginemen who venture on to tarmac roads with unrubbered wheels have found. The problem of moving on soggy ground was solved very simply by carrying loose on the engine a series of T-shaped projections which could be bolted to the hind wheels when necessary.

Another important feature of the traction engine, established in the early days, was a centrifugal ball governor to regulate the speed of the engine when it was running 'on the belt'. This was simply a miniature version of the governor applied to stationary steam engines. It was normally mounted on the cylinder block and connected to a butterfly valve in the passage between the boiler and steam chest. When the engine was on the road, the belt from the pulley on the crankshaft to the governor pulley was left off and the butterfly valve remained fully open. But when the engine was belted to a threshing drum, clover huller or other farm machine, the governor belt would be put on so that the engine speed would regulate itself, within limits, as the load varied.

Agricultural traction engines nearly all had a single cylinder and, to even out the motion, the crankshaft carried a large spoked flywheel at one end with a slightly convex rim for retaining the belt. This was in contrast to later portable engine practice where the drive pulley was separate from the flywheel. With two speed gears, crank, governor pulley, two eccentrics for working the valve gear and the necessity of ample-sized bearings, a traction engine crankshaft tended to become crowded enough without a drive pulley as well.

The general principles of transmitting drive through gears to the wheels also became established during the 1870s, but actual details varied widely between manu-

facturers. In general, engines were either three- of four-shaft machines, depending on whether there were one or two intermediate shafts between crankshaft and hind wheels. These intermediate shafts were arranged either above the boiler 'backhead' or just behind it—in a four-shaft engine, usually both. They were carried in bearings fixed to the 'hornplates' (described below) and the gears were fixed to one end of the shafts so that the whole train could be accommodated inside a neat casing down one side of the engine. Change-speed arrangements varied.

Clearly, in a four-shaft engine the crankshaft revolved in the opposite direction to the hind wheels and, in view of the extra friction and extra complication, it seems at first sight surprising that so many traction engines had this arrangement. The reason lies in the mechanical convenience of having an intermediate shaft close to the crankshaft which enabled the primary gears and gear change to be arranged wholly or partly between the bearings. Where the crankshaft pinion was overhung, there was always the risk of bending the crankshaft. For instance, should the engine get stuck in a tight corner, the driver might vigorously rock it forward and back on the reversing lever, using the energy in the flywheel and the backlash in the gears to precipitate motion. This sort of treatment, necessary at times, imposed great strains on the mechanical parts.

One of the biggest detail improvements in traction engine construction was Thomas Aveling's design of 'hornplates' which he introduced in 1870, in a small gear-drive engine. It was customary at that time for the crankshaft and intermediate shaft bearings to be carried on brackets bolted to the boiler, which used to cause problems when the bolts worked loose. Aveling hit on the simple plan of extending upwards and rearwards the outer side plates of the firebox, which, in a loco-type boiler, were of course flat. The shaft bearings were then fixed to openings in the hornplates. A 'spectacle plate' (so called because it had openings for the connecting rod, eccentric rods and governor belt to

Above: *a typical single-cylinder portable steam engine, intended to be towed from farm to farm by horses.*
Below: *Savage's first showman's engine, called 'Empress', was designed as a 'traction centre' for driving a roundabout.*

Above left: *a small fairground organ preserved in the USA. It is mounted on a circus-type road wagon and plays from perforated paper rolls.*

Above right: *a typical steam centre engine built by Savage of King's Lynn in 1923 for driving a merry-go-round or galloping horse roundabout. A centre engine is a type of small portable engine permanently mounted on the roundabout's centre truck and with the drive taken to the top gearing via bevel gears, and an upright shaft concealed within the decorative mirrored panelling.*

work through) was bolted across the front and a plain plate across the back, forming a highly rigid box construction. At first the arrangement was covered by patent but, because of its mechanical excellence, it later became standard. It was the last fundamental improvement in British traction engine construction.

Other features of Aveling's 1870 traction engine are worth noting, as they became more or less standard: the cylinder was steam-jacketed with the boiler safety valves and regulator valve in the top of the casing, well above the water level in the boiler and worked by a rod from the footplate. Four slide bars were used to guide the crosshead, and the slide valve was actuated by Stephenson's link motion, following normal locomotive and stationary engine practice. The reversing lever and quadrant were on the footplate.

DESIGN, LAYOUT AND REFINEMENTS

At this point it is worth summarizing the main features of traction engine design in its final form, combining the best ideas of the many early manufacturers. Water was carried in a deep tank which overhung the back of the engine, being rigidly bolted to the rear of the hornplates. The top of the tank acted as a footplate for the crew and the space behind was boxed in to form a coal space, the whole tender being of riveted steel plate construction.

Partly because of road legislation, it became customary to have two men on the footplate, although one could manage; rollers for example, with exactly the same controls but with perhaps less tendency for the steering wheel to be jerked out of one's hand, were always driven by one man. On a traction engine one man would steer—in later years a steersman's seat and footrest were provided—and the other would manage the regulator, reversing lever, handbrake and injector or feed pump, and look after the fire. This split of duties was quite different to railway practice and was made necessary by the steersman having to react constantly to the road ahead.

The handbrake was never more than a parking brake or gentle retarder for descending hills; usually it took the form of a band brake worked by screw and acting on either a driving hub on the hind axle or on one of the shafts. To stop suddenly, the engine would be put into reverse, causing the piston to act as an air compressor; or, for even greater effect, the regulator could be used to put a whiff of steam against the piston as well.

Feeding the boiler with water in accordance with the demand for steam was invariably by feed pump in the early days, but injectors also came into use, though never to the exclusion of the pump. Most engines had one of each. Injectors were also used on other self-moving steam equipment such as cranes and locomotives. On a portable engine, a feed pump was generally preferred because the rate of working was usually constant, which meant that the driver could adjust the pump to feed on every stroke and then forget about it. Similar remarks, of course, apply to a traction engine on belt work, and most makers always fitted a pump to traction engines.

Aveling's engine of 1870 lacked one important refinement which came later, namely, compensating gear or a 'differential' to enable the hind wheels to be driven at different speeds when cornering. Some makers fitted the compensating gear in the final drive to the back axle: others fitted it to the next intermediate shaft where it could be made smaller, as the torque was less, but it meant duplicating the final drive gears to serve each wheel independently. Aveling's 1870 engine had two driving drums keyed to the back axle which had a choice of positions for 'driving pins', for which there was provision in the wheel hubs. To get round tight corners, one pin could be withdrawn, freeing the wheel from the axle on that side. The driving pins did not disappear with the arrival of the differential, however, as it became usual to fit a winding drum and wire rope to one of the drive hubs, and to use the winch meant having the engine shafts revolving while the wheels remained stationary. Apart from the ability to pull down small trees or extract a threshing drum from thick mud, an engine could also use its winch to extricate itself from sticky positions, with the other end of the wire rope tied to a tree trunk or another suitable solid object.

No weather protection was provided until later in the 19th century, when full- or half-length awnings were fitted if the customer required. Full-length awnings extending to just behind the funnel were supposed to save wear on the motion-work by giving it partial protection from grit and soot, but many people considered that they spoilt the look of the engine. Another familiar piece of traction engine equipment was the wire-reinforced suction hose dangling from the tender. This was used in conjunction with a steam water-lifter on

Traction engines used on fairgrounds were equipped with special decoration and a dynamo in front of the chimney. The dynamo was belted to the engine's flywheel for generating power and light when the engine was not travelling. To move the equipment from site to site, the traction engine performed a haulage function: at 20–25 km per hour (8–10 miles per hour). This engine, called 'Supreme', was one of the very last built, in 1934.

the engine to replenish the water tank from ponds and streams. The whereabouts of all likely sources of water in the district was something with which the driver had to be intimate. No boss liked to hear that his engine was immobile far from base because it had run so seriously short of water that the driver had been obliged to throw the fire out.

When compounding by the use of a second cylinder became commonplace in stationary engines, and steam pressures had reached 1,034–1,379 kN per sq m (150–200 lb per sq in), compounding also became used in traction engines for the same reason—fuel economy. The normal arrangement was to have two cylinders side by side, of different diameters and the same stroke, the cranks being set at 90 degrees. To save pulling the flywheel round by hand if the engine stuck on dead centre at starting, a 'simpling' valve could be worked from the footplate which let live steam momentarily into the low-pressure cylinder. The double-crank compound arrangement produced an even more crowded crankshaft with its two cranks and sets of eccentrics, and gave the driver nearly twice as much to oil. Because of the added complication, compounding was used more on road locomotives (so as to save on the amount of water and coal to be carried) than on agricultural traction engines.

There were several other designs of compounding employed by various manufacturers. The most common, after the double-crank compound, was Burrell's single-crank compound which, as its name suggests, used only one set of motion. It did this by having the high-pressure piston rod joined to the same crosshead as the low-pressure one. This arrangement was cheaper but it still did not help get an engine off dead centre, and on later engines with front springs the large reciprocating mass could cause the front of the engine to jump up and down at critical speeds.

The most common sizes for portable and traction engines were 5, 6, 7, 8 and 10 nhp but some were smaller and many more, usually for export, were made larger, up to 20 nhp. To give an example of the amount of work a typical single-cylinder traction engine built at the turn of the century could do, here is a hill-climb test recorded in the journal *Engineering* on an 8 nhp engine built by Clayton & Shuttleworth for the Khedive of Egypt. Up a long hill near the works at Lincoln, with a gradient averaging 1 in $8\frac{1}{2}$, the engine hauled more than 26 tonnes (26 tons)—that is, more than twice its own weight—'at a good speed without a stoppage', steam blowing off at the safety valves all the way. The working pressure was 1,034 kN per sq m (150 lb per sq in), the road speeds of 3 and 6 km (2 and 4 miles) per hour corresponding to

This is a typical Continental fair organ built at the turn of the century by Gavioli in Germany and designed to occupy a narrow space in the middle of a roundabout. It is now mounted in a box truck for protection. In addition to wind and percussion effect, the miniature bandmaster in front moves his arms with the music, which is fed through the instrument in the form of a folding punched cardboard strip. Originally both the bellows drive and music transport would have been provided by belt from a small steam engine.

155 rpm of the crankshaft. The flywheel was 1,372 mm (4 ft 6 in) in diameter by a 165 mm (6½ in) face, hind wheels 1,905 mm (6 ft 3 in) by 406 mm (16 in) and front wheels 1,219 mm (4 ft) by 229 mm (9 in). The cylinder dimensions were 229 mm (9 in) bore by 305 mm (12 in) stroke, a common size for 8 nhp engines. Coal consumption is not recorded, but the fact that the engine was blowing off continuously (a very wasteful practice) indicates that it must have been high.

By the late 1870s, British traction engines had developed into fine, rugged machines with the minimum of 'frills'. Although speed on the road was limited by the legislation referred to in the steam carriage section, under the influence of Thomas Aveling a special variant, the steamroller, was introduced. Following the dramatic improvement in road surfaces caused by the use of these machines, another close relative evolved, the road locomotive. The refinements these engines brought with them included springing, solid rubber tyres and the use of a third, higher, speed, and they are more fully discussed on pages 134–8.

Agricultural traction engines continued to be built well into the 1930s, and the last example, turned out in 1938, was a 5 nhp single-cylinder engine by William Foster of Lincoln. It is now in private preservation. Until the early 1920s, harvest time in country districts was always enlivened by the sight of traction engines hauling threshing machine, living van and watercart. Manoeuvring the equipment through narrow gateways and lining up the engine to get the belt on demanded great skill on the part of the driver. Then, for the next two or three days, the farmyard would be animated by flying chaff, the smell of coal smoke and hot oil, and the gentle, rhythmic 'dufter-dufter' sound of the engine's exhaust as its flapping belt made the drum hum. Within a few short years, however, the tractor had almost completely taken over, and most traction engines stood silent in empty yards or corners of fields. The tractor was cheaper to buy, cheaper to run, and easier to manage.

The preservation movement began in 1952, when Oxfordshire farmer Arthur Napper challenged the Marquis of Bath to a traction engine race. The interest this aroused meant that most of the final survivors were saved. Now, every summer, on rally fields up and down the country, the peculiar sights, sounds and smells of steam on the farm are recreated.

An important facet of farm work is deep ploughing or cultivating, and a special breed of large traction engines was developed, from 1860 onwards, for this purpose, mainly by John Fowler at Leeds. His two-engine system was to have two traction engines facing in the same direction on each side of the field. Beneath the boiler of each was slung a large cable drum and, by a simple clutch device, the driver could change the engine from self-moving to drum driving in a few seconds.

Fowler built most kinds of steam engine at different times—this is a fine example of an 8-tonne (ton) compound steamroller by the Leeds firm.

Some steamrollers were of the tandem type, that is, they had a full-width roller back and front. This vertical-boilered example was built by Buffalo-Pitts in the USA and is now preserved.

The implement, usually the famous Fowler anti-balance plough, was towed from one engine to another, each engine alternately pulling the plough and then moving forward to the next position when its neighbour began pulling. Using double-crank compound engines of up to 16 nhp, six to ten furrows could be ploughed at once. The system had the great advantage that, since the cultivator was not towed directly by the engine, the ground being ploughed was not compressed by the weight of the engine first passing over it—as of course happens with modern tractor ploughing. The drawback of course was that the 'headland' on each side of the field on which the engines travelled was not ploughed, but in the very large fields common in East Anglia, the headlands were a small proportion of the total area.

A few sets of Fowler ploughing tackle remained at work until the 1950s. In West Germany, the last two-engine sets whose engines (by Magdeburg) were even bigger than Fowler's, and went up to 25 nhp, stopped work very recently. Many British-built ploughing engines went abroad to underdeveloped areas where they were in use for longer than in Britain, although under the ill-fated Government 'ground-nuts' scheme of about 1950, a number of derelict Fowler engines up and down Britain were hastily refurbished, which may have extended the steam ploughing era there by a few years.

A modification of the two-engine system used a single engine driving an endless rope, the loose pulley at the other end being carried on a mobile anchorage. A McLaren engine of 1881, said to have been used with this system, is in preservation. A number of traction engine builders also produced ploughing engines, mostly with underslung cable drums in the Fowler position, but some with side drums and one oddity, designed by Savory in 1861, which had the drum wrapped round the boiler. This was long before the days of regular boiler examinations!

No story of traction engine development would be complete without mentioning some of the attempts to turn the engine itself into a direct cultivating unit. Most of them failed because of the cumbersome nature of the equipment and the fundamental snag that the wheels of the prime mover tended to consolidate the ground in the path of the digging implements, thus causing extra work. The use of a traction engine to tow the implement behind it suffered a similar drawback.

The first steam digger actually built was James Usher's steam plough of 1849, used in Scotland. In this an undertype engine running on rollers drove a shaft in front with five circular discs carrying ploughshares and able to be raised for travelling and lowered for ploughing.

A joint effort by Canada and England was the Romaine-Crosskill digger, whose originator was a Canadian, Robert Romaine. A prototype built in each country in 1853–4 took the form of a horse-drawn portable engine driving a rear-mounted 1.95 m (6 ft 6 in) long hollow wrought-iron drum carrying a series of curved digging claws. Mr Crosskill of Beverley, Yorkshire, built a self-moving version which is reported to have ploughed 0.3–0.4 hectare ($\frac{3}{4}$–1 acre) per hour according to the ground, but no further developments seem to have occurred after 1858.

The earliest steam cultivator which enjoyed any success was the Darby broadside digger illustrated of which 30 examples were made by various traction engine firms between 1877 and 1898. Inventor Thomas Darby of Pleshy, Essex, designed a machine like a traction engine turned sideways. It was carried on four wheels on two axles in line and an outrigger at the rear carrying steering discs. The advantage of this strange arrangement was the enormous digging width of 6.2 m (20 ft 6 in). Three sets of tines like large combs were arranged parallel to the boiler and driven up and down by a cranked shaft running along its top. For travelling on the road, each axle was turned through 90 degrees and the steering outrigger unshipped. The principal manufacturer of the Darby digger was Frederick Savage of King's Lynn who brought in some modifications and used his own 8 nhp cylinder size of 232 mm ($9\frac{1}{8}$ in) by 305 mm (12 in).

By 1890 Davey Paxman was building a lighter version in which the engine faced the normal way; this was a modified single-cylinder traction engine with the dig-

NU 3867

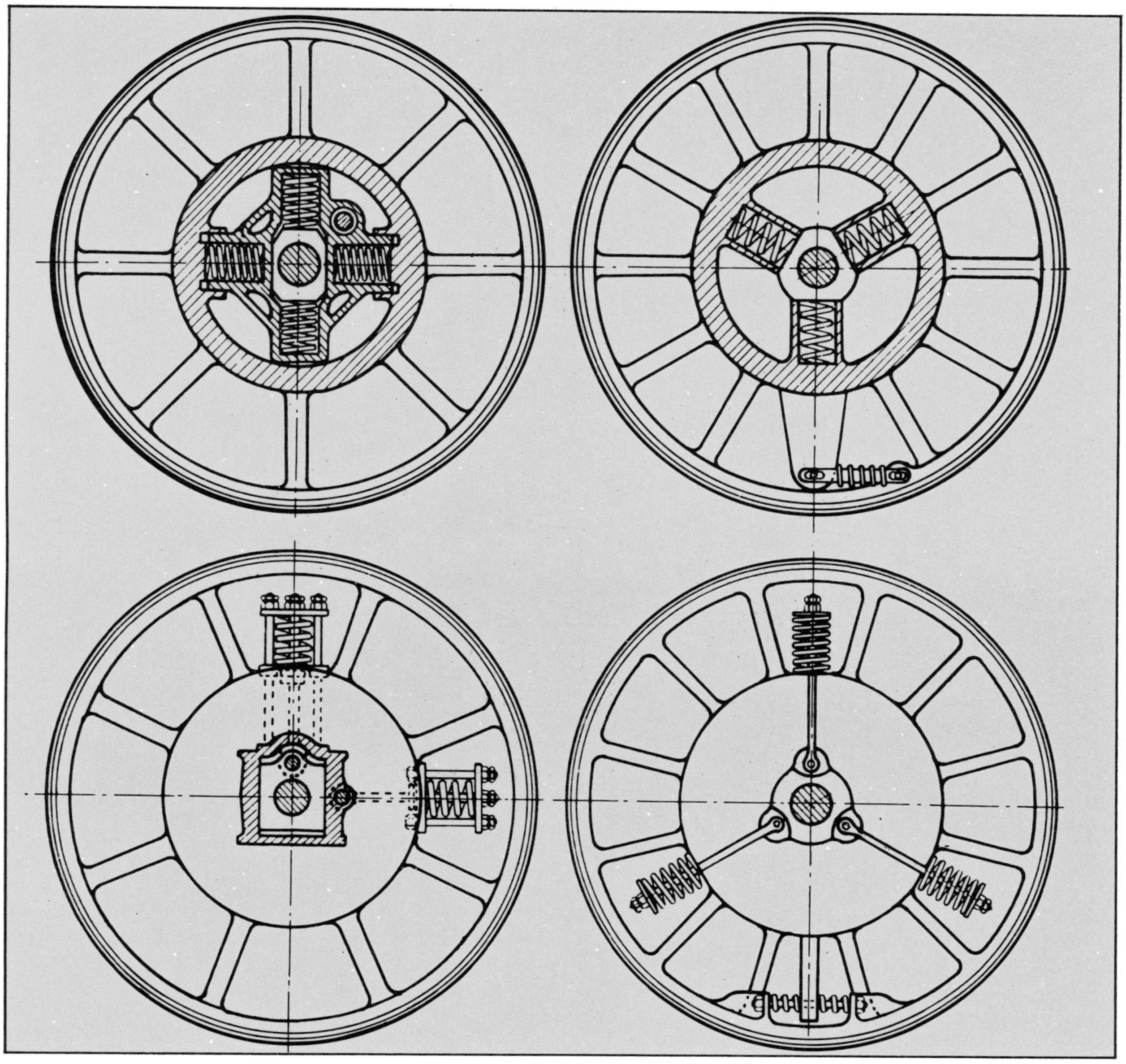

Rough roads were a problem to traction-engine builders: they jarred the machinery. Here are four ways of building springing into the hind wheels, which were patented by Charles Burrell of Thetford, Norfolk.

ging equipment, covering a 3.3 m (11 ft) width, fixed to the rear. The crankshaft, in the normal position over the firebox, carried two flywheels and a crankpin on each communicated motion to the tines via massive inclined levers. In the final version, the tines were replaced by a series of rotating forks projecting into the ground and taking their motion via a train of spur and bevel gears off the engine's second motion spur ring. Darby claimed that this digger could be fitted to any make of engine and among those so equipped were examples by Burrell, Fowler, Clayton & Shuttleworth, Wallis & Steevens, Ransomes, Sims & Jeffries and Ruston Proctor.

The Proctor digger also comprised a set of attachments which could be fixed to the rear of a normal traction engine. The equipment consisted of a series of fork arms projecting down from the back of the tender, actuated by a three-throw crankshaft running across the tender driven by spur gears from the engine's second motion shaft. One of the first Proctor diggers was fitted to a 4 nhp Burrell traction engine and went to South America in 1886. Further Burrell engines were so fitted, following improvements contained in a joint Frank Proctor-Frederick Burrell patent taken out in the same year.

The machine produced by the Cooper Steam Digger Company at King's Lynn was not an attachment but a purpose-built engine-digger unit. It used similar general principles to the final Darby design and one was tested at the 1900 Royal Show at York. The engine unit was a double-crank compound, at first arranged in the normal overtype position but later tried at the rear. The final design of 1903 had only three wheels and a very compact overtype engine arrangement with the driver in front. The engine was totally enclosed and forced lubricated and took steam at 2,068 kN per sq m (300 lb per sq in) from a return-tube boiler. Some of the earlier Cooper diggers were despatched to Egypt in pieces and assembled locally—an early example of 'ckd' practice.

This was the end of development of steam cultivating machinery in this country. How the modern combine harvester originated in the USA as an attachment to a traction engine is discussed later.

That relatively modern device, the crawler or the 'caterpillar' track, was also once fitted to a traction engine. The idea really originated with the Boydell wheel and a much later derivative, the 'Botrail', tried by Fowler. In these the shoes were not pinned to the wheels but fixed with steel cables to allow more flexibility. Short linked tracks passing over two wheels appeared in America in 1862 (see page 132), but the full-blown crawler track extending the length of the vehicle did not appear until Hornsby of Grantham built an oil-engined chain track tractor in about 1906.

An engineer in the Yukon, needing a cross-country tractor for hauling coal 64 km (40 miles) to Dawson City from a mine, then wrote to Hornsby asking for a steam caterpillar tractor. The result was a 41 tonne (40 ton) machine in which Foster of Lincoln supplied the double-crank compound engine and boiler, while Hornsby

Facing page: *Aveling & Porter of Rochester built more steamrollers than any other firm and exported them all over the world. Here is a typical 10 tonne (ton) piston-valve roller of 1921, now owned by one of the authors. Formerly it was used for rolling in road-base materials.*

Another problem for traction engines was soft ground which caused the wheels to lose grip and the engine to become bogged. This 'walking feet' idea was defeated by its complication—it was not patented by Diplock until 1909.

made the crawler undercarriage. Though the machine accidentally knocked down a house on its first trial run in England, it began work in the Yukon in 1910 and ran there successfully for many years. The undercarriage has recently been found intact and it is probable that the engine will be rebuilt for preservation. Sadly, Hornsby sold the track patent rights to an American firm who duly made a huge profit from it.

Although the USA was the other major centre of traction engine development, it is worth noting that there were also a number of engine builders in Europe. Large-scale production seems to have been confined to France and Germany, however, and, with the possible exception of the latter country, where firms tended to prefer sophisticated valve and cylinder arrangements, European development lagged behind that in Britain. Many British firms did a great deal of export business with south-east Asia, Russia and even Germany, and British steamrollers also did well in many parts of Europe. Subsidiary firms, such as Hofherr Schrantz, Clayton & Shuttleworth of Budapest and Vienna, and Fowler at Magdeburg in Germany, competed with the national manufacturers in the three countries.

The engine built by Hofherr Schrantz in Hungary in 1902, shown in the illustration, reflects in some respects what British firms had been doing 40 years earlier. The bevel-gear drive to a pinion shaft behind the hind axle is ingenious but note the single cylinder over the firebox driving forward as in portable engine practice, and the very shallow tender carrying only coal (water being held in a 'belly tank' slung under the boiler barrel).

Vertical-boiler tractors were being produced in France in the 1830s, particularly by Charles Dietz, but not until the trials of some Aveling products in 1863 was much attention paid to agricultural engines. Later, steam ploughing became fashionable in France, partly, no doubt, because the very large, gently undulating or level fields there made the use of these engines comparatively simple. Ploughing engines were built by *Societé Française de Materiel Agricole et Industriel* of Vierzon, L. Pécard of Nevers and other firms, while *Maison Albaret* of Rantigny produced both traction engines and steamrollers. The latter firm is well known for rubber-tyred vibrating rollers today.

In Germany, Kemna of Breslau, Wolf of Magdeburg and the Rheinmetal Company all produced ploughing engines with a resemblance to Fowlers, while Lanz, Maffei and Maschinen Fabrik Badenia built traction engines. Kemna also built what would today be called road locomotives and tractors. One of the latter was used by a British travelling showman, a very rare example of a Continental make being used in Britain.

Traction engines, and probably portables too, were also built in Sweden by Munktells, in Hungary by Elso Magyar Gazdasgi Gepgyar and in Poland by Cegielski.

AMERICAN TRACTION ENGINES

Traction engines became widely used in North America, as well as in Europe. Although their early development paralleled that of British and European engines, from the 1860s on their courses began to diverge. The ultimate solution adopted by most engine builders everywhere embodied the same overall principles (the locomotive boiler acting as the main frame and carrying an overtype horizontal engine transmitting its motion to the hind wheels through a gear train) but many of the actual details of the American versions were so different that it seemed almost that American engine builders were doing the opposite things just for the sake of it. Some firms were non-conformist in general layout right to the end; Avery, for example, insisted on putting the cylinders and motion underneath the boiler. He produced some fine, modern undertype machines which, because of the uncluttered boiler, bore a marked resemblance to railway locomotives. The firm of Best, founded by Daniel Best of San Leandro, California, favoured a vertical 'coffeepot' style of boiler, its chief asset being faster steam raising.

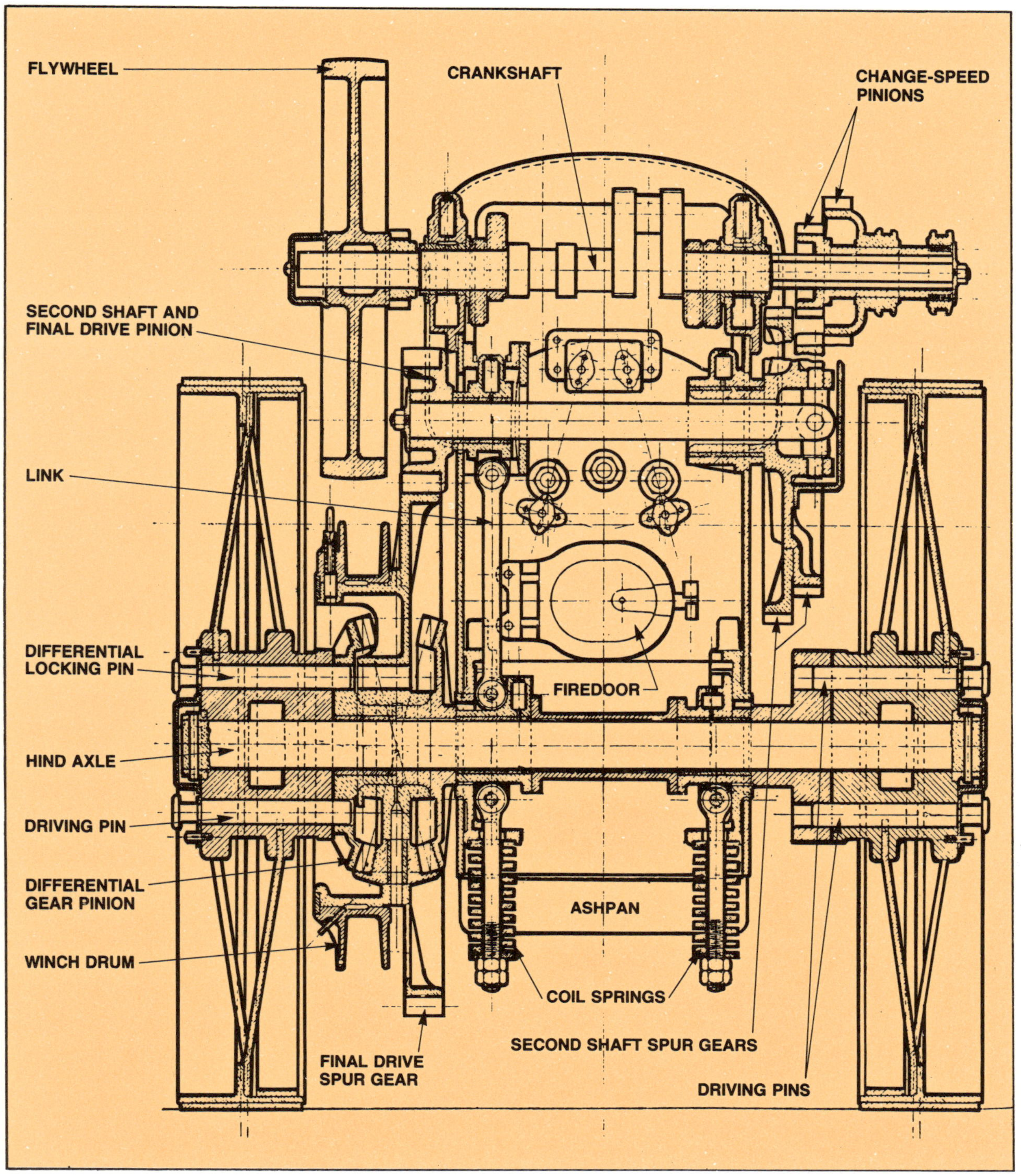

Hind-axle section through a Clayton & Shuttleworth traction engine, showing the drive gearing, compensating gear (differential) and springing.

Early American engines were equipped to burn wood or straw so that they could rely on local fuel supplies. (The only British engines with this feature were those built for export.) It meant that boilers had to have large heating surfaces to suit the lower calorific value compared with coal. At first, American traction engine practice was behind that in Britain; for example, portable engines shown in catalogues of 1880 still looked pretty primitive and had wooden wheels, while horse steering of traction engines was commonplace. The placing of the cylinder over the firebox also persisted for a long time.

American engines were also made in very large sizes, not only because of the fuel problem but because rural roads and farmyards were far more spacious than in Britain, and because the nature of the work to be performed demanded a high output. Thus the traction engine emulated the very large size of the American railway locomotives which was permitted by a generous loading gauge. It also, like its railway counterpart, tended to sprout a forest of ugly pipework and other external clutter.

Traction engines ranged in size up to 40 nominal horsepower. One of the biggest had hind wheel rims no less than 1.4 m (4 ft 6 in) wide. One engine over 40 nhp was the Gaar-Scott ploughing engine which had four cylinders in a double-tandem compound arrangement. Best's vertical boiler engines were actually the biggest self-moving road engines ever built. Their large logging engine had a chimney height of more than 6 m (20 ft) and weighed 66 tonnes (65 tons), compared with about 22 tonnes (22 tons) for the biggest British road locomotive (a showman's, with dynamo and special fairground trim).

There were also some traction engine manufacturers in Canada, and they closely followed US influence. Before discussing a few actual types, it is worth discussing in more detail the features by which American traction engines in general can be instantly distinguished from their European counterparts. Apart from size—

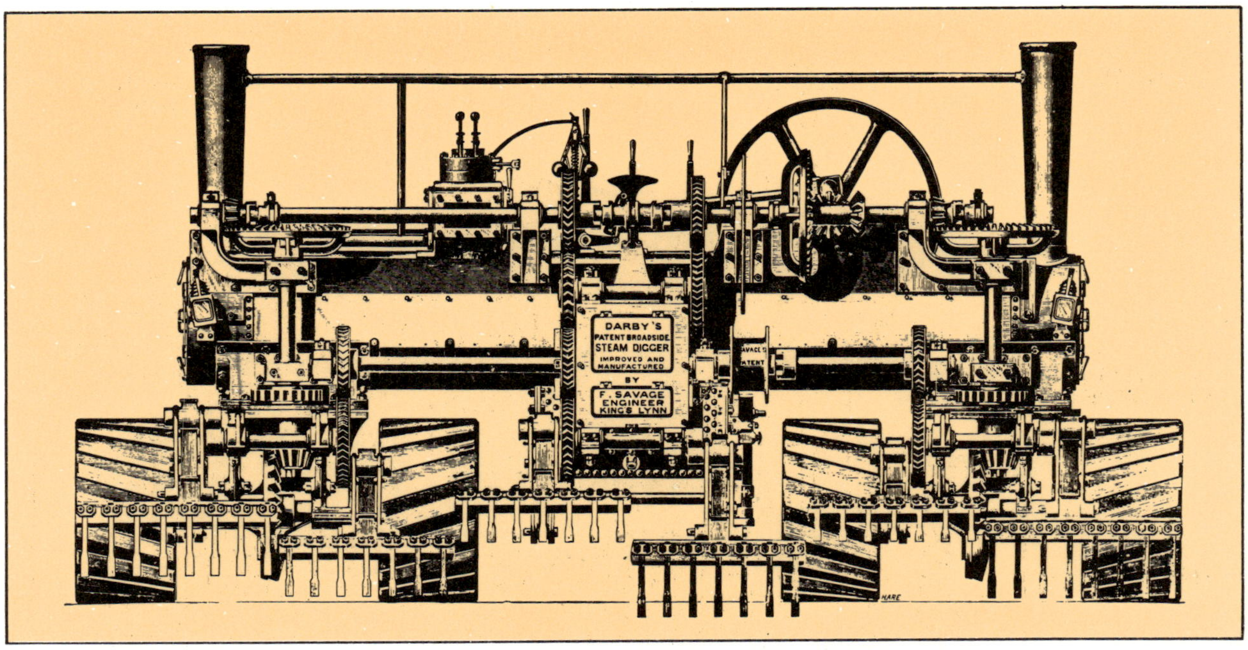

An early attempt at mechanical cultivation was the Darby transverse digger which moved sideways across the field, driving the digging forks up and down beneath it. This one was built by Savage of King's Lynn, in the 1880s.

and some firms did make small machines down to about 20 bhp, or 7 nhp—the boiler barrel was made long and slender and was entirely bereft of lagging, so that all lap joints and rivet heads in the plates could be seen. The front axle forecarriage pivot was fixed to the front of the boiler barrel instead of to the smokebox so the latter stuck out in front. This was done to reduce the turning circle on a long machine.

The motion, generally single cylinder but sometimes tandem compound, used a disc crank which was necessarily placed at one end of the crankshaft with the spoked flywheel at the other. It was normal to provide a cast engine frame incorporating a trunk guide for the crosshead, an arrangement which relieved the boiler shell of stresses due to piston rod thrust, in contrast to European practice. Because of the long boiler, the engine unit was much shorter and in large engines there would be a lot of boiler top not covered with motion. The engine unit was also offset to one side to miss the steam dome—another departure from European practice—and keep it close to the boiler centreline. The cylinder sometimes faced forward and sometimes backward, and the flywheel could be either on the offside or nearside, whereas in Britain it was almost invariably on the nearside.

The large steam dome was used to collect dry steam for the cylinder well above the water-line, and was connected to the cylinder block by a prominent external pipe in which the regulating valve was situated. This again was in direct contrast to English practice wherein steam was taken round past the cylinder and direct into the top of the block. The dome was necessary in view of the height to which the water level had to be maintained in such an attenuated boiler, to prevent the firebox crown from being uncovered when descending a hill. Live steam to feed the injector, water lifter, whistle and so on was also taken off the dome.

Another refinement was a friction clutch with which the driver could take up the load gradually. In Britain the gear shift could only be changed from the neutral position to the desired speed with the engine stopped, and to get the load rolling once the engine was in gear often meant momentarily reversing and using the backlash in the gears. On the debit side, US engines often had the gear wheels carried on stub shafts fixed to the boiler casing, instead of on shafts with two bearings, in the hornplates passing across the boiler.

However, perhaps the strangest feature of all to British eyes was the wheels. Unlike European engines which used heavy flat steel spokes usually cast into equally solid-looking hubs, most American engines used a multitude of spindly wrought-iron round spokes in combination with thin rims, almost like a bicycle wheel. Whether these wheels were supposed to be more resilient for cushioning the engine against road shocks or whether they were designed for lightness is not clear, but it was probably for both reasons. The 'no frills' approach to American engine design, already evident from the lack of boiler cladding, also led to gear trains being left exposed and ugly elbow bends instead of smooth curves in pipework.

There were, of course, exceptions to all these rules, from which one must assume there was intense competition among the 30 to 40 principal manufacturers. Where to keep water, for example, was a matter on which manufacturers varied. Apart from the conventional positions, some engines had ugly square tanks in front of or behind the chimney (or 'smokestack'), partly with the object of ensuring plenty of weight on the front wheels to prevent them from slipping sideways. The smokestack itself was rarely beautiful, having either a spark arrester or a plain stovepipe. Some later engines had awnings like their European counterparts, and a few even had all-round cabs like railway locomotives but the reduced all-round visibility from them must have made manoeuvring difficult.

Taking all these things into consideration, the average American traction engine was an ungainly and unsightly monster compared with the average British product. The grace which characterized the classic American 4-4-0 railway locomotives of the late 19th century was totally lacking, yet, in spite of this, they developed into functional and refined machines, fully capable of their many tasks and earning as great an affection among the men that worked with them as did the British Avelings, Fowlers, Burrells and the rest.

AMERICAN DESIGN EXCURSIONS

The development of the portable into the traction engine was accompanied (as in England) by some strange design excursions. One was a very early example of a prime mover on crawler tracks, built at Erie, Pennsylvania, in 1862. Rated at 45 nhp, the machine had a vertical boiler and was carried on three short sets of two-wheel crawlers, two of which were driven directly by an inclined cylinder and motion arranged outside the frames, as in railway locomotive practice. Even more like a locomotive was Charles Stratton's horizontal boiler undertype of 1893, with two sets of rear crawlers and a two-wheeled steering forecarriage which was worked from the footplate via a shaft.

In 1881, Atlas Engine Works of Indianapolis was using chain drive direct from a sprocket inside the flywheel. Much later Aultman & Tylor tried bevel drive. One way in which engines in America were more advanced than those in Britain at this time was in the use of a Pickering governor for belt work. In this type, the governor balls were carried on simple leaf springs so that, as speed of rotation increased, the weight of the balls deflected the springs outwards and pulled down the top yoke. This was linked to the butterfly valve in the steam pipe by a rod passing through the hollow spindle. Though much simpler mechanically than a Watt or other type of speed governor, and with less to go wrong, the Pickering did not find its way into Britain in any great number until after 1900; even so, this was a rare instance in the traction engine world of an idea crossing the Atlantic from west to east.

Daniel Best built vertical boiler engines, the original patent for which was held by D. L. Remington of Oregon, but Best himself is accredited with a number of inventions in agricultural machinery, the most important of which was his 'steam harvesting outfit' introduced in 1889. This consisted of a traction engine with a 'combined harvester' (sic) fixed to the rear. A contemporary illustration shows the engine carried on two large rear driving wheels with wide treads and a single smaller steering wheel in front. The 7.6 m (25 ft) long cutter was carried on a long cantilever arm projecting sideways from behind the offside rear wheel. Engine and steering were controlled by one operator in a cabin behind the vertical boiler with three other men tending the harvester.

The traction engine provided both support and locomotion and the whole assembly, in working order, weighed about 11 tonnes (11 tons). The cutting and cleaning machinery of the harvester was driven by an auxiliary engine taking steam from the boiler of the traction engine through a flexible pipe. The advantage claimed for this arrangement was that the speed of operation of the harvester could be controlled independently of the speed of motion. This suggests that other harvesting machinery had been tried which took its drive from the traction engine crankshaft, as happened in English steam diggers.

Best claimed that his harvester cut a swathe 7.6 m (25 ft) wide and, with an engine capable of moving at 5 km (3 miles) per hour, it could harvest between 26 and 40 hectares (65 and 100 acres) per day. The harvester apart, his engine had a towing bar to which a number of ploughs could be hitched as required. Best maintained that, with a dozen ploughs attached, the engine 'will do the work of 75 mules, and can be operated for the cost of the barley they would consume'.

In 1900 Robinson & Co. of Richmond, Indiana, produced a traction engine with tandem compound cylinders, and other firms who used this arrangement included Port Huron, J. I. Case and the Advance Threshers Company. Compensating gear and springing, also in use by that time, were likewise widely adopted. Reeves & Co. of Columbus, Indiana, produced a double-crank compound engine in which the disc crank was dispensed with in favour of a forged crankshaft of English form. A simpling valve was fitted to get the engine off dead centre or 'to give almost unlimited power for an extra heavy pull when on the road' and other refinements included cladding over the entire boiler, including the smokebox, and rocking firebars which the firm fitted to both its wood- and coal-burning engines.

Incidentally, the practice of 'simpling', or admitting live steam to the low-pressure cylinder when on the road with a compound engine, was used by some drivers to save changing gear when going up short hills. It was generally frowned upon by engineers in this country, not only because of the strain on the motion caused by a sudden trebling of the power output, but also because the extra blast on the fire could cause tubes to leak.

In 1905 the J. I. Case Threshing Machine Company of Racine, Wisconsin, built three 150 bhp engines which must have been close rivals in size to Best's logging engines. They had a single cylinder of 356 by 356 mm (14 by 14 in) bore and stroke and hind wheels of about 2,440 mm (8 ft) in diameter. A man could stand beneath

Heavy road haulage by traction engines continued until the 1930s. These three are seen in 1932 hauling a 60-tonne (ton) rotor on the last leg of its journey to the Dunalastair dam, part of the Tummel hydro-electric scheme in the Scottish Highlands.

the boiler barrel almost without stooping. These engines were used for ore hauling and heavy drawbar work and so they really count as road locomotives, although there were not the great differences between road and agricultural engines that there were in Britain. Engines in normal production did not generally exceed 30–32 nhp, that is, around 90 brake horsepower although, as mentioned earlier, some went up to 40 nhp. Ploughing engines were in fact large traction engines, since cable ploughing never came into vogue. American traction engines gradually became better looking after 1900, with a general switch from wood and straw to coal firing and the departure of the spark-arresting smokestack.

Americans are renowned for doing things in style, and 'traction engineering' was no exception. The 'Clarke School of Traction Engineering' in Madison, Wisconsin, offered correspondence courses to young men; these courses were claimed to fit them to become steam traction engineers (simply called drivers in Britain) after a few months' spare-time study. English drivers would have scorned such treatment!

With the arrival of diesel power, most American and Canadian traction engines disappeared from daily life even more quickly than their European counterparts. Some good examples have been preserved, however, thanks to the spread of the engine rally movement from Europe to America, where rallies are called 'threshermen's meets'. And in a few places traction engines are still in daily use. Driving along a freeway, one may sometimes spot a tall, smoking stack in the prairie. One example of such a machine is an 8 nhp single-cylinder Frick traction engine, used for soil sterilization, feeding steam through a network of pipes in the ground with the blower full on continuously to maintain pressure. The engine does this job regularly and travels from farm to farm under its own steam.

TRACTION ENGINE VARIANTS

Once the basic forms of the traction engine and portable engine had been firmly established (by 1870), production flourished and new developments from then on were mainly variants on the basic theme. Machines were designed for special purposes within the ever-broadening range of duties which required the use of steam power.

The first of these variants was the steamroller, which was ultimately produced by most traction engine builders in large numbers. It enabled much higher standards of road construction and maintenance to be achieved. Some rollers were conversions of wheeled types of traction engine. Then came the road locomotive, designed specifically for heavy haulage as opposed to farm work. A peculiarly British version of the road locomotive was specially equipped and finished for fairground work. Soon after 1900 came the road locomotive's small brother, the steam tractor, designed to beat a system of taxation by weight. Initially steam tractors were used for direct ploughing as well as haulage because they were so light.

The portable engine was also the basis of numerous variants. Among these were the attractive, diminutive engines of the fairground (affectionately remembered by many today). They were decorated with a great deal of polished brass, and drove the big roundabouts, steam switchbacks and mechanical organs of the late Victorian and Edwardian eras. Other types of engine used in fairgrounds powered lights or drove the 'steam yachts'—large steam swings—the latter having crankshafts which never managed to turn a complete circle. Also associated with polished brass was the horse-drawn steam fire engine and its American variant, the self-propelled fire engine. In these, a squat vertical boiler for fast steam-raising was combined with a two-cylinder engine direct-coupled to high-pressure pumps. Ingenious design, for compactness and low weight, was a feature of these machines.

Other types of small engine, which for convenience may be classed under the heading 'portable', were attached to less glamorous movable equipment such as hoists, pile drivers, concrete mixers, stone crushers, and even lawnmowers. Some of the more specialized forms had vertical boilers separated from the engine unit and so owe more of their origin to stationary than to portable practice. Two simple high-pressure cylinders were the rule to avoid dead centring. Space does not permit more than a mention of most of these special types, the working principles of which were in no way different from 'mainstream' types. (Steam cranes and excavators are regarded as production machines and are discussed in other chapters.)

Another variant, though not an engine in the true sense (because it produced sound energy rather than mechanical work), was the steam 'calliope' (pronounced cal-eye-o-py), peculiar to North America. It comprised a towed portable boiler carrying an array of whistles on which tunes could be played by means of keys. Calliopes were paraded through the streets in full song, to herald the arrival of fairs and circuses, and their raucous voice carried for miles. The term 'steam organ', applied to mechanical fairground organs in Europe, is a loose one since air was the operating medium, not steam as in the calliope. Steam was, however, present to some degree since bellows and keyframe drives were provided by a small, separate steam engine.

The principal variants of the traction engine described in more detail below are steamrollers, road locomotives and tractors, and fairground steam engines.

STEAMROLLERS

Without the steamroller and the dramatic improvements in road-making techniques which resulted from it, it is doubtful if the development of other vehicles, particularly lighter ones, could have proceeded at the pace it did. From 1870, when Thomas Aveling introduced his extended hornplates and other improvements, until the end of construction in the late 1940s, the general form of the steamroller was a two-speed unsprung traction engine, but mounted on rollers instead of wheels. The front axle, held in a massive fork, carried a two-part roller, and the hind wheels had thick, smooth treads so they also acted as rollers. This basic type is still used, though today's version is diesel powered.

When the vehicle moved in a straight line, the overlap between the front and hind rollers was sufficient to produce a smooth, rolled track equal in width to the distance across the hind wheels, that is, generally between 1.2 and 2 m (4 and 7 ft). The worm-driven chain-barrel steering followed traction engine practice but was lower geared, as it required anything up to 50 turns lock-to-lock to provide the greater effort needed to get the front rollers to swing round.

Another type of steamroller was the tandem roller. It was produced in several guises in the 20th century when machines were built lighter to work on tarmac. In this type there were two full-width rollers, back and front, but it was never numerous compared with the overtype three-wheeler.

Aveling could not claim to have invented the steamroller, as a Birmingham engineer called Batho had had one made for work in Calcutta in 1863. Aveling's first machine of 1865 was merely a traction engine towing a 15 tonne (15 ton) cast-iron cylinder made for him by Easton, Amos & Anderson of Erith, a firm better known for beam engines. Adoption of the three-wheel principle took place after he and Batho started to work together, and the outcome of their cooperation was a 22 tonne

One of the best-known systems for negotiating soft ground was Boydell's so-called 'endless railway' introduced in 1856, by which the weight was taken on large, hinged pads. This is the final form of the Burrell-Boydell engine as shown at a London exhibition in 1862.

(22 ton) machine built for Liverpool in 1867. This was a back-to-front design in that the driving wheels were in front, chain-driven from the crankshaft, and the steering roller was at the rear. Boiler and engine dimensions were the same as Aveling's 12 nhp traction engine.

A number of such rollers were built in various sizes up to 30.5 tonnes (30 tons). Half of them were exported to France, India and the USA. The heaviest sizes tended to overdo their effect by crushing the roadstone instead of consolidating it, and so their production was halted. Batho tried a two-cylinder engine to increase the low power-to-weight ratio for hill-climbing, and fitted springs to enable more rapid travelling between jobs. Aveling, however, wanted simplicity, and by changing the design to the same form as his traction engine, he got it.

The 'breakthrough' was in the use of a massive cast-iron steering head projecting in front of the smokebox to carry the roller pivot, and so restore the steering to the front. The rest of the machine was simply a traction engine on smooth treads. At first the front pivot was carried down to axle level by tapering the rollers but this soon gave way to a hollow cast-iron fork embracing parallel rollers and carrying a straight axle, and this became universally adopted.

As the years passed, the policy of rugged simplicity and realistic pricing proved so successful that the firm of Aveling & Porter was able to concentrate primarily on steamrollers. In all, more than 8,000 were made and the brass rampant horse of Kent which each machine displayed on its steering head came to be recognized all over the world. To obtain the maximum effect of deadweight, these Rochester-built steamrollers had no springing. Nor was there any differential gear, winding drum or governor for belt work, as there was on a traction engine. Instead, there were simply two driving bosses keyed to the axle with, on the non-gear side, a band brake. Either, both or neither hind wheel could be driven by the axle as it rotated, by inserting driving pins as appropriate, just as Aveling used in his early traction engines. Side motion covers and disc flywheels concealed moving parts from nervous horses. Standard sizes were 6, 8, 10, 12 or 15 tonnes (tons) corresponding to 4, 5, 6, 7 or 8 nhp traction engines; all-up weights on the road with fuel and water were, however, considerably higher than the ratings.

Most of the world's traction engine firms added steamrollers to their ranges as time went on, as there was a big market for them. Designs were very close to those for traction engines, and some manufacturers fitted differentials. Optional equipment included an awning, a tender-mounted scarifier for digging up the road, water sprays for damping the rollers when working on tarmac, and even tarspraying gear for surfacing. Some steamrollers were given compound cylinders for greater economy; and some were converted from traction engines, especially after 1920 when their work in agriculture was declining. A few traction engines were deliberately built as convertibles, with ingenious ideas for altering the traction-type smokebox to make it suitable for the roller steering head. Also from 1920 onwards, the demand for the heaviest machines declined and several lighter steamrollers purpose-designed for work on tarmac appeared. Of these, the Wallis & Steevens 'Advance' roller fully lived up to its name and in some respects, for example, its balanced weight distribution and quick reverse facility, anticipated features of today's diesel rollers. It was in production until just before the Second World War.

Avelings were always the most numerous steamrollers; some contractors owned large fleets of them to the virtual exclusion of other makes, which enabled spares to be standardized. One of the best known of these contractors was Eddison of Dorchester, which had some 400 machines.

All over the south of England, until the late 1950s, one could come across an 8 or 10 tonne (ton) single-cylinder Aveling with living van and watercart in tow; their drivers, living on the job, thought nothing of an 80 or 100 km (50 or 60 mile) trek from one job to the next, taking two to three days over it. The drivers would do a route survey before attempting a journey, noting such things as water pick-up points, steep hills, awkward turns, upstanding manhole lids, points of congestion and most important, places where a roller could pull off the road. Eddison also used to tell its drivers to seek police permission before entering a large town. Going down a steep hill was always hazardous with a steamroller, particularly with a driver unfamiliar with the road. It was always essential to change into bottom gear and ensure both pins were in before beginning a descent. But on an icy or muddy surface or on melting tar, the smooth treads on the hind wheels could suddenly start to slide, and prompt action would be needed in order to avoid having a runaway vehicle.

A handful of steamrollers remained at work in Britain in the late 1960s; in some parts of the world they are still used but, as boilermaking skills vanish, their future must be uncertain. Today steamrollers are collector's pieces although, perhaps because so many have survived, rally organizers assign them to a lower class than engines on wheels. The beauty of a steamroller in preservation is that it can still be used for the work for which it was designed, and it is to be hoped that the active preservation movement will spread.

EARLY ROAD STEAMERS

A number of early engines designed primarily for road haulage, and sometimes termed 'road steamers', were tried with varying success. One of these was the Bray traction engine which had driving wheels in which projecting teeth moved in and out as the wheel revolved. An eccentric device controlled from the footplate decided the point in the revolution where the projection was maximum; in other words, the engine could adjust its feet to the softness or hardness of the surface on which it ran. Later Bray engines had undertype engines and leaf springs; one is on record as having hauled a heavy boiler through London in 1861. Examples sent to Turkey a few years later looked just like 2-2-0 railway locomotives with uncluttered boiler, long side tanks, 1.5 m (5 ft) driving wheels and no flywheel.

Three-wheeled Thompson road steamers enjoyed a few years' popularity between 1871 and 1873. The two-cylinder engine unit geared to the hind wheels was arranged vertically; early engines had a vertical boiler but later examples used a horizontal one. R. W. Thompson's chief claim to fame is that he pioneered the use of solid rubber tyres for the wheels of steam vehicles. To improve grip as well as to cushion the ride he used rubber blocks which (because bonding techniques were unknown) were fixed to the wheel rim by steel clasps. His engines managed 15–16 km (9–10 miles) per hour, hauling a wagon or omnibus.

From 1860, a few 'ice locomotives' were built to British design for winter haulage on Russia's frozen lakes and rivers. They looked like railway locomotives with plate frames, springs, two outside cylinders and side-rod drive to a crankpin on each hind wheel. The rims had metal projections to provide grip and there were no front wheels, only a steerable sledge device.

ROAD LOCOMOTIVES AND TRACTORS

Because the final form of the heavy road haulage engine is so similar to the agricultural traction engine—one being a more refined and expensive version of the other—and because manufacturers often embodied road loco features in their traction engines, it is difficult to draw a distinct line between the two. The following may be regarded as road locomotive features, though by no means every road locomotive had all of them:

- double-crank compound cylinders;
- three speeds instead of two to give a higher top speed;
- generally heavier construction;
- springs to both front and rear axles;
- extra water-carrying capacity in the form of a 'belly tank' slung under the boiler barrel;
- disc instead of spoked flywheel and side plates to conceal the motion, as used on steamrollers;
- a higher standard of painting and embellishment;
- a steersman's seat;
- an awning;
- solid rubber tyres—a legal requirement from 1930.

The question of springing requires explanation; the problem of how to cushion the machinery (rather than the driver, one suspects) against road shocks occupied much time and attention by traction engine designers from the early days. The fundamental dilemma was how to build resilience into the system while also keeping the gears in the drive to the back axle in constant mesh. Early solutions avoided the problem: the springing was built into the hind wheel itself. The disadvantage of this was the heavy wear on the springs caused by mud, dust and grit.

Two distinct springing designs ultimately came to be used in engines of normal overtype traction engine form. One design used leaf or coil springs to allow vertical movement of the hind axle limited to about 25 mm (1 in), and the teeth on the final drive gears were made extra long so that they remained in mesh. The more exotic form, normally employed in road locomotives, allowed movement of 50 mm (2 in) or more, by arranging the second intermediate shaft to articulate, as shown in the drawing. The intermediate shaft carried its driven and driving gears on opposite ends. Normally the final drive was on the offside end, in which case the shaft bearing in the hornplate could move up and down. Since it was constrained by a link to do so in unison with the hind axle bearing, the final drive stayed in constant mesh. The hind axle bearings were normally fixed to leaf springs arranged under the axle, to avoid cluttering the footplate.

The first true road locomotives began appearing in the 1880s, after John Fowler had introduced the double-crank compound traction engine in 1881. By 1900 some British firms were making road locomotives in considerable numbers for both home and overseas. The 'generally heavier construction' referred to earlier included shafts of larger diameter, wider gearing and larger wheels. While the third speed gear in theory permitted a top speed of 10–13 km (6–8 miles) per hour, in practice, with an engine on solid rubbers, this was often doubled. Road locomotives had a winch which was often used in conjunction with 'snatch blocks' for pulling the load on or off its trailer.

Among the finest of all road locomotives was the Fowler 'Big Lion' type. Though rated at 8 nhp, it could sustain 70–100 bhp for a short time. Cylinders were 171 and 292 mm ($6\frac{3}{4}$ and $11\frac{1}{2}$ in) bore by 305 mm (12 in) stroke, working pressure was 1,379 kN per sq m (200 lb per sq in), the all-up weight was 17 tonnes ($16\frac{3}{4}$ tons) and the length 6 m (19 ft 9 in). Its normal load of 41 tonnes (40 tons) was often greatly exceeded. Like other road locomotives, the Big Lion was often turned out in showman's trim for fairground work. For generating at the fair, a dynamo with its belt pulled in line with the flywheel was mounted on a 'dynamo shelf' protruding in front of the chimney to protect the dynamo. Showman's engines were also made by Burrell and Foster, and some of them had a 'feast crane' fixed to the tender for lifting the cars on or off the tracks of scenic railways. Some travelled very long distances with three or even more loads behind: for example, Jacob Studt's Burrell 'General French' regularly ran between Cardiff and

Most steam fire-engines were horse-drawn but one of the best-known manufacturers, Merryweather, won a prize for this self-propelled engine which they called 'The Motor Fire King'.

the London area in only two days. In summer, when showmen had to catch as many fairs as possible, the fires in their engines scarcely ever went out.

With their twisted brass fluting to the canopy uprights, brass stars and rings on the flywheel and motion covers, bright colours and elaborate lining, the gaudy showman's engines were very attractive. Today, in preservation, they are very expensive to buy.

In 1905, when new legislation permitted one-man operation and a lower tax on road engines weighing less than 5 tonnes (5 tons), several manufacturers began turning out the road locomotive's small brother, the 'steam motor tractor'. This was simply a miniaturized road locomotive, and many of them were very attractive. One or two manufacturers had clever ways of saving weight: Tasker of Andover, for instance, used chain drive; and Foden used a design based on the firm's steam wagon to produce its highly successful timber tractor. To save the weight and complication of fitting a third speed, some makers clung to two, but when a Foster 'Wellington' tractor, for example, was running at 16 km (10 miles) per hour, which it could easily do,

Thomas Savery

the crankshaft was revolving at 500 rpm, which threw a lot of oil away from where it was needed. Wallis & Steevens tackled this problem by enclosing the whole motion in an oil-bath, anticipating modern ic-engine practice.

Steam tractors were popular in about the same period as the overtype steam lorry. They were used for moving trailer loads beyond the capacity of a lorry-type vehicle, such as vegetables, furniture, bricks, roadstone and the like. Some were equipped with fairground fittings exactly like road locomotives.

To ensure that a new design of steam tractor weighed less than 5 tonnes (tons) for the official weigh-in, the manufacturer would attach wooden funnels, wooden firebars and even wooden wheels. These components would be duly changed for the metal variety once the Ministry inspector had departed. Even so, an all-up weight of 6–7 tonnes (tons) including fuel and water, for a steam tractor easily capable of pulling 20 tonnes (20 tons), represented quite an engineering achievement, especially as they did not have the benefit of the light alloys available today.

FAIRGROUND PORTABLE ENGINES

Portable steam engines used on the fairground fell neatly into two classes: those which stood on their own wheels and those which were mounted on a separate wheeled piece of equipment. The best-known, the tiny centre engines which powered roundabouts, came into the latter category. It is not known who was the first to produce one, but it is known that in about 1870 Henry Soame of Marsham, Norfolk, applied steam drive to a roundabout, using a belt from a portable engine. Frederick Savage of King's Lynn, renowned for inventive work in agricultural machinery, then turned his attention to fairground machines. St Nicholas Ironworks was soon turning out not only purpose-designed steam engines for powering roundabouts but the rides themselves, several of the ideas for which he held patents.

Centre engines took two forms. One resembled a normal portable but had the addition of a 'cheese wheel' around the chimney (or centre pole) with bevel gearing from the crankshaft to rotate it. The roundabout (or switchback) would be assembled around the engine at the fairground, the long radial arms (or swifts) which formed the framework of the rotating top being inserted into the slots in the cheesewheel which gave it its name. This was the 'end-on' centre engine.

In about 1890, when horse roundabouts began to have the well-known galloping action provided by overhead rotating crankshafts, the entire weight of the machine plus riders came on to the centre. This meant that the centre pole had to be much more massively built and supported. To allow this, the engine was moved to one side or, more correctly, put across one end of a 'centre truck' which now formed the base of the machine. This was the 'transverse' centre engine. It resembled a small portable engine but with neither wheels nor, apparently, a chimney. (Actually, this went out sideways from the smokebox in a flue incorporating the blast pipe and leading to the centre pole which still acted as the chimney. Being on the side remote from the riders, the exhaust flue could not be seen.)

Savage's transverse centre engine was a masterpiece of compactness, being scarcely more than 1.3 m (4 ft) long and 1.3 m (4 ft) high, and yet, with two high-pressure cylinders and steam at 552 kN per sq m (80 lb per sq in) it could develop some 10–12 bhp. The water tank was tucked under the boiler barrel, while the cylinder block over the firebox carried the regulator, spring balance safety valve and siren. To keep down the height, the slide valves were arranged between the cylinders so the balanced cranks and balance weights did not come over the highest part of the boiler. Feed pump and injector took their supply from the tank and a water lifter enabled the tank to be replenished from other tanks placed on the ground while the fair was open.

A small vertical organ engine on the smokebox, always called the 'model', had a stepped belt pulley which lined up with a corresponding pulley on the back of the organ when the ride was built up. Steam supply to the model was controlled from the 'footplate'. Normal fuel was coke which was kept in a box on the narrow platform which served as the driver's footplate.

Many of these engines wore out and had to be replaced, partly because of the use of light boiler plates and partly because their small size made boiler cleaning difficult. The last ones were built in the early 1920s, by which time fairground drives were going electric—though a new firm which took over Savage's drawings and patterns will build one today to order. Cylinder dimensions of Savage's 'No. 6' size, the biggest, were 102 by 203 mm (4 by 8 in). Centre engines were also turned out by two other merry-go-round builders: Robert Tidman of Norwich and Walker Brothers of Tewkesbury. Savage also built a few electric light engines which were simply portable engines with a dynamo attached. Most fairground generating requirements were, of course, met by the road locomotives.

An ingenious extension of the centre engine principle, of which Savage built a few examples, was the 'traction centre' in which the functions of showman's road locomotive and centre engine were combined. It was really a traction engine with the addition of a cheesewheel assembly carried on a frame formed by upward extension of the hornplates. In view of the considerable difference in power output when driving the machine, as compared with hauling the loads, to say nothing of the difficulty in finding time to do any repairs, the traction centre never became popular.

The oddity of all portable engines, the steam swing or 'steam yacht' centre engine must also be mentioned. Despite their specialized application, these engines demonstrated the versatility of steam power, in that the pistons were of variable stroke. The engine, which from a distance looked like a normal farm portable, was placed in between two pairs of A-frames. Each A-frame carried a shaft at its apex from which was suspended a large swing boat seating between 20 and 25 people. The apex shaft carried a chain wheel at its inner (engine) end. Two high-pressure cylinders over the engine firebox drove forward to two half-crankshafts, each carrying a chain wheel identical to the one on the apex shaft and linked to it.

The driver worked the slide valve for each cylinder by a long lever, called a 'pump', each side of the firebox. When the ride stopped, both swings were at the bottom of their travel and the pistons in mid-position. To start, the driver opened the stop valve and then pumped the valves. As the swing boats moved through an increasing arc, so the travel of the pistons increased. When the arc of swing approached 180 degrees, that is, just before the hapless passengers felt they would be tipped out, a trip gear came into operation worked by a tappet on the engine crosshead which automatically moved the slide valve just before each piston dead-centre position. The whole arrangement was patented by William Cartwright in 1894, the trip gear being known as 'rabbit gear' owing to the shape of the yoke which the tappet engaged. Savage built about a dozen sets between 1894 and 1928.

The last fairground machine associated with steam was the Tunnel Railway, in which a small four-wheeled railway locomotive with carriage ran on a circular track, half of which was in a tunnel. The only departure from normal practice on the locomotives was that their axles were set radially to suit the continuous curve.

Boilers and Accessories

The history of the boiler obviously goes back further than that of the steam engine, even if we use the word boiler only to describe a vessel normally closed; by a lid, plug, cork or other device. We can safely assume that the first attempts at obtaining mechanical power from steam were inspired by the apparent violence with which steam issued from confinement, or even by the explosion of a closed vessel which was closed too well. Any lidded water-boiling vessel could have demonstrated this effect, especially if it had been made of pottery.

However, the earliest boilers (those belonging to the prehistory of the steam engine) were made of copper or bronze. From all accounts and all reconstructions, Hero's aeolipile was mounted on top of a boiler in the shape of a lidded, round-bottomed, cooking pot. From experience with reconstructions of this device, one knows that it required a pressure of at least half an atmosphere or 48 kN per sq m (7 lb per sq in) to work at all convincingly, and it works much better at twice that pressure. Provided that it is not made too large, a cooking-pot boiler should be able to withstand such a pressure, even if made in Roman times.

Two remarkable Roman proto-boilers were discovered at Pompeii and are now in Naples Museum. They are clearly not pressure boilers, but water heaters, with hinged lids and elegantly shaped handles—proto-samovars, perhaps. But their construction is very sound and their shapes would be quite well adapted to withstanding internal pressure. Their truly remarkable feature is that they have water-tube fireboxes inside them. The water tubes form the grate at the bottom of the box. The box is fully water-jacketed and has a fire door in the side. One of these water heaters has a firebox which tapers upwards and curves to the side, to give a proper chimney effect, but the other, though it has a domed firebox crown, is simply provided with a small opening high at the back of the firebox, which would need to be connected to a long rising pipe to generate any natural draught. The skill evidenced in their construction makes it plain that a Roman pressure vessel was perfectly possible, provided that it was not too large.

During the revival of ideas of using steam to provide power, in the 16th, 17th and early 18th centuries, the pressure of confined steam is still the first thought in the minds of inventors such as Branca, who conceived an impulse turbine supplied from a cooking-pot type of vessel, and even the nearly successful Savery, who hoped that steam pressure would do most of the water raising in his 'Miner's Friend'. However, when it came to doing real work, pressure vessels proved unsatisfactory, because they could not be made large enough without leaking, deforming, or even bursting. For most of the first century of practical steam power the boiler was a kettle, a copper, a brewer's vat, or some other form of familiar water-heating vessel which produced steam at atmospheric pressure only, or just sufficiently above that pressure to make it flow in the desired directions. It did not even have to be made of metal.

Savery's boilers, which were intended to withstand a pressure of several atmospheres, were based on the design of the brewer's copper. They were made of copper, externally fired by a fire beneath them, and housed in brickwork which followed brewery practice in providing some sort of encircling flue through which the gases passed on their way to the chimney. This flue might even, according to one description, be spiral. A point of particular interest was Savery's use of boilers in pairs: a small boiler was arranged to be able to refill the larger main boiler whenever necessary, and for this purpose had to develop a higher pressure than that of the main boiler. Obviously, this should have been done when the main boiler pressure was rather low, and the fire damped down, so that no excessive pressure was required in the small boiler. There was always a danger of the small boiler 'blowing up' the large one if the operator was careless and there was any overheating caused by low water in the large boiler. The system was ingenious (as was the whole Savery engine) but, with the state of technology at the time, clearly contained the apparatus of its own destruction.

Newcomen's engine was quite different and only used steam to produce a vacuum. The height to which it could raise water was not directly determined by the pressure of the steam, but by the relative sizes of piston and pump, and the number of stages in the lift. The steam was kettle steam at the pressure of the atmosphere—96 kN per sq m (14 lb per sq in)—or very slightly above, and the boiler from which it came had only to provide a large water capacity and effective transfer of heat. These boilers were rather like Savery's, in that they were circular vessels encased in brickwork with an encircling flue. The bottom and sides, exposed to the fire, were generally of copper in the early years, and the top was sometimes of lead. The cylinder of the engine was above this, and secured in crossbeams of the engine house, but undoubtedly there must have been some relative movement between cylinder and boiler as the engine operated. As the two components were connected by a steam passage and valve, the use of relatively flexible lead for the boiler top probably prevented fracture, and this is to some extent borne out by the later practice, after the whole structure had become firmer, of making the lower part of the boiler of iron, but the top of copper.

Newcomen's boilers were the shape of an enormous, short, round-headed rivet. The top was hemispherical, and below was a parallel part of smaller diameter. This provided the boiler with an internal shelf, of no particular benefit except that it enabled the top of the boiler to be borne directly on the brickwork. As there was no appreciable internal pressure, the boiler could be replenished directly from a tank at a higher level. The bottom of the boiler was slightly concave.

This basic form was soon being modified by various engineers. As early as 1717 Henry Beighton flattened the hemispherical top to something more like the domed lid of a cooking pot, and turned the parallel and rather deep lower part into a shallower cooking-pan

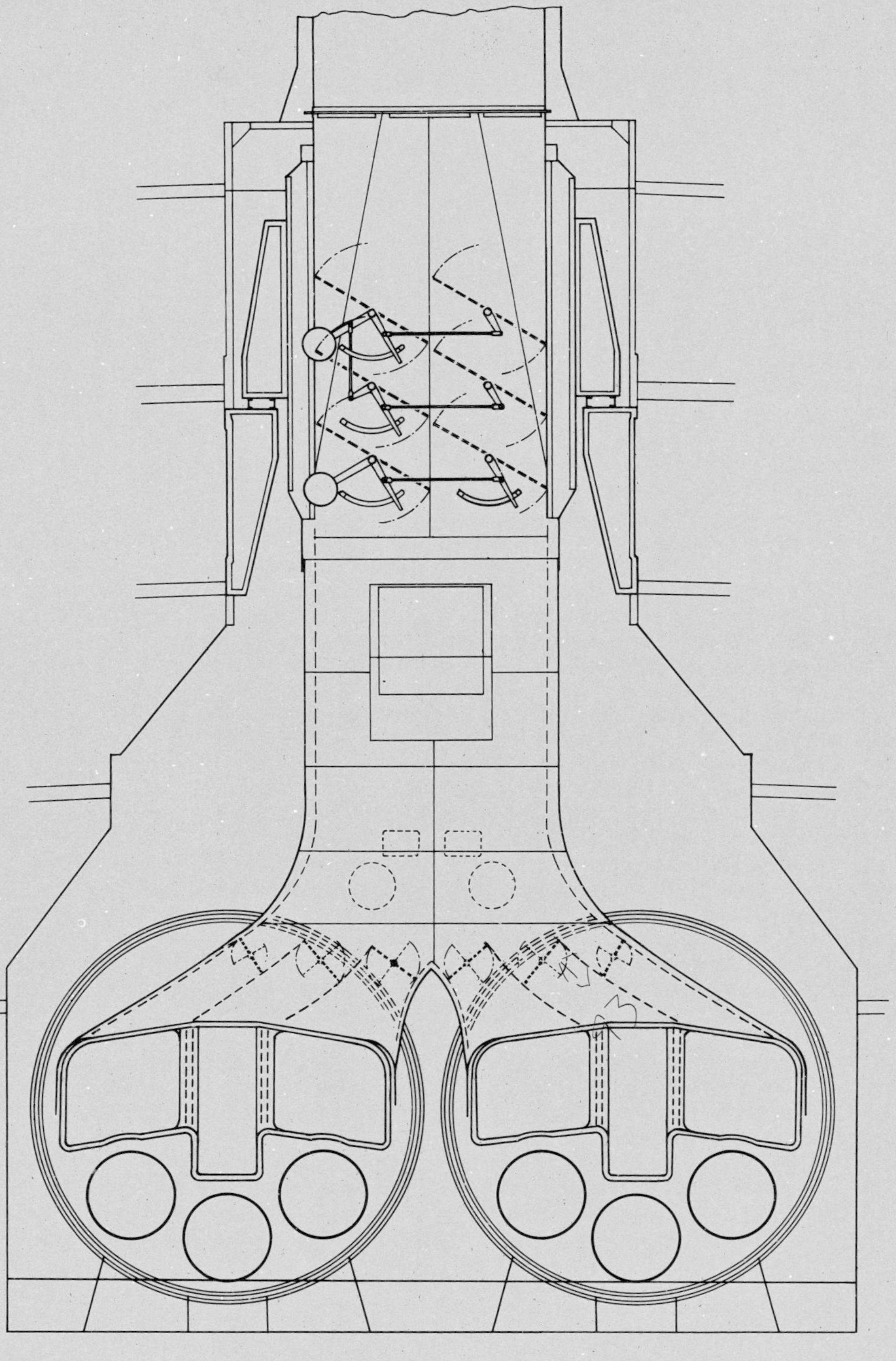

Layout of boiler uptakes, flue dampers, etc. for a large steamer. Each boiler has three furnaces, with individual dampers for each furnace, in addition to dampers at the base of the funnel. The curved tops of the banks of tubes are to ensure that the tubes are covered by water when the vessel rolls.

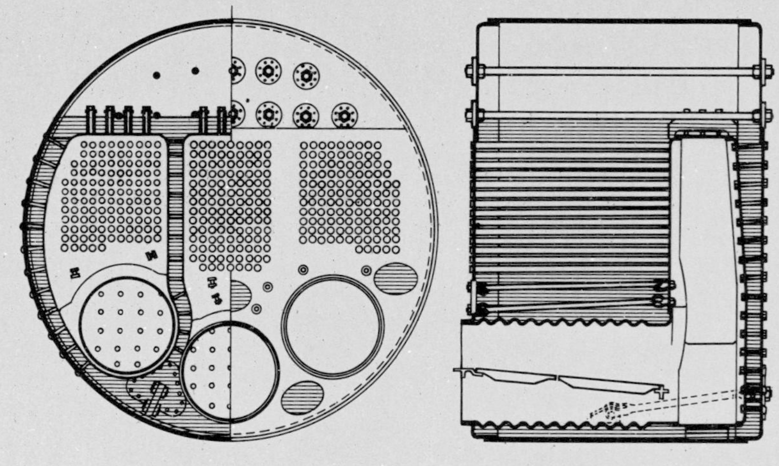

Scotch boiler with corrugated furnaces, for a lake steamer.

shape with flared sides and a more sharply domed bottom. The shelf was reduced in width, but was exposed to heat in the encircling flue. This boiler must have been very sensitive to changes in water level, because the wetted, heated area would drop sharply, as would the area of the water surface, if the water dropped below the shelf; but there was insufficient height above the shelf to allow much water there. Again, as the water level rose above the shelf, the area of the water surface diminished sharply and in practice this would have hindered the production of steam.

After 50 years of experience, boiler makers were using iron plates and making boilers which had highly domed tops and no shelf; they tapered down to a base ring somewhat smaller than the maximum diameter at the bottom of the dome. The lower taper might be concave, and the boiler bottom was nearly flat, being supported by internal stays. Cast-iron was also beginning to be used, and boilers were assembled from sections bolted together.

In the later part of the 18th century, James Watt was using the wagon-type boiler. Although closely identified with Watt, it was not developed by him and had earlier antecedents. Its name arose from its general resemblance in shape to a covered wagon, as it was a rectangular box with a rounded top. The flat sides and bottom were in fact slightly concave, and tied together with a few internal stays, but the ends were actually flat. The half-cylindrical top was not stayed, all the internal stays joined the more-or-less flat surfaces. These boilers were assembled by riveting quite small wrought-iron plates, large plates not being available until well into the next century.

Water feed into wagon boilers was again by gravity, because Watt was still using the near-atmospheric pressure of the Newcomen engine. An internal float could be used to open a valve in the gravity feed-pipe and thereby maintain a constant water level. This type of boiler was not particularly susceptible to changes in water level, because there was no internal shelf which would be exposed; neither did the area of the water surface change much at different heights, unless the boiler was overfilled. But there was an encircling flue reaching to about half the height of the boiler, which made it undesirable to let the water fall too low.

Plain cylindrical horizontal boilers, with no internal flue and arranged in brickwork with the fire underneath and an encircling flue (or a pair of return flues in parallel) were used for low pressures, notably in America, during the period of the popularity of the wagon boiler. This type was better adapted for higher pressures, but only became common when fitted with one or more internal flues. As it needed no stays, except perhaps to join the flat ends, it was clearly superior to the wagon type, but probably more expensive. With the adoption of higher pressures it became more attractive and was developed in two quite different ways: by the addition of internal flues, giving rise to the Cornish and Lancashire boilers with their several variants; and by the addition of extra drums, joined to the main barrel by large diameter pipes. This last type, developed especially by Arthur Woolf and exported by him to the European continent, became known as the 'Elephant' or 'French' boiler. It consisted of a relatively slim main drum, with two slimmer ones beneath. The whole was set in brickwork and the grate was beneath the lower drums at one end. The flames passed along the lower drums and then rose into a higher flue in which they passed along the upper drum. These boilers were long used and very successful, provided that the arrangement of flues allowed sufficient temperature differences to ensure a good circulation of water within the boiler.

Internal flues, and internal fireboxes, were tried many times during the 18th century, but did not become common until the 19th. Their outstanding advocate was Richard Trevithick, but he had several predecessors, notably William Smeaton, the great improver of the Newcomen engine, who produced a small, nearly spherical, boiler with an internal firebox in 1765; and Nicholas Cugnot, whose artillery tractor of 1770–1 (an example of which is preserved in Paris) incorporated a preserving-pan-shaped boiler with a firebox at the bottom and rising flues terminating in two short chimneys. This boiler worked a non-condensing engine so it must have had a working pressure of not less than 48 kN per sq m (7 lb per sq in).

Trevithick produced many boiler designs suitable for the high pressures he advocated, and the most important of these has already been described in the section dealing with the first locomotives. Whether he could be regarded as the originator of the 'Cornish' boiler is open to doubt, though this type has frequently been attributed to him. A Cornish boiler is a horizontal cylindrical boiler with a single internal flue along its length, the grate being in one end of this flue. Hot gases emerging from the other end pass beneath the boiler and return along the sides, before ascending the chimney. (The great height of factory chimneys was commonly needed to produce an adequate draught to draw the fire through arrangements such as this.) The weakness of the design lay in limited water capacity, because the flue had to be large and its upper surface was not far from the top of the boiler. Water level had to be held within narrow limits: too high a level would reduce the water surface and restrict liberation of steam from it, and too low a level would expose the flue crown and risk its burning.

LANCASHIRE BOILERS

Lancashire boilers have two flues of smaller diameter, offering a greater heating surface and a lower flue crown, though at the expense of reduced flue cross-section, of little significance in their usual applications. The two flues were sometimes united at the firebox end, providing a large grate area, but complicating the design and construction, requiring the use of stays. Both flues led the gases to external flues in the brick setting of the boiler, as in the Cornish type. The Lancashire type was patented by William Fairbairn in 1844, and has remained in use ever since. Both these types of boiler could be fitted with water tubes within the flues, and this practice was common with later Cornish boilers. The Lancashire boiler formed the basis of some short-lived freak designs, such as one with the two flues united into a single flattened shape, with vertical water tubes acting as stays. This sort of thing made it far more difficult to withdraw the flues for overhaul, and even to climb inside them for inspection.

Though the cylindrical boiler appeared early and was in use for a long period, it must be remembered that a great number of the early internally fired boilers were operated at very low pressures and so could be rectangular boxes with labyrinthine flues inside them. Such boilers were very common on land, and almost universal at sea for the first half of the 19th century. They had large rectangular flues through which a cleaner could pass, and these turned back and forth to fill a large part of the space inside the boiler shell. The shell itself was not exposed to the fire anywhere, so it could be made of flat plates caulked with lead, and if it leaked at the seams it did not matter, because scale would soon build up and seal the gaps. The shell had only to be strong enough to contain the weight of water, and it did not have to be made of iron, or indeed of metal at all. Wooden boxes served the purpose on occasion, notably in America at the turn of the 18th and 19th centuries, and there were also a few boilers made of stone in England.

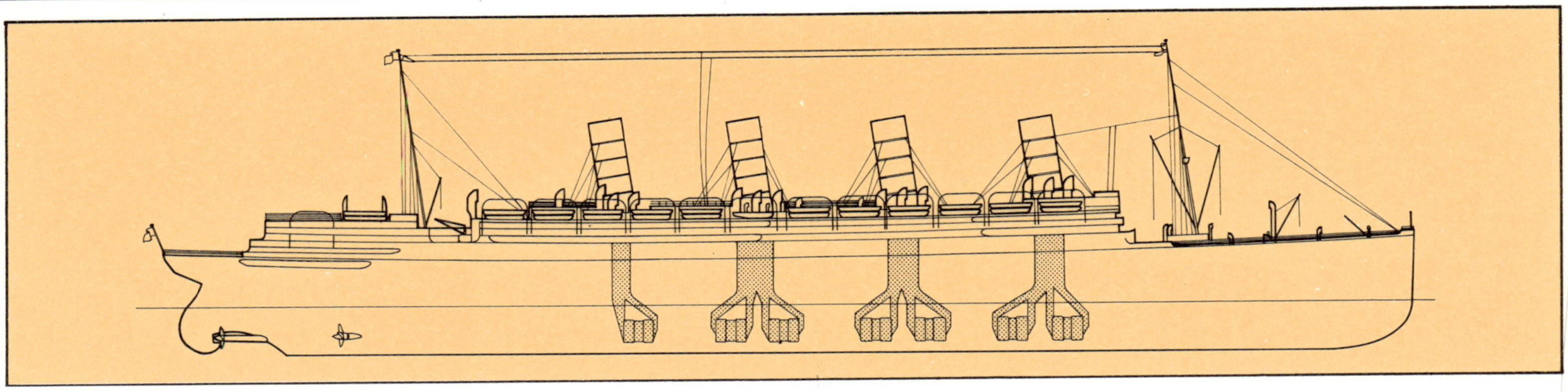

The arrangement of double-ended Scotch boilers on the transatlantic liner Aquitania *of 1914. There were seven transverse sets of five boilers each, originally coal fired. The* Aquitania, *of 53,850 tonnes, was the last large four-funnelled liner in service, and lasted until 1950—a record among Atlantic liners.*

Marine boilers were made on this principle until the 1860s, when double expansion involving higher pressures ruled them out. However, by this time the labyrinthine rectangular flues had given way to arrangements comprising fireboxes, smokeboxes and multiple fire tubes. The box boilers of the *Great Eastern*, of 1857, had six double furnaces in each, with a box-shaped rising chamber above the middle of each one, from which fire tubes of small diameter ran to each side of the boiler where they terminated in smokeboxes, still within the main boiler shell. All this required an enormous number of stays, because there were so many flat surfaces; even the furnace chambers had flat sides. All the same, these boilers had only to be made circular, and fitted with cylindrical furnaces, and the whole thing would be a double-ended Scotch boiler, the first examples of which appeared in 1862. They had two cylindrical furnaces, with a grate at each end, and from the middle of each rose a rectangular chamber with tube plates on opposite walls, from which straight fire tubes ran parallel with the furnaces out to the front and back of the boiler. The smokeboxes were external to the boiler shell, and, as also in the 'Great Eastern' and other fire-tube box boilers, smokebox doors gave access to the tubes for cleaning. Above all, the shell was cylindrical.

The Scotch boiler remained associated with marine steam plant until very recently. It was a coal-fired boiler in conception, but was successfully used with oil fuel after conversion, and was actually built for oil firing in the 1920s and 30s. It was made double- or single-ended and presented a large circular 'front' to the stokehold, with a number of furnace doors at the bottom, placed to conform to the shell shape, so that the central ones were lower than those at the side. Each furnace door was divided at the level of the grate, with firing above and air admission below. Above each furnace door was its associated smokebox, the doors leaning forward, and above these were the funnel uptakes. In a large ship, the boilers were arranged in rows across the vessel, so that the stokers worked in a transverse passage with fire doors on both sides. Coal was carried in bunkers at the side of the ship, as a rule, and was brought to the stokers in barrows.

A large double-ended Scotch boiler might have a diameter of 5 m (16½ ft) and a length about the same, with five furnace flues and ten grates totalling 12 sq m (about 135 sq ft) in area. The pressure would be about 138 kN per sq m (200 lb per sq in) and it might contain 50 tonnes (tons) of water. Until near the end of the 19th century, tall funnels were relied upon to ensure a good draught, but a boiler of the size described would probably be found in a large ship of the early 20th century, and would have forced draught. The entire stokehold would be pressurized by means of large fans driven by small auxiliary steam engines. These would be mounted above a pressure deck passing over the boilers, and would force air downward into the stokeholds. Staff entering or leaving the pressurized area would do so through double-door air locks. This is still normal practice in modern oil-fired ships with water-tube boilers: the pressure is low and not injurious to those working in it. Similar forced draught arrangements were applied to water-tube boilers almost from the beginning.

LOCOMOTIVE-TYPE BOILERS

The locomotive-type boiler is also characterized by a firebox within the shell and straight fire tubes. Some mention of its form has already been made in the sections dealing with locomotives in this book, but it is important to realize that this type of boiler was extensively used away from railways. This was because, in relation to its bulk and weight, the locomotive boiler was a prodigious steam raiser (able in extreme cases to produce something like half its own weight of steam every hour) without being unduly 'sensitive', and possessing sufficient reserve to tolerate water-level and furnace heat fluctuations, or to meet excess demand over a short period. Any boiler works best under constant conditions, but the locomotive type is the best adapted to variation.

The shape of the locomotive boiler, basically a horizontal cylinder with a downward-extending firebox beneath one end, has also made it suitable, as a structure, to serve as the foundation of a machine. This happened with some types of railway locomotive, such as early American 4-4-0s and the *Rocket* itself, but was the rule with road locomotives, traction engines, agricultural engines and steam rollers. Locomotive-type boilers were also fitted in steam launches, fast torpedo boats, some sternwheel river steamers, many steam lorries, some early electric generating plant, stationary engines with the cylinders beneath ('undertype' engines), similar machines with the cylinders on top ('overtype' engines), in factory boilerhouses and, because of their compact and portable nature, on construction sites to provide steam for capstans, winches or any other purpose. However, it was on the railway that it reached its highest performance.

The barrel of the locomotive boiler was at first made cylindrical, in theory, but an element of taper might be introduced by the need to lap the joints of the rings of plate of which it was made. Because of the relatively small diameter, each ring could usually be made of one plate, as was not possible with marine fire-tube boilers. At first, most barrels were made of three rings, the central one often being smaller in diameter than the other two, so that it could fit inside them for riveting. The best arrangement would seem to be to have all three rings diminishing in size towards the front, but this was not necessarily a preferred practice. As larger plates became available, short boiler barrels were made of two rings, and occasionally of one only. These variations account for the various positions of the steam dome on locomotives: a two-ring barrel could not have it in the centre.

Tapered boiler barrels, with conically rolled rings, appeared very early in America, at about the time when the American 4-4-0 ceased to look like the Norris 4-2-0 and its cylinders were placed horizontally. These boilers had one conical ring just ahead of the firebox, and were sometimes known as 'Wagon Top' boilers. The arrangement gave a large steam space over the firebox, which was further augmented by a large steam dome. The main virtue of a tapered barrel, however, was that it provided for a greater gas area through the tubes and greater freedom of circulation for the water around the firebox tube plate. The firebox tube plate needs to be the same size as the smokebox tube plate, but whereas the latter is a flat plate, in effect, closing the front of the barrel and needing only to be large enough to allow water to flow outside the tubes as well as between them, the firebox tube plate needs width at the sides of the tubes (and above them) to enable a smoothly rounded flange of generous radius to be provided for joining to the wrapper of the inner firebox. This inner firebox has therefore to be as wide as the minimum width of the front of the barrel, but outside it come the water walls of the firebox, which should be about 100 mm (4 in) wide. The firebox outer wrapper is therefore much wider than the minimum size of the front of the barrel, and a parallel barrel means either that the firebox is cramped or that extra and unnecessary weight is being carried at the front end. The latter condition does not matter greatly until a locomotive is designed which is at the maximum weight limit and has to provide the maximum power. An example of this in practice was to be found on the former London and North Eastern Railway, where the late Sir Nigel Gresley's engines all had parallel boiler barrels except for the very largest, the Pacifics, which had a pronounced taper.

The early American locomotives of the mid-19th century, and G. J. Churchward's first locomotives for the Great Western Railway in Britain (which owed a great deal to American practice) had one tapered ring just ahead of the firebox, and parallel rings ahead of it. Later Great Western boilers, like American boilers before them, had the whole barrel tapered. In Britain, the Great Western style was eventually copied by the LMS Railway in the 1930s, when it acquired an ex-Great Western chief mechanical engineer. In continental Europe, the tapered barrel has only been found where weight limitations and maximum power made it obligatory, as on the LNER. The outstanding exponent of it was Karl Gölsdorf of the Austro-Hungarian Railways, who, in the early 20th century, built some notable large locomotives with restricted axle loads. A moderate taper was also a feature of the boilers of the Chapelon 4-8-0s of the Paris–Orléans Railway and the SNCF, boilers which showed an evaporative capacity in proportion to weight and size which has never been bettered.

The shape of the firebox has undergone far more variation than that of the barrel. In the *Rocket* it was a water-jacketed box attached to the rear of the barrel and linked to its water circulation system by copper pipes of quite large diameter. In the *Northumbrian*, only a year or so later, the firebox was fitted inside the rear of the barrel, which was modified underneath to accommodate the deep sides. The basis of this construction, which persisted ever after (apart from a few experimental or very small types), was the foundation ring, a bar bent into a circle, or a rectangle or some other appropriate shape, which formed the bottom of the water walls (or water legs, as they are sometimes called) of the firebox. The inner firebox was fitted inside this 'ring' and the outer one outside it. At the front, the inner firebox wall rose from it to become the tube plate, and at the back, the firehole door or doors joined inner and outer boxes with firehole rings. The inner box had a more or less flat top, called the crown, about three-quarters of the height of the barrel. The tops of the tubes were necessarily lower, and the water level necessarily higher than this crown. Locomotives used on steep gradients had to have short boilers, otherwise, when they tilted nose downward at the start of a descent, the water would expose the crown sheet of the firebox as it surged towards the front of the barrel. This was an effect less marked, of course, if the barrel was tapered.

The early locomotives of Bury in England and Norris in America had fireboxes which were circular in plan, except for a flattened front, where the tube plate was situated. This circular form extended upwards beyond the top of the barrel to provide a steam dome the full diameter of the firebox. When, in later engines, the plan became square, this steam dome was also squared in plan, but became 'Gothic' in elevation, and was sometimes called a 'Haystack' firebox. Because of the prevalence of the 2-2-2 wheel arrangement, it was often possible to fit a firebox somewhat wider than the barrel, there being no large-diameter wheels to be accommodated outside it. This allowed a construction in which the firebox top was made of a half-cylinder plate, curved to a radius greater than that of the barrel and consequently raised in profile. Eventually, this profile was preserved even in boxes no wider than the barrel, simply by raising the centre of curvature. All this was intended to give greater steam space over the crown sheet, and, at first, to provide for the collection of dry steam from a high position above the water level. However, raised firebox tops persisted long after the adoption of separate steam domes on the boiler barrels.

At some time in the early 1860s, boilers with square-topped fireboxes made their appearance. This construction did enlarge the steam space over the crown sheet, but, more importantly, it resolved the rather awkward (at the time) problem of how to arrange internal stays to support the crown sheet. These could now simply join two parallel surfaces, and were not in essence different from those required in the water legs. Similarly, transverse stays between the flat vertical sides of the box, above the crown sheet, could prevent them bulging, either under the pressure of the steam or as a result of deformation of the top. The inventor of the square-topped firebox is usually said to be Alfred Belpaire, of the Belgian Railways, where coal of very low calorific value and high ash content led to some very large fireboxes for their time. It is possible that Jules Petiet of the French Northern Railway arrived at the same solution independently. Whatever their advantages, square-topped boxes were not universally adopted, but they were to be found in the best practice of many countries right up to the end of the steam locomotive's effective existence. They were very common in stationary and marine applications of the locomotive boiler, but they never superseded the round-topped type flush with barrel, known as the 'Crampton' type. For the very largest locomotives, square-topped boxes had the disadvantage that they blocked too much of the driver's view from the cab.

The locomotive boiler as applied to purposes other than locomotives usually had a flat-sided firebox very slightly wider than the barrel, and this design was found on Beyer–Garratt locomotives as well. However, most locomotives could not take such a boiler, and had to be fitted with boilers having wide or narrow fireboxes. A wide firebox was one which was not flanked by wheels and so could extend if necessary to the full width allowable for the locomotive, though its upper part conformed generally to the size of the barrel. An exception to this last point was made in the case of some of Belpaire's Belgian engines, which had wide, square-topped fireboxes much wider than the barrels: the enginemen's outlook was over the top of the firebox, not round the sides. The most acute 'spread' of round-topped wide fireboxes was that of the Wootten fireboxes of some late 19th century American wood-burning locomotives. Such wide boxes were made shallow, because carrying

Sir Nigel Gresley

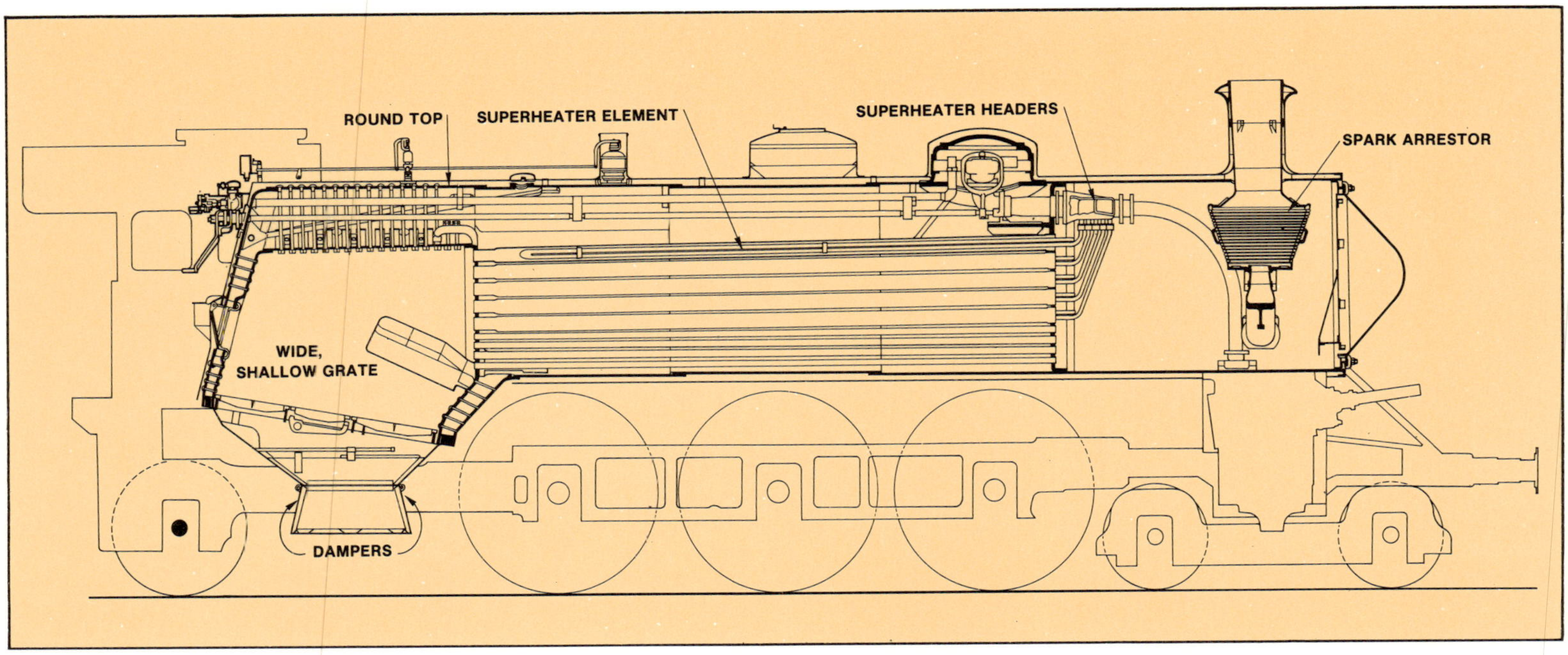

Superheated boiler with wide firebox, as fitted to the four-cylinder compound Pacific locomotives of the former Bavarian State Railway, built from 1908.

wheels had to be accommodated beneath them, and they required more than one firedoor (and sometimes more than one fireman, as well). Larger coal-fired American engines with wide fireboxes were mostly fitted with mechanical stokers from the second decade of the 20th century, but in Europe this practice was adopted rather later, and was only common after 1945. Even then, only a minority of locomotives, those with at least 3.5 sq m (40 sq ft) of grate area, were so fitted.

The usual wide firebox had a grate which sloped slightly towards the front, and its ashpan was arranged to clear trailing wheels beneath it at the rear. Because the grate could not be low down, such boxes tended to have a low volume in relation to grate area, a condition not conducive to full and efficient combustion. It therefore became common to provide a forward extension of the inner box into the barrel, known as a combustion chamber, instead of allowing the tube plate to rise vertically from the foundation ring, or the top of the inner throatplate (the lower part of the tube plate which has no tubes but forms part of the front water 'leg'). This also reduced the length of the tubes, which could be an advantage in an elongated form of locomotive with a long barrel.

The narrow firebox, with plate frames in European practice, fitted between the frames which were, of course, inside the wheels. As it could have driving wheels at the sides, it could be longer than a wide box for any given overhang beyond the coupled wheelbase, and it could be deeper, because the ashpan had only axles to clear, not the tops of carrying wheels. Although it was most commonly applied to locomotives having no carrying wheels at the rear, this was by no means an invariable rule. The low axle of a carrying wheel allowed a very deep narrow firebox giving excellent combustion conditions, and some very successful express locomotives of the 4-6-2 and 4-4-2 wheel arrangements were built everywhere, but perhaps the most outstanding were the Atlantics, designed in 1899 by du Bousquet for the Northern Railway of France, and the Super Pacifics of the same railway, designed under Bréville in 1921. In fact, most European Atlantics had narrow boxes, the important exceptions being the British Great Northern engines built by H. A. Ivatt from 1902, those of the Bavarian and Baden Railways built shortly afterwards by Maffei and the Prussian examples of von Borries' design and their larger successors. However, most Pacifics everywhere had wide boxes, while there were more than 1,000 French engines of the Pacific, 2-8-2 and 2-10-0 types which had a compromise type of box, narrow and deep at the front, and wide and shallow at the rear, which gave an eight-sided grate. These boilers were completely successful and easy to fire.

Very many British locomotives were built, especially in the 19th century, with deep narrow fireboxes and low horizontal grates. Because these boxes had to fit between axles, the grate area was necessarily small, about 2.2 sq m (24 sq ft) being the limit. Such fireboxes were very good for the excellent coal usually supplied for hard work in those days. On the European continent (as in Britain later) larger grate areas were often required and so grates were somewhat higher, and usually sloped upwards towards the back to clear the rear axle, over which they had to pass. In America, the use of bar frames meant that a true narrow box had to be very narrow indeed, so by the end of the 19th century most narrow boxes were made to pass between the backs of the wheels, but to sit on top of the bar frames (which were never as high as plate frames). They were thus wider than European narrow boxes, but also shallower. On the other hand, the greater height allowable to American locomotives reduced the consequences of this. It was also common in America, but exceptional elsewhere, to fit wide fireboxes passing above the driving wheels, not only of freight engines but also of mixed traffic machines of the 4-6-0 type, with quite large wheels: 1.75 m (5 ft 9 in) in diameter.

Steam locomotives burned coke or wood in the first half of the 19th century, but coke was mostly replaced by coal after that time. However, the burning of coal presented some problems, which were solved by the invention of the brick arch in the firebox, devised by Charles Markham under the guidance of Matthew Kirtley of the British Midland Railway. Many other devices were tried, to ensure full combustion of coal, but the brick arch, together with the introduction of slightly more air above the grate (usually via the firehole door) was really all that was needed. This brick arch extends from the tube plate, just below the bottom row of tubes, backwards and upwards towards the rear of the box, angled to a line above the firehole. It stops short of the back of the box by an amount (which was often too great) between a half and a third of the box length. It has the effect of equalizing the paths of the gases as between the upper and lower tubes, of taking the flames right round the box walls, and of mixing the secondary air thoroughly. In American practice of recent years, when the fireboxes were sometimes of enormous size—grate areas of 15 sq m (150 sq ft) were by no means the largest—the brick arch was often supported on water

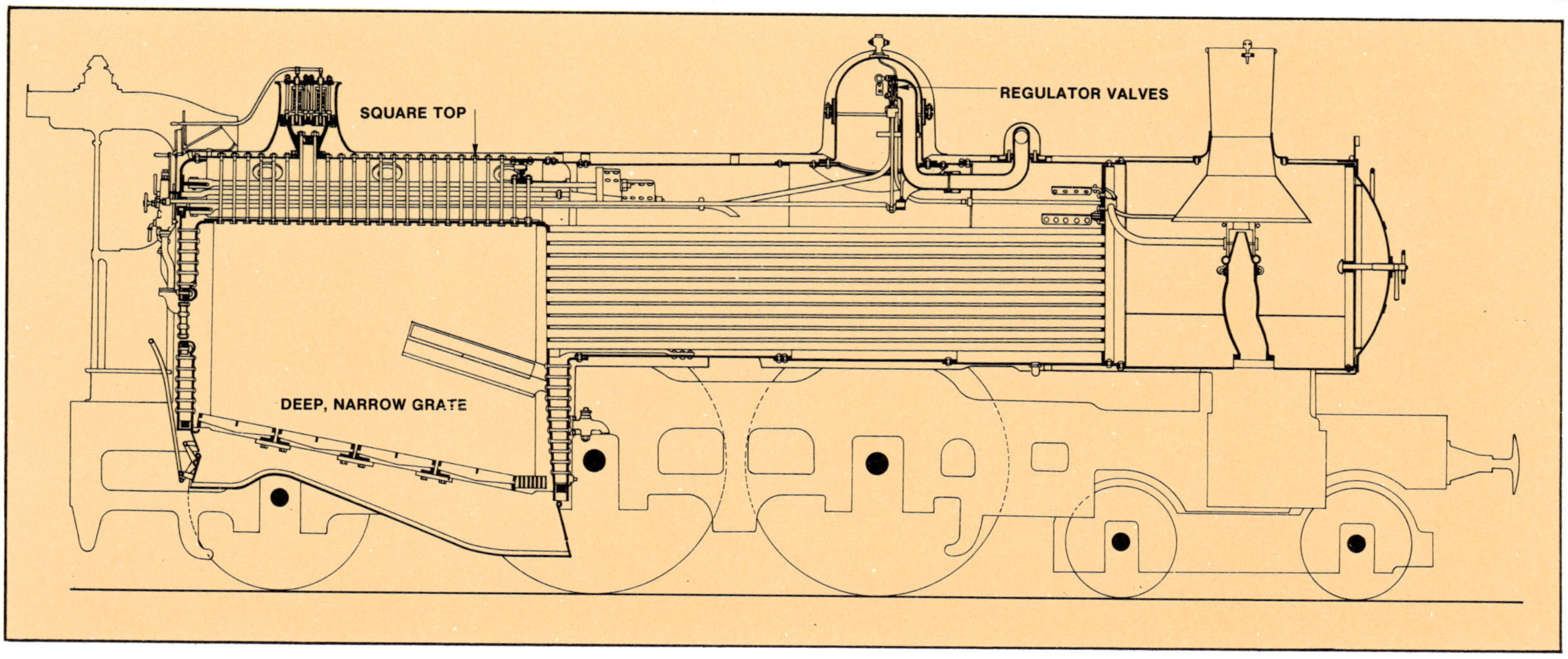

Saturated boiler with long, narrow, square-topped Belpaire firebox, as fitted to the four-cylinder compound Atlantic locomotives of the French Nord railway, built from 1899.

tubes, which also added to the evaporative heating surface. Similar in function was the thermic siphon, a water tube coming from the lower front of the box and opening out to form a water leg inside the box, which served also to keep the crown sheet supported and flooded with water.

The boilers of large locomotives presented problems of draughting. As the barrel diameter was limited by the need to pass the moving load gauge of the railway, it had to be increased in length to provide adequate water capacity and heating surface. The tube length could be kept within limits by the fitting of a combustion chamber, but nothing could be done about the barrel diameter and consequently about the free gas area through the tubes, except to increase the draught. As large locomotives passed a great deal of exhaust steam, the only real difficulty was to make that steam induce a good vacuum in the smokebox without too much back pressure on the pistons. The problem was made worse by the restricted height available for the blast pipe and chimney arrangement in large locomotives.

Much work had been done on the ideal proportions of blast pipes and chimneys, and success had been achieved in solving the problem posed by the need to keep the fire burning well, whether the locomotive was working hard and using much steam, or working easily and using little. William Adams, of the London and South Western and other railways, adopted an annular blast nozzle, the inside of the annulus being in communication mainly with the lower tubes of the boiler. This produced an even draught over all the tubes and was much copied. There were a number of devices to vary the effective orifice of a plain nozzle: the movable internal cone (invented by Koechlin, works manager of the French Northern Railway) which reduced the orifice as it rose; the automatically 'jumping' blastpipe cap of the Great Western, which increased the orifice in response to increased pressure within the blast pipe; the Macallan cap, which was simply a supplementary nozzle ring lowered over the main one by a driver's control; and the hinged vanes just above the nozzle, again under driver's control, which were a common feature of French locomotives at the beginning of the century. But of these, only the first had any potential as a device for coping effectively with large steam flow within a restricted height.

The solution was to fit multiple nozzles. Legein, of the Belgian State Railways, greatly improved some large four-cylinder simple expansion Pacifics just by fitting a double chimney and blastpipe—the first modern instance of this practice—in the early 1920s. Kylala of Finland split the jet from the nozzle into four streams which filled an enlarged circular chimney, and Chapelon improved this with specially designed intermediate 'petticoat' pipes, designed to improve the entraining efficiency and draw in the furnace gases at several levels and across a greater width of the tube plate. This 'Kylchap' exhaust was commonly fitted in double form, and once in triple form. Lemaître of the Nord Belge Railway fitted very large diameter chimneys, filled by steam from a ring of nozzles with a central adjustable Koechlin nozzle in the middle and, more recently, Dr Giesl-Gieslingen of the Austrian Railways has invented an oblong ejector in which a row of nozzles, set in a line fore-and-aft in the smokebox, discharge upwards through a long and narrow chimney. All these devices greatly improved the effective capacity of locomotive boilers and reduced back pressure on the pistons. They represent the results of scientific study of gas flow in fire-tube boilers, and there can be little doubt that, had the steam locomotive continued in general use, equally thorough attention would have been given to the circulation of the water. In American practice, arch tubes and thermic siphons were considered essential on really large boilers, and in Europe and America attempts had been made, from the 1890s onwards, to replace the water-leg firebox with one built like a water-tube boiler, but these had all failed in the long run because the shaking to which a locomotive is subjected had proved detrimental to water-, steam- and airtightness.

WATER-TUBE BOILERS

The water-tube boiler, as now universally used in large marine plant and power stations, is a type in which the flow of water and of furnace gases is very carefully designed from the outset. It has had little evolution in its basic design because it did not make its appearance in the days of primitive technology and virtually no science in engineering: what has happened is that detail design has improved and that the boiler has had to develop as a structure as it has grown larger. Its design has also varied a little according to the type of fuel used, whether it be a long flame coal, a short flame coal, pulverized coal, bunker oil of normal grade or some economically available oil residue, or gas. Fuels other than coal (most commonly oil) were burned in locomotive boilers, of course, but the boiler had to remain the same whatever the fuel used. The applications of water-tube boilers are different in this respect, and the actual shape of the boilers can vary greatly according to application and fuel type.

A basic industrial water-tube boiler may, like the 'Babcock' type, consist of a top drum, containing mainly steam, with a battery of sloping water tubes running between 'headers' underneath. When working, water is pumped in continuously, is carried up the sloping tubes by convection as it heats up, and gives up its steam in the drum at the top. Such a boiler is likely to be set in brickwork, and may have a hand-fired grate, or a moving chain grate, if coal fired. If oil fired, there will be an atomizing oil burner just inside the fire door, in which the oil may be atomized mechanically into droplets, probably centrifugally, or it may be atomized by a steam jet.

Box-like headers are not suitable for very high pressures, and by the 1880s Thorneycrofts were already developing boilers for naval use with extra drums in place of them. A problem was the need to clean tubes out: box headers could have screwed plugs opposite the tubes to permit the use of cleaning rods, the tubes being straight. Such cleaning was essential in industrial applications where dirty feed water was usual but, at sea, where the water was in effect distilled as it came from the condenser, this was scarcely necessary. The tubes could therefore be curved, and so were less likely to disturb their fixings as they expanded and contracted, and this in turn made different boiler layouts possible.

Remarkably, this form of water-tube boiler, with a large number of curved tubes running between drums, had been made in the late 1820s and early 1830s by Goldsworthy Gurney, for use in his technically very successful steam carriages. In many respects Gurney's ideas were extraordinarily advanced, and the design of his boilers anticipated those of the end of the century. He no doubt devised these boilers because of the need to keep down the weight of a road vehicle. The lack of reserve, which is an inherent quality of the water-tube boiler, was of little importance and the concomitant rapidity of steam-raising from cold would have been advantageous. Unfortunately, his steam carriages aroused great hostility and they were never allowed to be used for any great length of service. Had they been, it is probable that the boilers would have proved unsatisfactory because of the quality of the water.

The major development of water-tube boilers started in high-speed naval craft. The first Thorneycroft boilers had two bottom drums, and a single, slightly larger, top steam drum, with the grate between the bottom ones. A nest of curved tubes formed a sort of arch over the fire, with the steam drum at the apex. The drums extended outside the boiler casing, where they were joined together by large diameter pipes away from the fire, thus ensuring circulation of the water. The water level was about half way up the steam drum. Contemporary Yarrow boilers were built with straight tubes: from two bottom drums these rose diagonally and crossed each other on the way up to two top drums placed close together. The drums were in fact of semicircular section, with flat tube plates. These boilers were used in fast torpedo boats.

For large warships, the Royal Navy made extensive and rather unsatisfactory trials of Belleville boilers in the later 1890s. There was nothing wrong with these boilers, which were of French design, provided that full training was given to those who had to operate them, but they were undoubtedly complex in design and unsuitable as a first introduction to water-tube boilers for many of the naval personnel confronted with them. They were assembled of straight tubes, fixed into arrays of return bends in such a way that they formed a set of waterways zigzagging upwards within a rectangular furnace box, with the fire-grate at the bottom of it. Above the main evaporative tube bank was a similar one which served as a feed-water heater or economizer. The space between the two tube banks acted as a secondary combustion chamber, and was provided with a supply of warmed air. In some respects this was more like a modern power station boiler than the marine type, in that it appeared to have been designed as a furnace first, and then provided with the most efficient means of extracting the heat into the water and steam, whereas the Yarrow and Thorneycroft types had all the appearance of being designed as water circulation systems first, and then having the furnace added in the best position. The difference may appear academic, but it is worth pointing out that there are two different approaches to a similar conclusion and it is usually fruitful to examine a design problem from two opposite viewpoints.

For reasons of space, the marine boiler has not generally followed the Belleville style, while the power station boiler has. Marine boilers have acquired more drums, and baffles to guide the furnace gases through the various nests of tubes. Thorneycrofts adopted early three bottom drums, and two grates, the gas flow being vertical; but other makers, such as Stirling and Clarke Chapman, arranged a diagonal gas flow and disposed the tube banks accordingly, between several top and bottom drums. The adoption of oil firing removed the constraint of providing for a large, level, grate which could be fired manually from a number of fire doors, and the Yarrow boilers of the *Queen Mary*, of 1936, showed the possibility of 'tailoring' an oil flame to the precise needs of the boiler. The main gas (and flame) path was divided, one part passing through a large tube bank between the largest bottom drum and the steam drum, while the other part passed through smaller banks and a superheater. Both flows, before joining up again at the top of the boiler, passed through heat exchangers to preheat the combustion air. The cleanliness of the oil flame made this last provision a better proposition than it would have been with coal firing.

The Babcock type of boiler, with straight tubes and box-like headers (or downcomers, as they depended from the top drum) continued in marine use, especially in America, where it was much used by the US Navy in the earlier part of this century. In recent years, however, power station technology (though not the whole morphology of the boiler) has had great influence at sea, and those aspects of design which derived from the days of solid fuel have mostly disappeared. The rise in steam pressures and temperatures has already been instanced in the case of Atlantic liners. Information on the steam pressure of the *United States* is not available, but the more modern though less powerful *Queen Elizabeth 2* uses steam at some 60 atmospheres, or 6,200 kN per sq m (900 lb per sq in). Even this is still well below the pressures used in power stations, which can exceed 200 atmospheres, or 20,000 kN per sq m (3,000 lb per sq in). Such pressures, and temperatures sometimes over 500°C (916°F), are obviously only possible as a result of the development of alloy steels, and welding techniques capable of dealing reliably with them. Early water-tube boilers were wholly assembled by screwing pipes, expanding them into holes, and bolting joints together. Even they depended mainly on steel, but in Gurney's time there was only wrought-iron or the more reliable but softer copper. (In telling the story of the progress of an engineer, it must be emphasized that his thought processes are conditioned by materials available and the machinery of manufacture, and his triumphs often owe as much to the metallurgist as to himself.)

It is difficult to give any adequate account of the growth in size and power of water-tube boilers in power stations. These boilers are hardly units in the ordinary sense, but buildings with forests of pipes in them. The whole assembly may be capable of producing 1,000 tonnes (1,000 tons) of steam per hour, but it may also be

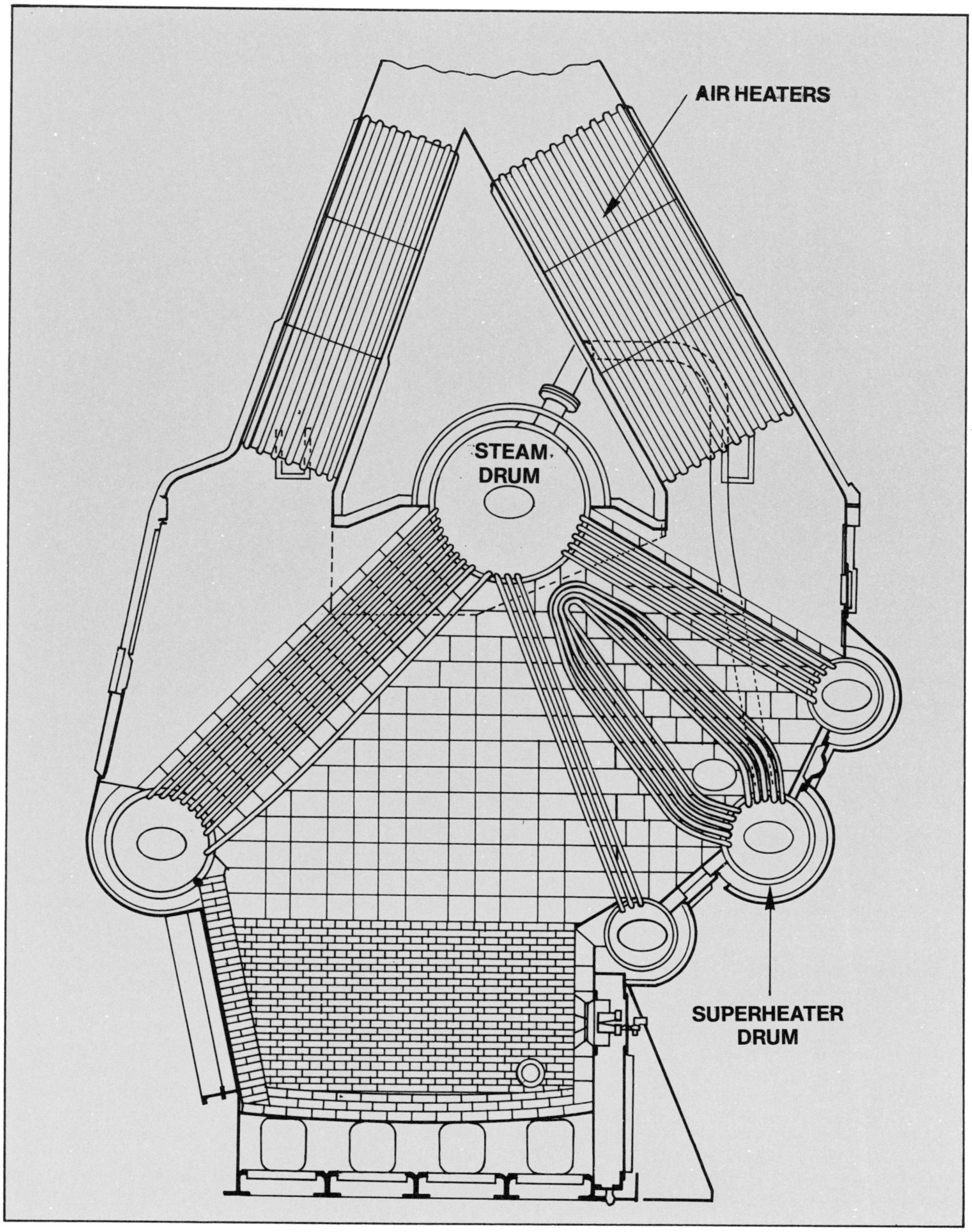

One of the 24 Yarrow water-tube boilers of the transatlantic liner Queen Mary *of 1936.*

capable of being operated in part only, at lower output. The evaporative capacity, as quoted above, is also no basis for comparison with other types of boiler, because the very high superheat represents another way of using the fuel, and temperature is gained at the expense of steam tonnage. The outputs of power stations are dealt with elsewhere, but it may be pointed out here that the only real measure of a power station boiler is its capacity for turning the heat of the fuel into kilowatt-hours.

With marine boilers, however, one can still talk of units. While a single large Scotch boiler might produce steam equivalent to 1,237 kW (3,000 horsepower) at the propeller, one of the *Queen Mary*'s water-tube boilers produced about 5,592 kW (7,500 hp), and one of the three fitted in the *Queen Elizabeth 2* produces nearly 30,000 kW (40,000 hp). In terms of output in relation to the volume occupied by the boiler, the locomotive type compares surprisingly well with the marine water-tube boiler, provided that its induced draught is produced by one of the improved draughting systems described earlier. Something like 10,000 indicated horsepower has been achieved in the largest locomotives in the USA, and the thermal efficiency of the engine part could only be about half that of a marine turbine set. The boiler output must therefore be seen as equivalent to 20,000 in marine terms: this is indeed a remarkable figure for a boiler working on induced draught, coal fired and built to pass the moving load gauge of a railway.

A brief reference must be made to the small vertical

Facing page: *Babcock and Wilcox sectional-headed boiler, arranged for oil firing. A second boiler was set back-to-back, to the right.*

boilers used in cranes and for so many other low-power applications of the steam engine, if only because many are still in industrial use and many have been preserved in various applications, from old fire engines to steam launches. These boilers have cylindrical shells, usually with flat tops but occasionally with domed ones. Cylindrical fireboxes with water walls (which require no staying) are located in the bottoms of these boilers, and their crowns serve as tube plates for a nest of fire tubes which rise to a small, domed smokebox at the base of the chimney. Alternatively, there may be a single large flue, with cross water tubes in it. Launch boilers were built low, and had the multi-tubular arrangement, as they were required to maintain considerable outputs in relation to their size. Crane boilers, being subjected to intermittent loading, often had centre flues, which had the advantage of better natural draught and could thus be fitted with a short chimney. A special type of vertical boiler is the Cochran, which has enjoyed a long period of success in all sorts of small industrial applications. Its peculiarity is that its fire tubes are nearly horizontal, and pass from a combustion chamber above the firebox on one side to a smokebox on the other. It is almost a single-ended Scotch boiler of small size, arranged in a shell set vertically instead of horizontally.

SUPERHEATING

As already indicated, superheating has greatly influenced boiler proportions, because it involves applying a part of the heat of the furnace in a different way. In a water-tube boiler, the addition of a nest of superheater tubes presents no particular problem, provided that suitable materials and designs are available. The superheater simply takes its place somewhere in the flow of hot gases—not usually in the hottest part, for fear of its burning, but not far from it, on the general principle, adopted whenever possible in boiler design, that the flow of water or steam runs against that of the furnace gases, so that feed water enters the coolest part of the boiler and steam is taken from the hottest. As water-tube boilers are generally confined to stationary or marine applications, where they are not subjected to vibration or random shocks, the mechanical design of the superheater is not particularly difficult. It is very different on a locomotive.

The modern locomotive superheater was devised by Dr Wilhelm Schmidt at the beginning of this century, but during the same period there were a number of other attempts at locomotive superheaters, none of which was able to produce much increase in temperature. These were mostly smokebox devices: nests of tubes, or drums with extensions of the main fire tubes passing through them. They were unsuccessful because they ignored the principle stated above, and were situated in the coolest part of the boiler, only really suitable for installing a feed-water heater. Schmidt's superheater consisted of small bore pipes folded back on themselves to form 'elements' which were inserted in enlarged fire tubes in the upper part of the tube bank. Usually the small tube made four passes in the fire-tube, or superheater flue. There were many such elements arranged in parallel, providing a path between saturated and superheated steam manifolds or 'headers' fitted in the smokebox.

The advantage of Schmidt's layout was that it brought the steam close to the firebox, so it could gain a great deal of heat. It also had the excellent feature that the elements were well supported within the flues. However, the steaming of the locomotive depended very much on there being no steam leaks in the smokebox, to destroy the vacuum; and it was also important to be able to withdraw elements easily for flue cleaning. These requirements led to many other versions of the Schmidt superheater, all basically the same in concept, but showing much variation in constructional detail.

The changes in the nature of the tube bank raised delicate questions of proportion. When dampers were applied to the superheater flues, as they usually were in the early days, it was possible to increase the draught through the lower, ordinary, evaporative tubes at the expense of the flow through the flues. This was sometimes necessary, to raise boiler pressure, at the expense of superheat. Eventually dampers disappeared and, fortunately, at about the same time satisfactory proportions of flue and small tubes were generally achieved. The advantages of superheating have been already described in connection with locomotives, and the same principles apply in other applications of steam power.

The other main contributor to economy which may be found incorporated in the boiler assembly is a feed-water heater or economizer. In marine practice this goes back to the days of paddle steamers, when a feed-water tank was sometimes placed round the base of the funnel (one such exploded aboard the *Great Eastern*). One original water colour appears to show such a device around the funnel of Stephenson's first locomotive, the *Blucher*, and even before that the idea was found in stationary steam plant. All these early feed-water heaters were at atmospheric pressure—in theory—but with the advent of water-tube boilers at sea and on land, continuously pumped feed began to be circulated through pipes within the boiler uptakes, on its way between the pump and the boiler.

Feed-water heaters applied to locomotives have almost always used the latent heat of part of the exhaust steam to heat the feed. Two such devices stand out owing to their widespread use: the exhaust steam injector and the mixing tank apparatus with steam feed pump. The exhaust injector is an extension of the principle of the live steam injector, first invented by Henri Giffard in the mid-19th century. This device has been used extensively in all types of steam plant and consists of an assembly of cones, in which steam from the boiler is condensed in cold feed water. The paths of steam and water and their combination, through contracting and expanding cones, are designed so that the energy given up by the steam to the mixture is sufficient to overcome the pressure within the boiler, and a stream of warmed water enters the boiler via a check valve. The exhaust injector simply uses exhaust steam for the same purpose, and is therefore more economical. However, it is more complicated, partly because more cones are required to make exhaust steam do the job, and partly because, in a locomotive, there has to be an automatic change-over to live steam operation every time the regulator is closed. This device is therefore best suited to long runs without stops, as frequent stopping or slowing causes the change-over mechanism to wear out rapidly.

Of the mixing tank devices, the best known are the German Knorr apparatus and the French ACFI one. The latter has a double-sided pump, for feeding cold water to the apparatus and hot water to the boiler. The mixing tanks are placed on top of the locomotive boiler. These feed-water heaters, like the exhaust injector, require oil separators to prevent contamination of the feed water. Their use can produce economies of the order of 10 per cent of fuel.

There are three main types of condenser. The oldest is the water jet device, as found in rudimentary form in the Newcomen engine, but later much refined and long used at sea for paddle engines. In stationary plant, it has sometimes been combined in a unit with the boiler feed pump. In later marine installations and in almost all large stationary steam plant, the surface condenser has replaced it. This is basically a multi-tubular heat exchanger, though in recent years the actual form of the tubes has sometimes ceased to resemble those found

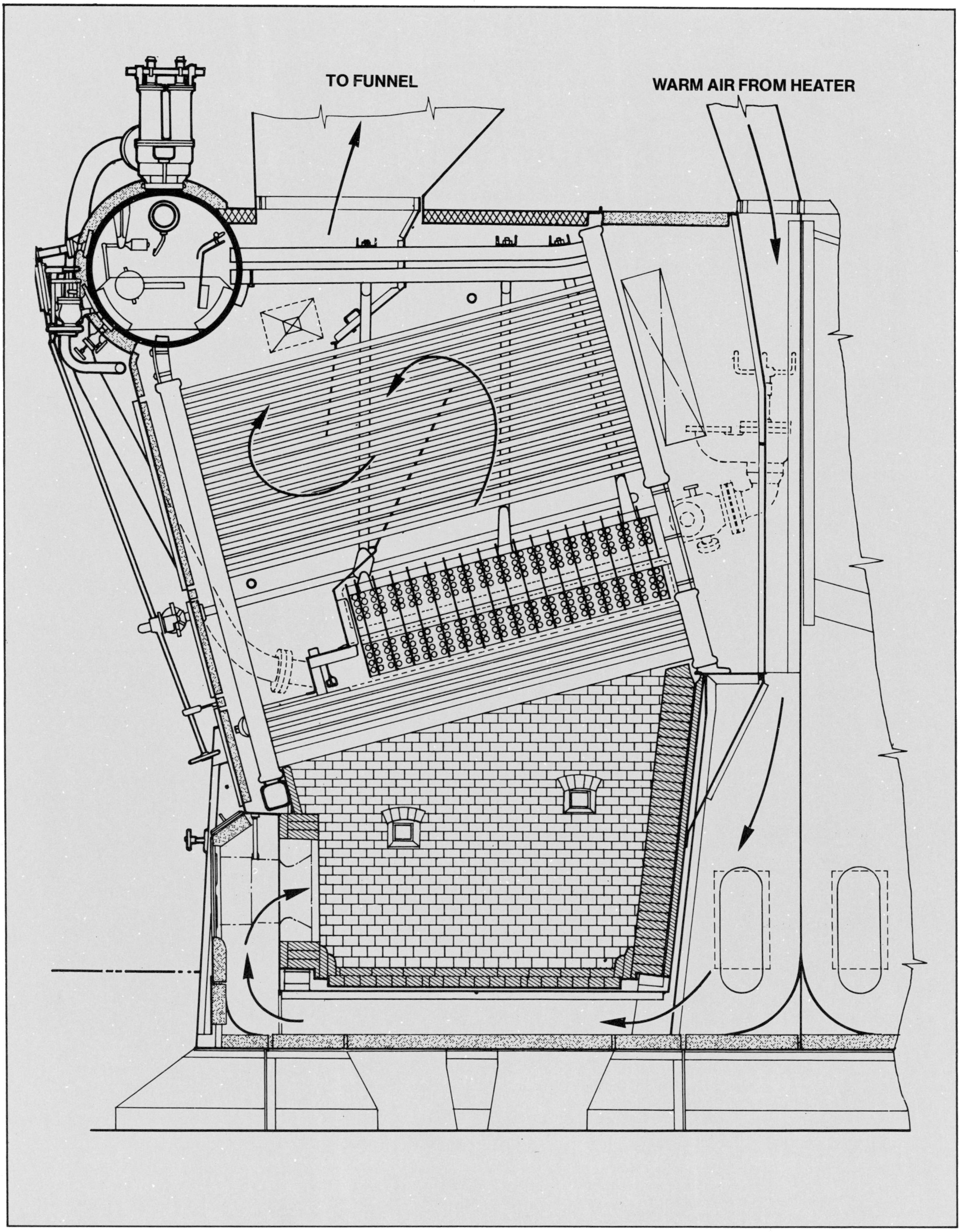
TO FUNNEL
WARM AIR FROM HEATER

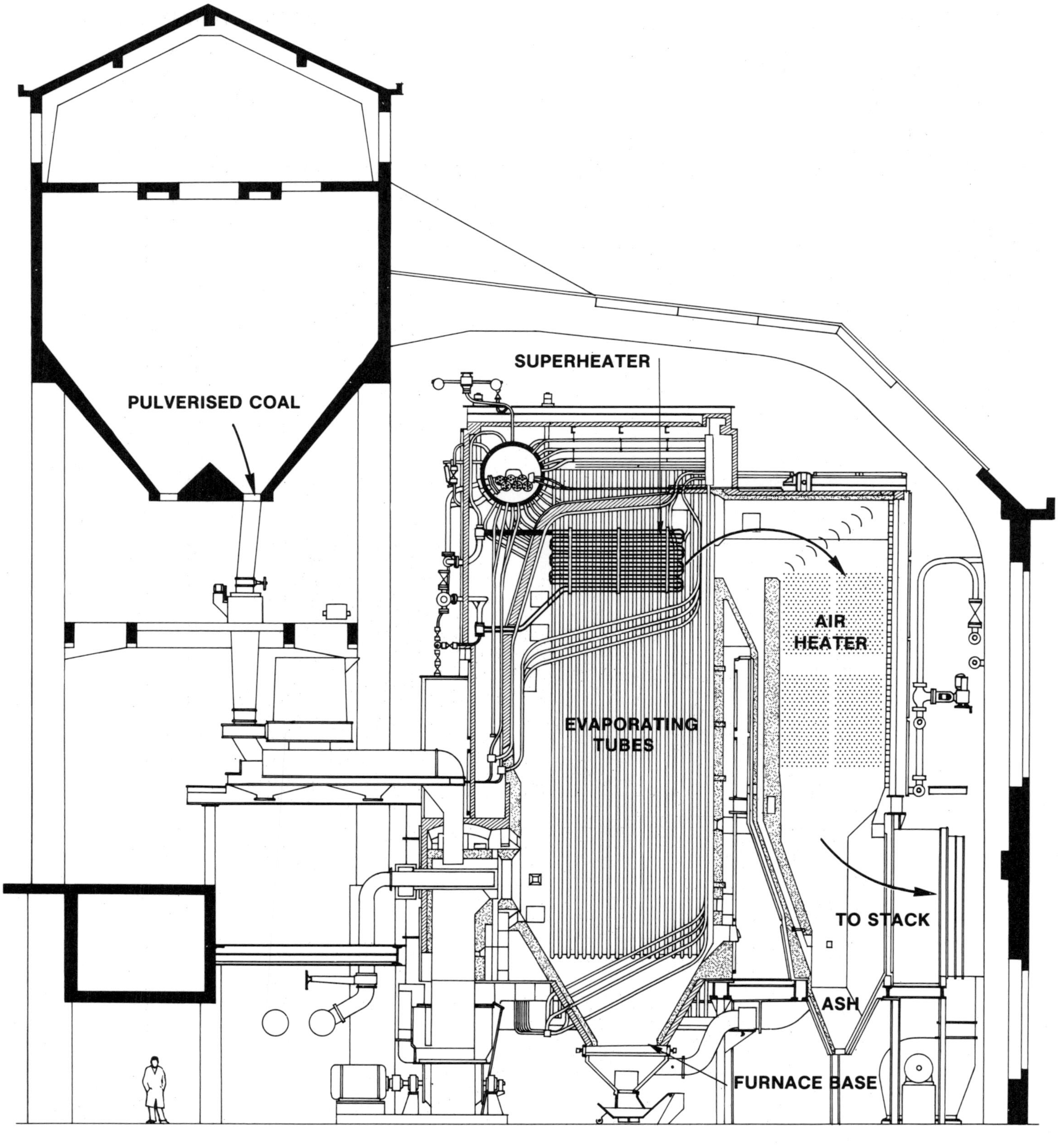

Layout of a power station boiler, c. 1940. No longer a recognizable pressure vessel, this type of water-tube boiler can be designed to fit the natural path of flames and hot gases.

in a fire-tube boiler. These condensers need cooling water to be pumped through them, in a circuit separate from the steam/water circuit. At sea, sea-water is of course used, but on land river water is most commonly used unless the plant is a very small one. This may need to be cooled in turn, before its return to the river, hence the large cooling towers to be seen at power stations. Ejector condensers, sometimes used in small plant, use the cooling water to provide a vacuum directly, after the manner of a filter pump, by passing it through a set of jets in series, so that it entrains the steam, condensing it and passing warmed water to a hot well.

The construction of some of the gauges and valves used on steam boilers is visible in the illustrations in this section. There are many minor variations in the products of different manufacturers, which can affect the convenience and life-span of these fittings when in use. However, the illustrations here are confined to the main types, the minor details being omitted for the sake of clarity.

Electricity Generation

HIGH-SPEED ENGINES

The high-speed reciprocating steam engine and the steam turbine are both inseparably related to the generating of electricity; indeed the steam turbine still satisfies the great bulk of the world's demand for power and is likely to do so for a long time into the future. However, the first true high-speed engine came into being several years before the first electric lighting and the consequent sudden demand for an efficient, constant-speed, fixed power source.

An early pioneer was Peter Brotherhood, whose engine was first exhibited in 1872 in London. Despite its revolutionary design, it rapidly gained acceptance. The design consisted of three single-acting cylinders arranged radially and driving on to a common crankpin. There were no piston rods, the connecting rods being fitted directly into the pistons—this feature and the totally enclosed crankcase to keep oil in and dirt out anticipated later internal combustion engine practice. Steam was distributed by balanced piston valves driven from the crankpin. The governor was of the spring-loaded type arranged on a horizontal axis and fitted directly on the crankshaft, which was fully balanced. These engines were small and compact, the biggest developing 55 bhp at 500 rpm with cylinders of 178 mm (7 in) bore by 152 mm (6 in) stroke. In engines at the small end of the range the crankshaft revolved at 1,000 rpm. This engine design was later used as an air motor for torpedoes and was also adapted as a hydraulic motor and an air compressor.

Though Michael Faraday discovered in 1831 that electric current could be produced by mechanical means, and Sir Humphrey Davy had demonstrated the principle of the carbon arc lamp long before that, it was not until after 1870, when a Frenchman, Gramme, produced a self-excited dynamo with a ring-wound armature, that electric lighting became a practical proposition. Before then, only a few lighthouses on both sides of the English Channel were lit by electricity. In 1875 electric lighting supplied by Gramme machines came to Paris; by 1878 there were several hundred of these machines, usually installed in basements and belt-driven from a 7–15 kW steam engine. Some of these produced alternating current.

Electric light came to London in the winter of 1878 when the Société Générale d'Electricité de Paris put in several schemes. A typical one, with 20 lamps along the Victoria Embankment, was powered by a 60 hp Ransomes, Sims & Head semi-portable engine, belted via a countershaft to a 650 rpm Gramme dynamo. It was housed in a wooden shed close to Charing Cross Bridge, where it ran for almost five years, by which time the number of lamps had risen to 50 and the mains extended for nearly 2 km (more than a mile) along the Embankment in each direction.

From 1880 onwards, the first somewhat uncertain electric lighting (by arc lamps) began to make its appearance. There were several attempts to adapt various existing types of slow-running horizontal engine to drive the dynamo by belting and in so doing, multiply the speed. Unfortunately the engines' governors could not exercise a sufficiently accurate speed control to prevent the lamps flickering, and the plants also took up far too much valuable space in urban areas. There was still an urgent need for a quick-revolution engine that could be coupled direct to the dynamo. Brotherhood's engine did help to fill the gap, but meanwhile there was a rash of inventions to meet the dynamo's demands on the steam engine, which were far more exacting than those of any of its other applications. Apart from close governing, a fresh look had to be taken at improved lubrication, balance of reciprocating and rotating parts (with their conflicting requirements) and the use of better materials so as to reduce weight, particularly of moving parts.

In Britain, nearly all the engines developed for this new need were of the inverted vertical double-acting type, side-by-side or tandem compounds. The most successful, and one that has never ceased production, was the engine designed by George Edward Bellis of Birmingham who had had 20 years' experience in the marine engine field. The success of this engine was largely due to its revolutionary approach to lubrication wherein oil was forced under pressure direct to the surfaces requiring it. The system was patented in 1890 and 1892 by Albert Pain, a draughtsman in the Bellis works, and it has since become a basic feature of every kind of high-speed engine and machine.

The first Bellis engine was built in 1890 and it worked for nearly 30 years, supplying electricity at the works. When it stopped in 1919 the journals and bearing surfaces showed little wear, though the crankshaft must have made more than 4,000 million revolutions. It ran at 625 rpm and developed 15 kW; it has since been preserved.

The most popular form of the high-speed 'self-lubricating' engine produced by Bellis & Morcom, as the firm became known, was a two-cylinder side-by-side inverted vertical compound, in which only one eccentric and valve rod worked both high- and low-pressure piston valves between the cylinders. An oscillating pump working from the eccentric supplied oil at 70–200 kN per sq m (10–30 lb per sq in) pressure. These engines were not very exciting to watch in action: the only moving parts to be seen were the solid flywheel and short lengths of piston and valve rods visible between the main glands and oil seals at the top of the crankcase. An enclosed shaft governor connected to a sensitive stop valve through a bell-crank linkage completed the assembly. Later the firm manufactured small turbines as well as their high-speed reciprocating engines.

Many other makers entered the field with similar types of engine, including Allen of Bedford, Ashworth & Parker of Bury, Tangye of Birmingham, Browett & Lindley of Patricroft, Manchester, and Westinghouse of Schenectady in the USA. For very small drives including power generation, single-cylinder engines with forced feed lubrication were produced by firms such as Bumsted & Chandler of Hednesford, E. Reader of Nott-

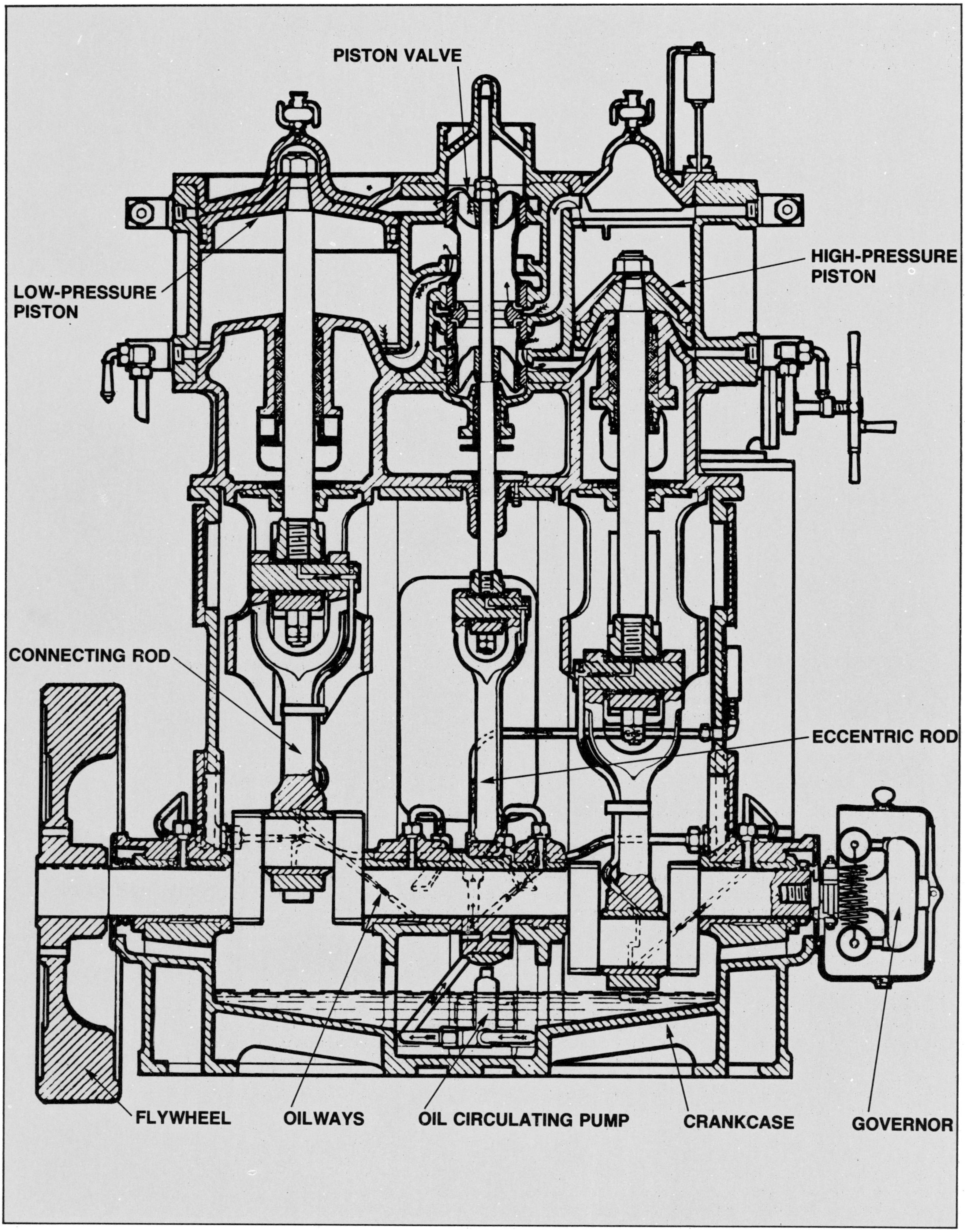
PISTON VALVE
LOW-PRESSURE PISTON
HIGH-PRESSURE PISTON
CONNECTING ROD
ECCENTRIC ROD
FLYWHEEL
OILWAYS
OIL CIRCULATING PUMP
CRANKCASE
GOVERNOR

ingham, Robey of Lincoln and Sissons of Gloucester. Bumsted & Chandler's speciality was the 'Silent' engine which used a single-acting cylinder, or cylinders, and this avoided stress reversals in the motion. It really was almost as quiet to run as the company claimed.

A type of vertical engine of entirely original design, and much used in the early days of area power generation, was the Willans central-valve engine, patented by Peter Willans from 1884 onwards. This was an entirely original form of inverted vertical tandem compound engine with one, two or three pairs of single-acting cylinders with their pistons acting on a one-, two- or three-throw crankshaft. The guide or crosshead worked in a trunk which acted as an air cushion. Its chief innovation was the steam distribution which was done by a long piston valve working inside a slim cylinder where the piston rod would normally be and to which the pistons were attached. Thus the valve travelled with the two pistons and received its motion relative to them from an eccentric on the crankpin. The crank chamber was enclosed to retain an oil bath and the speed was regulated by a spring-loaded shaft governor.

These engines were made in large numbers in compound and triple-expansion forms up to about 500 hp, and ultimately 2,500 hp. A typical Willans engine installation which was considered a model of its kind at the time, was at a power station near Regents Park in London. It was commissioned in 1891 by the St Pancras local authority. Nine main generating sets each consisted of a 120 hp two-crank triple-expansion Willans engine direct-coupled to a 90 kW Kapp dynamo, giving an aggregate output capable of supplying 10,000 incandescent lamps of 16 candlepower. There were also two 51 kW Willans-driven sets used for supplying arc lamps. Steam at 1,172 kN per sq m (170 lb per sq in) was provided by five single-drum Babcock & Wilcox watertube boilers, and the engine exhaust passed either to jet condensers or to atmosphere. Three more 90 kW Willans sets and three more boilers were added to the station later.

As a point of interest, St Pancras' second station at Kings Road was equipped with two-crank compound Bellis engines, each one driving two 65kW dynamos. Thus the authority used the two most popular types of high-speed reciprocating engine for power generation. Willans engines, being single-acting, were very quiet when running and, in small sizes, much used for driving auxiliary machinery such as centrifugal pumps on board ship. A floating crane in London docks with Willans auxiliary engines survived into the 1950s.

The USA differed in its approach to high-speed engines and favoured the horizontal arrangement. The best known was that introduced by P. Armington and Winfield Scott Sims in 1888. This had piston valves and a shaft governor. Another type was the Ames horizontal high-speed engine, examples of which may still be found driving generators today.

Curiously, the power generation requirement which gave birth to engines which were among the smallest ever built in relation to power output, also gave rise to some of the largest. These were the four-cylinder compound 'Manhattan-type' engines with the cylinders in pairs, the high-pressure horizontal and the low-pressure vertical both driving on to a common crankpin with the generator in the middle. These 'megatheriums of the engine world', as they have been described, were installed in two New York power stations in 1903–4, one in Manhattan and the other to serve a section of the city's rapid-transit railway network.

The perspective sketch shows one of the 6,000 kW (8,000 hp) Manhattan engines which took steam at 1,034 kN per sq m (150 lb per sq in). For maintenance, either set of cylinders could be uncoupled at the crankpin and the engine run as a single set. The cranks of the two half engines were arranged at 45 degrees instead of the customary 90, thus giving eight impulses per revolution. This enabled the 152 tonne (150 ton) weight of the generator's rotating parts to act in place of a flywheel. Cylinder diameters were 1,118 and 2,235 mm (44 and 88 in) by 1.5 m (5 ft) stroke and the normal working speed was 75 rpm. Corliss valves were used throughout. The incoming steam was superheated to about 214°C (450°F) and the temperature of the steam entering the low-pressure cylinders was given a boost during its passage from the high-pressure cylinder via reheating receivers. Condensers were of the surface type which were standard in power station practice. These engines could be worked up to 12,000 hp on overload if required.

Some idea of how enormous the parts of these engines were can be gained from the size of the main shaft bearings. The journals were 1.5 m (5 ft) long with a shaft diameter of 864 mm (34 in); the crankpin itself was 457 mm (18 in) in diameter and 457 mm (18 in) long. The rapid-transit engines differed slightly in detail, having fractionally smaller cylinders in view of a higher steam pressure of 1,206 kN per sq m (175 lb per sq in), and drop valves instead of Corliss valves on the high-pressure cylinders, to avoid problems caused by the superheat temperature.

Apart from problems of the manufacture, erection and maintenance of such large engines, they took up an enormous space compared with the turbine and may be said to represent the practical limit in size achieved by a land-based reciprocating engine. In view of this, it is somewhat surprising that as late as 1906 four similar but smaller engines were installed in the London County Council Tramways power station at Greenwich. These had cylinders of 851 and 1,676 mm (33½ and 66 in) by 1.2 m (4 ft) stroke; they took steam at 1,241 kN per sq m (180 lb per sq in), ran at 94 rpm and developed a normal output of 3,500 kW (4,700 hp) or 4,375 kW (5,900 hp) on overload. With the object of improving drainage, the New York arrangement of cylinders was reversed, that is, the low-pressure cylinder was horizontal and the high-pressure vertical. These engines were built by one of the principal mill-engine builders, John Musgrave of Bolton. The last four of these for which the station was arranged were never built, and their place was taken by 5,000 kW turbo-alternators. By 1922 all four reciprocating engines had been scrapped and their place had been taken by turbines. Since 4,000 kW Parsons turbines had been operating satisfactorily on Tyneside for nearly two years before the Greenwich station was opened, at a steam consumption of only 7 kg (15.4 lb) per kW-hour, it must be conceded that the English Manhattans were out of date before they were even built.

However, while the turbine was rapidly taking over all central power generating duties from the reciprocating engine, for smaller individual factory installations, process plants, laundries and on board ship, the single-cylinder or compound high-speed enclosed engine driving a generator, a pump or a fan enjoyed more than half a century of popularity. The form varied little, whether it was built in Britain, Europe or in North America. In a generating rôle, a totally enclosed, force-lubricated engine with a shaft governor was direct-coupled to the dynamo; a flywheel with holes in the rim was between them, so it could be barred round to get the crank(s) in the best position for starting. Engine and generator were mounted on a common baseplate. On board ship, the exhaust was normally led to the main engine condenser. On land these sets were used for works, pumping and power station lighting in the days when the Grid was less reliable. They also provided emergency power for such places as hospitals. Some engines spent virtually their whole lives on hot standby with the cylinder drains open and the throttle valve just 'cracked', to pass enough steam to keep the

Facing page: *a cross-section through a Bellis & Morcom compound self-lubricating engine, showing the single piston valve between the cylinders. This type was probably the most popular of all high-speed reciprocating steam engines and was much used in conjunction with a dynamo for local or auxiliary power generation.*

Charles Parsons

cylinders warm ready for instant start-up. (Since water is incompressible, a cylinder could easily be smashed during a start from cold if condensate had been allowed to accumulate inside.) Another important rôle was generating for process plants, where the exhaust could be used as process steam, the engine acting merely as a reducing valve in the system and providing power almost as a bonus.

The advances made in packings for pistons, valve rods and piston rods must also be mentioned in connection with reciprocating engines. Hemp rope with vegetable oil and tallow lubricant had proved adequate until steam pressures began to exceed 550 kN per sq m (80 lb per sq in), when they were liable to form acids which attacked the metal or became carbonized and useless. Lubricating oils based on the distillation of petroleum began to be available in 1860 (and are still in use); asbestos became increasingly tractable, and by interweaving it with brass wire into a fabric impregnated with vulcanized india-rubber (and now graphite), and making it up into rope coils or blocks, it stands up to the highest temperatures.

Metallic packing for pistons began with Barton's segmental wedge-packing pressed outwards by springs and patented as early as 1816. Many variations on this theme were used at different times and by different builders, but the simple pair of rings pegged to prevent rotation and sprung into grooves in the piston (used in automotive practice today) was widely used in steam engines. The design was patented in 1852 by John Ramsbottom for locomotive cylinders. For piston and valve rod packing, the system marketed by the United States Metallic Packing Company, combining the efforts of several inventors, was widely used in stationary and marine engines. It has the advantage that it can follow small lateral and angular displacements of the rod, relative to the gland, without leaking. It consists of a series of blocks and springs acted on by steam bled from the cylinder to provide just enough sealing force without causing excessive wear.

THE HEAT ENGINE

The turbine is, of course, the most fundamental change in the form of the steam engine since Newcomen's day. As we have seen, the evolution of the reciprocating engine was the result of the work of practical men whose aim was to use the pressure of, first, the atmosphere, then of steam, to act on a piston.

When the ideas of theoreticians such as Carnot, Joule and Rankine came together in the third quarter of the 19th century, it was recognized that the reciprocating steam engine was a particular form of heat engine. Heat is let down from a high level of temperature to a lower one, and the heat which disappears does so through performing mechanical work. Clearly, there must be ways of achieving the same end other than that of simply using the pressure of steam.

When a given volume of water is turned into steam at atmospheric pressure, it multiplies its volume 1,642 times. The opposite happens when steam is turned back into water, and it is this tremendous contraction which makes the vacuum form in the condenser.

In 1915 Professor W. E. Dalby published a diagram showing the heat–energy stream in a condensing engine and boiler. It reveals the overall thermal efficiency of the whole installation to be only 14.9 per cent—the engine's efficiency being 18.9 per cent. No less than 60 per cent of the original heat is lost in the condenser cooling water alone, and in a non-condensing engine, because of the lack of any heat recovery (in the absence of a feed-water heater), the efficiency is appreciably worse, perhaps as low as 5–7 per cent.

Today a modern coal- or oil-fired power station can at best reach an overall efficiency of around 40 per cent. It is mainly the turbine which has made this improvement possible, in that it has enabled the upper limit of temperature to be pushed higher and higher by increase in pressure and by superheat. In a turbine there are none of the condensation losses which occur in a reciprocating engine because of the alternate heating and cooling of metal. A turbine also permits expansion down to the lowest vacuum that the condensing water can produce—expansion that a reciprocating engine could not handle, because of the huge low-pressure cylinder that would be needed to contain the large volume of steam. Some early marine turbines in transatlantic liners were used to expand the steam from the low-pressure cylinder of a triple-expansion engine down to the highest vacuum attainable in the condenser.

In the same way that a water turbine may be compared with a waterwheel, a turbine uses the kinetic energy in the steam rather than the pressure used by a reciprocating engine. There are two types: a 'reaction turbine', where steam issuing from a jet tangential to the direction of rotation pushes the jet in the opposite direction to the jet stream, as in Hero's aeolipile described in chapter 1; and an 'impulse turbine', where a jet of steam strikes vanes on a wheel, causing it to revolve. The method for the second type was first postulated in 1629 by Giovanni Branca in Italy.

PARSONS AND THE TURBINE

The Hon. Charles Parsons achieved his first breakthrough with a turbo-generator in 1884 and pursued development so rapidly beyond it that he ranks in stature with Newcomen, Trevithick and Watt. The following is an extract from a lecture he gave in 1922:

> In commencing to work on the steam turbine in 1884, it became clear to me that in view of the fact that the laws for the flow of steam through orifices, under small differences of head, were known to correspond closely with those for the flow of water, and that the efficiency of water turbines was known to be from 70 to 80 per cent, the safest course to follow was to adopt the water turbine as the basis of design for the steam turbine. In other words, it seemed to me to be reasonable to suppose that if the total drop of pressure in a steam turbine were to be divided up into a large number of small stages, and an elemental turbine like a water turbine were placed at each stage (which, as far as it was concerned, would be virtually working in an incompressible fluid) then each individual turbine of the series ought to give an efficiency similar to that of the water turbine, and that a high efficiency for the whole aggregate turbine would result; further, that only a moderate speed of revolution would be necessary to reach the maximum efficiency.

From the start, the idea of the turbine was that it should be able to drive directly the newly introduced dynamo or electric generator which ran at about 1,200 rpm, a speed which even Brotherhood's high-speed reciprocating engine only barely approached. But this speed was only a tenth of what the turbine was capable. So Parsons' famous patent for 'Improvements in rotary motors actuated by elastic fluid pressure and applicable also as pumps', dated April 1884, was accompanied by another, 'Improvements in electric generators and in working them by fluid pressure'. In other words, he redesigned the generator to run at the much higher speed of his turbine, and so well that the proportions he established were used for more than 25 years.

In view of the number of practical problems Parsons had to solve before he could build a successful working machine, the short time he took to do so is an impressive

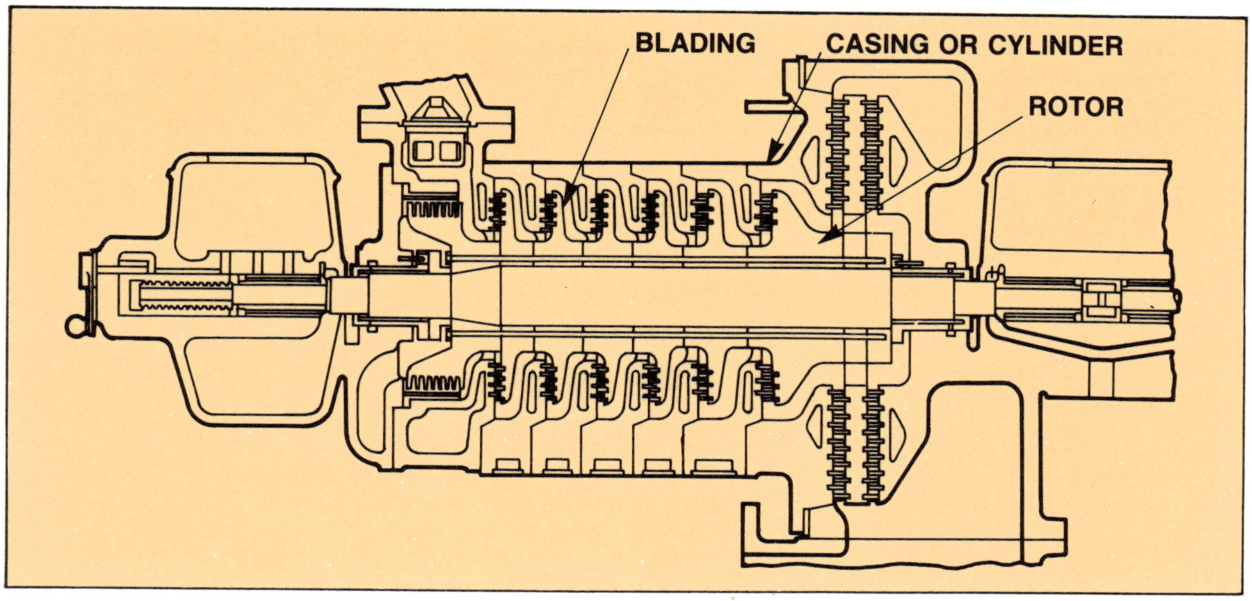

A simple form of axial-flow turbine based on Parson's early units of the 1880s, in which the steam flowed through the blading parallel to the rotor.

achievement. Not only had suitable blading to be invented and manufacturing methods devised, but the design generally had to meet conditions completely outside the range of normal engineering practice. The rotational speed he adopted, for example, was more than 50 times that of the fastest reciprocating engine of the day. Means had to be provided for continuous lubrication of the bearings and a totally new method of controlling the speed had to be found. His generator, too, demanded of him an almost equally revolutionary approach.

It was in January 1884, only three months before he filed his famous patents, and at the age of 29, that Parsons joined Clarke-Chapman & Co. of Gateshead as junior partner in charge of the electrical department. The company was anxious to produce a small, lightweight, steam-driven lighting set for use on board ship and Parsons was given the task of developing it. This he lost no time in doing, and the first turbo-generator on his new plan appeared before the end of the same year.

The reaction turbine part of the machine consisted of a 'rotor'—like a thick shaft—free to rotate inside a steam-tight casing or cylinder. Steam flowing axially (that is, parallel to the rotor) through the annular space between it and the cylinder impinged on a series of blades, alternate sets of which were fixed to the static cylinder and moving rotor, with the clearance between them kept small. The blades consisted of slots cut on the edge of gunmetal rings, the slots being angled at 45 degrees to impose the direction of rotation on the rotor blades. The cylinder could be split into top and bottom halves for internal access. To avoid end thrust, the machine was of the 'double-flow' type, that is, the steam entered at the middle of the cylinder and flowed both ways towards the ends, passing through 14 sets of blades as it did so (the total being 28). The small pressure drop across each set of blades caused a progressive reduction in pressure and therefore an increase in volume of the steam on its way through the turbine, so the pitch and length of the blades were increased towards the exhaust end.

Although this pioneer machine leaked, it did work: taking saturated steam at 551 kN per sq m (80 lb per sq in) and exhausting at atmospheric pressure, it developed 7.5 kW (10 hp) at 100 volts and a speed of 18,000 rpm. Oil was fed continuously to the bearings by a pump. Steam consumption was, of course, high: about 59 kg (130 lb) per kW-hour compared with 4 kg (8.5 lb) 40 years later and even less in the best units of today. Nevertheless, the machine was used for 16 years. It is now in London's Science Museum.

Parsons' turbines quickly became established for ships' lighting sets and by 1888 about 360 had been sold. In the same year one was installed for the first time in a power station, and was the first of four 75 kW (100 hp) turbo-alternators for the Newcastle and District Lighting Company's Forth Banks power station. These were the first generators for supplying alternating current, all those previously supplied being for direct current. At about this time Parsons left Clarke-Chapman and started his own company, C. A. Parsons & Co. of Heaton, Newcastle, in 1889. Since he had relinquished his patent rights on the axial-flow machine, he courageously began work on a modified form of turbine in which the flow of steam was radial, that is, at right angles to the rotor axis both inwards and outwards. Though this type was constructionally more awkward, in 1891 Parsons succeeded in producing a 100 kW (134 hp) machine which went to a power station in Cambridge. This machine was the world's first condensing turbine and, when it was tested by Sir Alfred Ewing, it was found to have a steam economy equal to the best equivalent reciprocating engine. The rotor had seven wheels and ran at 4,800 rpm; at a steam pressure of 965 kN per sq m (140 lb per sq in) and on full load, steam consumption was 18 kg (39 lb) per kW-h; and the whole unit was 4 m (14 ft) long and weighed (without condenser) just 4 tonnes (4 tons). The alternator produced single-phase, 80-cycle current at 2,000 volts. This machine is also at the Science Museum.

The biggest radial-flow turbine, which was a 150 kW machine, was built for Portsmouth in 1893 but the next year Parsons bought back his original patents and was able to resume manufacture of his preferred axial-flow machines. One of the first was a 350 kW machine built for the Metropolitan Company's Manchester Square power station in London, where Willans sets were creating such a nuisance because of vibration that an injunction had been taken out against the company. The new machine represented a colossal jump in output, but worked so well that more sets were ordered immediately. The turbine incorporated two important advances: to keep the blading of uniform height, namely 12 mm (1 in), the rotor was made of increasing diameter towards the exhaust end; and to keep down the overall length and halve the blade-tip leakage, Parsons made the flow unidirectional. He also took care of the end thrust by the use of 'dummy' pistons, one for each of the five diameters of the rotor. The stepped rotor persists to this day.

Parsons managed to have his patents extended for five years, until 1903, because of the years lost to him while they belonged to Clarke-Chapman, and develop-

Assembling the rotor of a 500 Megawatt steam turbine for Fawley power station in Parsons' works at Newcastle. By comparison with the two men just visible in the centre of the picture, the huge size of a modern power station becomes apparent.

ments continued apace. In 1896 he patented an important advance devising independent blades made of drawn strip metal. This enabled the blades to be made crescent-shaped which meant that their curves could match accurately the flow requirements, instead of being restricted by having to be made out of one solid piece. Since the blading was now closer to the ideal shape, pressure loss was reduced. The crucial factor now was how to fix the blades, and this was done by cutting a groove in the rotor (or cylinder), inserting a piece of strip of the right blade length, fitting a correctly shaped distance piece between it and its neighbour and caulking it up tight, and so on until the whole assembly was complete. The blade shape and spacing Parsons established have, in principle, been used ever since, though there have been improvements in detail such as sharper edges and better finish. In the words of his life-long friend, Dr Gerald Stoney, 'the prescience of Parsons seems the more remarkable when it is remembered that in 1896 the present knowledge of aerofoil shapes and of aerodynamics was non-existent.'

To increase the rigidity of long blades, Parsons introduced one or more circumferential lacing strips soldered into notches in the blades. In 1899 came another patent, jointly with Dr Stoney and D. F. Fullagar, covering methods of making the blades up in segments to cut out some of the tedium of fitting blades by hand. Meanwhile the size of unit continued to be increased. In 1900 Parsons built the first 1,000 kW set, with twin cylinders in tandem, for Elberfeld in Germany. In this set, steam consumption was down to 9 kg (20.2 lb) per kW-h.

In 1904, a comparative trial was carried out at Neptune Bank power station, Newcastle, where both four-cylinder inverted vertical triple-expansion reciprocating engines driving generators, and a 1,500 kW turbo-generator, had recently been installed. It was found that, at full load, the turbine consumed little more than three-quarters the amount of fuel per kilowatt-hour used by the reciprocator, and even less when on overload.

In 1902 Parsons had patented another important invention, the vacuum 'augmentor' or 'intensifier', to assist the air pumps in maintaining a high degree of vacuum. It worked on the principle of a steam jet air pump. After its introduction a maintained vacuum of 710 mm (28 in) became commonplace, and by 1918 the figure was up to 735 mm (29 in), which was the highest practical. Further progress since then has invariably been towards higher pressures.

All alternators now produce three-phase current, usually at 50 cycles per second but 60 cycles in America and in places under US influence. The first three-phase alternator was a small set of 150 kW, which Parsons supplied to a colliery in 1900. Another important milestone was in 1912 when Parsons built a 25,000 kW machine which was four times the output of anything built previously, for Fisk Street power station in Chicago. The design introduced the use of two cylinders in tandem, the low-pressure one being double-flow, an arrangement which became almost universally adopted in large axial-flow turbines and which gave rise to the three- and even four-cylinder machine. The Fisk Street machine was very large indeed, 22 m (72 ft) long, 6.7 m (22 ft) wide and 9 m (30 ft) high from the basement; had double-flow not been reintroduced its length would have been doubled.

TURBINE PIONEERS OUTSIDE BRITAIN

Parsons was not the only engineer in the world who began work on the turbine at this time. A French inventor, Carl de Laval, tackled the impulse turbine, where the big stumbling block to efficiency was in the nozzle, by which steam is directed on to blades on the wheel rim. His breakthrough came in 1889 when he discovered that providing his nozzle with an expanding orifice made the conversion of steam pressure into kinetic energy so complete that high efficiency became attainable.

His other problem was that, while the turbine rotor would find its own speed at full power, with rigid bearings destructive vibrations would be set up while getting up to speed, because of the practical impossibility of achieving perfect dynamic balance in the rotor. He solved this problem by fixing his wheel in the middle of a long, flexible shaft with a spherically mounted bearing at one end. A de Laval turbine runs at an extremely high speed, anywhere between 10,000 and 40,000 rpm, so to make his machine suitable for driving a dynamo or cream separator (also his invention) he reduced the speed of the drive shaft in the ratio 10:1 by helical reduction gearing. Since then the de Laval turbine has been widely adopted for all manner of main and auxiliary drives in the 1–225 kW range; it is still made in Sweden and Britain.

Combining the principles of the impulse (de Laval) and reaction (Parsons) turbine is the machine of

Auguste Rateau of Paris. This first saw the light of day in 1898. It can be described as the pressure-compounded impulse or multi-stage type. Here a series of de Laval-type wheels is interspersed with fixed guide diaphragms, each containing a set of nozzles. Since the pressure drop in the steam as it passes through each stage is only a fraction of the total through the whole machine, the inconveniently high velocity of the single-wheel de Laval turbine is avoided. Le Société Rateau opened works in Paris and Maysen, Belgium, in 1903; in England Fraser & Chalmers of Erith and Metropolitan-Vickers manufactured the Rateau turbine under licence.

Another approach to reducing the speed of the de Laval turbine was made by Charles Curtis of New York who, in about 1900, introduced the velocity-compounded turbine. Here the steam leaving the nozzle at a velocity too high to be dealt with efficiently by a single row of moving blades is made to act upon two, three, or in marine practice even four, rows of moving blades on the same wheel, being reversed in direction by fixed guide blades between successive rows. Although low blade speeds can be obtained with a small number of wheels, the high friction losses in the steam as it is turned this way and that reduce the possible efficiency compared with other methods of compounding. The Curtis turbine was produced by the General Electric Company, Schenectady, and was also taken up by AEG in Berlin and the British Thomson-Houston Company in Rugby, both associated with GEC.

Another inventor who was inspired by the de Laval turbine was Birger Ljungström of Stockholm. In 1910 he produced a 373 kW (500 hp) double-rotation turbine using the de Laval nozzle and having the blades which were previously fixed rotating in the opposite direction. The blading was arranged on overhanging discs facing each other, the two contra-rotating shafts not being mechanically coupled but kept uniform in speed by the generators (to which they were coupled) being in parallel and mutually balancing. A bigger 1,000 kW machine, when it was tested in 1912, gave an efficiency of 77 per cent which put the design into the front rank. This same set worked at Willesden power station, supplying current to London's tramways, for 12 years.

After the First World War, the Ljungström turbine was being built by a company called STAL of Finspong in Sweden, and by the Brush Electrical Engineering Company of Loughborough in England. It became common practice to add axial-flow exhaust stages, to which the turbine layout lent itself; this anticipated the present-day situation, whereby no large turbines are of one single type. Two 37,500 kW Brush–Ljungström turbo-alternators which were supplied to Southwick power station, Brighton, just before the Second World War, showed a steam consumption of 3.8 kg (8.4 lb) per kW-h, an efficiency of about 92 per cent.

LATER DEVELOPMENTS

After the First World War developments began again, nothing being so striking as the escalating size of the unit. In 1928 two huge sets were installed, the 168,000 kW reaction set built by Brown-Boverie for Hell Gate station in New York, and a 165,000 kW four-cylinder impulse-reaction set built by American Westinghouse for the same station. A year later came the colossal 208,000 kW impulse machine for the State Line generating station, Illinois. The tendency to obtain maximum output at moderate speeds is conditioned by the tip speed of the last row of moving blades, the USA being more generous in this respect with a limit of some 1,320 km per hour (1,200 ft per second), compared with 1,100 km per hour (1,000 ft per second) in Britain.

Since the Second World War, the increase in world demand for power has been fairly steady, at around 8 per cent per year. The trend towards larger units and higher steam conditions has continued, both for greater efficiency and to cut the cost of power station construction per kilowatt installed.

Among many technical changes that bigger and bigger sets have made necessary are the unit approach, reheat, and large single-shaft machines. Making each boiler-turbine an independent unit, instead of having the pipework arranged on a ring-main system, offers a major saving in pipework and valves which, as steam conditions have risen, have to be made of costly alloys.

While it is good for efficiency when the temperature at the exhaust end of a turbine drops very low, if condensation develops and starts eroding the blading, it means expensive outages from service and repairs. With a reheat cycle the steam, after passing through the high-pressure stage of the turbine, returns to the boiler for a boost in temperature before entering the intermediate-pressure stage. Thus its temperature remains higher throughout its passage. When reheat was first introduced in the USA, the large volume of steam in circuit was found to be a drawback. If in an emergency the generator suddenly lost its load and the governor shut the turbine stop-valve, there was still enough steam in the turbine and reheat elements to accelerate the turbine rapidly to a catastrophic overspeed. To meet this eventuality, reheat turbines have additional governor-controlled interceptor valves in the system. In recent years the USA has introduced double reheat, that is, with the steam going through a second reheat between the intermediate and low-pressure stages, but the Central Electricity Generating Board in Britain considers the efficiency benefit insufficient to justify the extra complication and cost.

In 1960 came twin-shaft machines for outputs above about 600 MW, because the sizes of rotors, casings and generators had reached their practical limits. Two turbines operating as cross-compounds, that is, with the steam passing through each in turn, can conveniently have the low-pressure rotor (where the diameter across the blades is greatest) running at half the speed. Since each shaft is direct-coupled to its own generator which is electrically coupled to the other, the rotor speeds are kept in step without having to be geared together.

A typical twin-shaft cross-compound is a 900 MW set commissioned at Bull Run in the USA in 1965. Steam conditions are 24,130 kN per sq m (3,500 lb per sq in) and 538°C (1,000°F), with reheat to the same temperature, and shaft speeds are 3,600 and 1,800 rpm. Incidentally, at pressures above about 22,000 kN per sq m (3,200 lb per sq in) the steam goes 'super-critical', which means that its density begins to exceed that of the water from which it started, and this has an effect on the volume ratios in the turbine's high-pressure stage.

Today, with further advances in materials technology, single-shaft machines are judged to be practicable for ratings up to 1,300 MW. Nothing of that size has yet been built: for example, Britain's newest power station, Drax B, now under construction in Yorkshire, is having 660 MW single-shaft machines.

Nuclear power stations may seem to represent the latest technological development, but the steam turbo-generator is hampered by the low steam conditions as yet obtainable from the various types of reactor boiler. This means not only that the turbines are of excessive size in relation to output, but it reintroduces the problem of wetness in the exhaust. Despite its huge size, however, a modern power station turbine is a sophisticated, precision-built machine. At 3,000 rpm, the tip speed of a long low-pressure blade may exceed 1,930 km (1,200 miles) per hour, the centrifugal pull on it may amount to 110 tonnes (tons) and yet there is only a minute amount of space between it and the neighbouring rows of fixed blading. The fact that turbines with a design life of 20 years often run day in, day out for 30

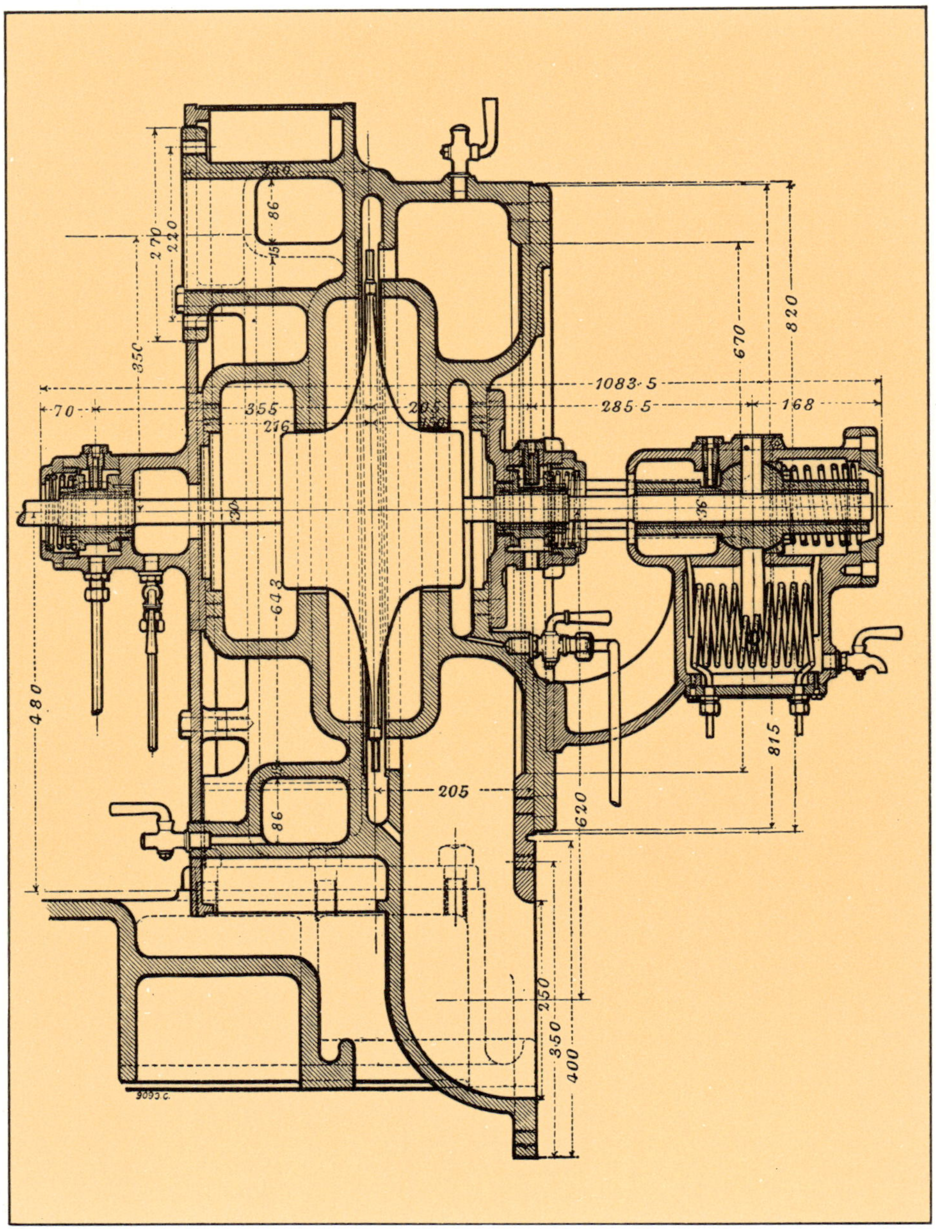

In the de Laval impulse turbine, steam at high velocity is directed through nozzels on to blades fixed to a wheel-shaped rotor. The long, thin rotor shaft with spherical bearing at one end (right) eliminates vibration problems.

or even 40 years without major trouble is a tribute to Parsons and his successors.

Because of the physical limitations mentioned, it looks as though the virtual standstill on unit size and steam conditions of about 24,130 kN per sq m (3,500 lb per sq in) and 548°C (1,050°F) will continue for the next few years. Now that everything in the modern home that can conveniently be motorized already has been, right down to the electric toothbrush, the demand for power should also level off.

Just as reciprocating engines varied enormously in size, from huge marine and stationary units to tiny engines no higher than a table, so turbines also exhibited a great range of sizes. They still do today, though the very small turbine for auxiliary drives has largely been displaced by the electric motor because of bulk electricity tariff incentives, even in situations where there is a ready steam supply available.

Among the smallest turbines still found are those used to generate current for headlamps and cab fittings on steam railway locomotives in a few parts of the world. The whole turbo-generator is no bigger than 304 mm (1 ft) in diameter and 762 mm (2 ft 6 in) long, a truly compact power unit. Other small turbines still in use are de Lavals no bigger than 610 mm (2 ft) across, which may be found driving fans and other auxiliaries in the paper and process industries.

Next in size are those which drive centrifugal pumps, also now largely displaced by motors taking current from the Grid. The turbine was either direct-coupled to the pump, as in land drainage or some water supply applications, or it drove a generator feeding current to an electrically driven pump at the bottom of the borehole. Medium-sized generating and air-compressor plants, serving collieries and other industrial premises, have also been major applications for turbines which were sold in direct competition with high-speed reciprocating engines.

It is not, perhaps, generally known that until current frequencies and voltages became standard many large office blocks, department stores and hotels had their own generating plant in the basement. In New York, particularly, these plants were often dual, one engine driving generators, the other providing hydraulic pressure for lifts and elevators. Some elevators, incidentally, were directly steam-driven. A number of working plants have survived into recent years, the generating/hydraulic machinery being a mixture of slow-speed Corliss engines, Unaflows, high-speed reciprocating engines and turbines.

However, the turbine's dominant rôle is still in bulk power generation. The USA was a major influence in pushing up machine size and steam conditions from the 1920s to the present, and Parsons himself was involved in the early American large sets. As with all steam engines, each stage in the development of the turbine

Because large reciprocating engines could not run as fast as the dynamo required, it was common practice in early power stations for the speed to be multiplied by using rope or belt drive from the engine flywheel. This shows three rope-driven machines at London's Richmond Road power station, Brompton, installed in 1889.

has been prompted by a need for improvement to meet a new demand, and, as with other types of steam engine, a practical visionary has emerged, able to recognize the problems of the situation and point to the solutions.

Parsons, like Newcomen, Watt, Trevithick and Stephenson before him, was unquestionably one of the great men associated with the history of steam power and, like them, was dependent on the circumstances in which he found himself. He arrived just at the moment when the new demand for electricity began to place demands on the reciprocating engine that it could not hope to meet. However, it would be wrong to conclude from this that, without Parsons, the turbine would not have happened. De Laval and Ljungström, for example, pursued their early developments quite independently. Without Parsons, however, the turbine would have taken much longer to assume an established form, both for bulk power generation and in geared-down form for marine propulsion. (The stir caused by his historic craft *Turbinia* with direct drive to three screws has been covered in chapter 5.) His subsequent development of single- and double-reduction gearing for passenger ships put the turbine into the forefront of marine propulsion machinery, a position it held for a quarter of a century. Even today, despite the steady development of the large diesel engine, it plays a vital rôle in many of the world's biggest oil tankers and containers.

One of the most remarkable facts of all is that, even if Parsons had done nothing in the marine field and had been content with his work on turbo-alternators, he would still have been one of technology's most prolific and influential innovators. Some 96 per cent of all the world's power is produced by steam turbine today, with either a fossil-fuel-fired or nuclear heat source.

It is tempting to consider possible future developments of the steam engine. The turbine will certainly become still bigger, in answer to the need for greater efficiency to combat depletion of the world's resources, and as better engineering materials become available. In an attempt to beat the chronic nuisance from vehicle exhaust fumes in many of the world's large cities, engineers in Sweden are at this moment actively reconsidering the steam engine as a vehicle-propelling unit. There is also some research going on in Britain into the possibility of using small reciprocating steam engines fed with steam from solar-heated boilers, to provide basic energy needs in developing countries.

In the more wealthy industrialized nations there is increasing interest in the amenity rôle—the preservation of large reciprocating and self-moving steam engines simply because they are so enjoyable to see, hear and smell. As the years pass, other areas of opportunity may well occur as the world becomes more socially aware and environment-conscious. The expression 'steam engine' might sound old-fashioned, but the subject is very far from dead.

Steam in the Factory

Throughout the 19th century and early into the 20th, before the advent of electric power, the stationary reciprocating steam engine was almost universally employed for providing power to mills, factories, breweries, brickworks, process plants and other industrial premises. Its only real rival, until the advent of the gas engine in about 1890, was the waterwheel. The steam engine also wound up coal and ore from mines, pumped water (as discussed in chapter 3) and drove the rolling mills and workshop machinery.

Unlike the steam engine in its land transport rôles, the stationary engine remained apart from daily life, concealed in an imposing engine-house or lost within a sprawling industrial maze. Few people ever saw it or were even aware of its existence, and most people still think of the steam engine only in its locomotive forms. The stationary engine, however, was entirely separate from its boiler, or boilers. When boilers wore out or became outdated by the need for higher-pressure steam, they could be replaced without disturbing the engine. Moreover, the dust and dirt associated with coal burning were confined to the boilerhouse.

For reasons of fuel economy, the great majority of stationary steam engines had condensers and did not throw their exhaust into the atmosphere. Industrial premises were usually located close to a river so there was no cooling water problem and the condenser and its auxiliary pumps could be tucked away beneath the engine-room floor. The principal exceptions to this rule were colliery winding engines (where there was less incentive to save fuel since the boilers could be fed with unsaleable low-grade coal) and large rolling mill engines. In both cases, the intermittent working cycle would have presented problems in maintaining a steady vacuum in the condenser.

The drive from the engine to the machinery was normally transmitted by gearing or by belt, or a combination of the two. However, direct drive was used in some applications, notably mine winding and pumping where the engine speed was about right. The low speed of a beam engine, commonly about 30 rpm and never higher than 50, meant that it was usual to provide step-up gearing to the system of shafting that served the factory. When beam engines gave way to multi-cylinder horizontal and vertical engines and crankshaft speeds increased, multi-rope drive using V-grooved pulleys became commonplace, particularly in the textile mills of Yorkshire and Lancashire. Here the system had the advantage that a series of driven shafts at different levels serving the various floors could be rope-driven from a single flywheel-pulley on the engine crankshaft. Also, failure of one rope would not immobilize the mill.

The length of revolving line shafting in a large cotton mill, for example, could be several kilometres (miles)

Facing page: *a Murray vertical cycloidal engine of 1801, in which a pinion and internal gear was used in place of a crank. This example is preserved in the Birmingham Museum of Science and Technology.*

Left: *non-rotative beam pumping engines were also used to force air into blast furnaces, when they were termed 'blowing engines'. This example, preserved in the open in Birmingham, was built by Boulton & Watt in 1827 and worked until the 1950s.*

Right: *Richard Trevithick's experimental model road locomotive of 1798 as represented by a model in the Science Museum. The upright cylinder is sunk into the boiler, an early example of steam jacketing to save heat and thereby improve efficiency. The crankshaft is beneath and drove the hind wheels via step-up gearing.*

Far right: *one side of a 2,500 hp cross-compound Corliss valve engine in a Lancashire cotton mill. Rope drive was taken from the flywheel pulley in the background, direct to countershaft pulleys serving the various floors.*

arranged overhead, carrying flat belt pulleys serving individual machines and all driven by one large engine which probably also drove a generator for the lights. It says much for the engineers of the past that breakdowns and consequent total loss of production were infrequent. The rumble of the shafting all through the mill was a characteristic facet of the Steam Age.

EARLY DEVELOPMENTS

As already discussed on pages 26–9, the rotative condensing single-cylinder beam engines of Boulton & Watt sprang from the firm's low-pressure pumping engines. By the early 1800s, rotative engines were already far more numerous than pumping engines. The expiry of Watt's restrictive patent led to other firms entering the manufacturing field, and Trevithick's high-pressure 'puffer' type exhausting to atmosphere began to satisfy demands for self-moving and semi-portable types of engine. It was therefore inevitable that the condensing principle and use of high-pressure steam would soon be incorporated in the same machines. This produced much greater efficiency in the Cornish pumping engine from 1815 onwards, culminating in the famous trial of Austen's 2,032 mm (80 in) engine at Fowey Consols in 1835. Less well documented was a comparison made at Wheal Alfred in 1824 when Arthur Woolf was asked to set up two pumping engines, one having two compounded cylinders to his design and using his boilers, the other having a single cylinder and Cornish boilers. When they began working, Woolf's double-cylinder engine gave a duty of 40 millions, while the single-cylinder engine bettered it with 42. As Lean said of double-cylinder engines at the time, 'The greater expense of their erection and the want of simplicity in their construction were objections to their general use', and thereafter they were discontinued.

However, with rotative engines things were different. Double-acting engines in which steam acted alternately on both sides of the piston before passing to the condenser were found to be more efficient when provided with a second, low-pressure cylinder as in the Woolf arrangement. Indeed, compound rotative beam pumping engines were for a time the only type which could equal the Cornish engine for efficiency. Under the influence of other major engine builders of the early 19th century, such as Matthew Murray, John Penn and Henry Maudslay, the rotative compound engine with two cylinders in line on the same side of the beam fulcrum became firmly established, along with the single-cylinder engine, in a wide range of sizes.

Small beam engines were often self-contained machines on a 'tank' frame containing the condenser and having the beam supported on a simple A-frame. Another popular arrangement for small engines was to have a flat bed and the beam supported on a central pillar. Larger beam engines used A-frames or pillar frame while the biggest 'house-built' engines had the beam trunnions supported on one of the engine-house walls or on a massive cast-iron entablature built into the house. Large engines had the beam in two cast halves separated by spacers with the connecting rod top bearing and auxiliary pump offtakes in between, but the parallel motion links remained outside.

The entablature and supporting pillars were often elaborately moulded and decorated, in an effort to disguise the function of the engine beneath it. Beam engine cylinder sizes ranged from about 203 mm (8 in) bore and 533 mm (1 ft 9 in) stroke, to 1,524 mm (5 ft) bore and 2,134 mm (7 ft) stroke; it all depended on the power required and whether the engine was single-cylinder or compound. Slide or drop valves, worked from an eccentric on the crankshaft, were the rule.

Other forms of overhead beam engine, apart from the conventional one, included the half-beam or 'grasshopper' engine. This was popular in small sizes; its short length was due to the cylinder and flywheel coming close together. The arrangement, first used in America by Oliver Evans in 1804, was such that one end of the beam was pivoted to the top of the piston rod and constrained by links or guides to move up and down with it. The other end was anchored to a pair of tall back links which swung to and fro as the beam's angularity changed; this motion was likened to that of a grasshopper's hind legs. The fork-ended connecting rod was pinned to the beam as close to the cylinder as the size of the crank allowed.

A typical American 'Unaflow' engine built by the Skinner Engine Company and capable of delivering 900 hp from a single cylinder.

The extremely long life of many beam engines, extended by such devices as McNaughting and fitting new cylinders and valves to take higher pressure steam, helps to explain why the horizontal engine took so long to become established. Fear of excessive cylinder wear due to the weight of the piston was another deterrent, as described in chapter 3, and it was not until 1850 that the horizontal began to be built in large numbers. However, a London firm, Taylor & Martineau, had begun building them for factory purposes in 1825. Their engines took the form of a cast-iron box-girder bedplate supporting the cylinder at one end and a single crankshaft bearing at the other, with roller or slide crosshead guides between them. The condenser was underneath, with the air and feed pumps driven by an angled lever from the crosshead. The flywheel revolved in a pit at one side of the bed and the second crankshaft bearing was built into the wall close behind the flywheel. The latter's proximity to the wall was no accident: as with beam engines, it was common to anchor a toothed rack to the wall by which a bar placed through the flywheel spokes could be used to lever the engine off dead centre at starting. Some drivers were more adept than others at stopping with the crank in the right position!

As the horizontal engine became more popular, so did compounding systems, and which was used often depended on the shape of the space available. Tandem compound engines had the cylinders in line with one crank and the drive to the valves duplicated; cross compounds had two cylinders with the cranks at right angles, each half engine having its own bedplate and valve gear, and with the flywheel and drive pulley or gear pinion in between. As the scope of their applications increased, so did the need for engines able to be run in either direction, and the reversing gear ultimately used was the same as that employed in railway locomotives and steam road vehicles, the ubiquitous link motion.

This of course had its origins in stationary engine practice. Murdock's famous slide valve, described in chapter 3, was activated by his equally famous eccentric fixed to the crankshaft. The eccentric was really a small-throw crank, and had the same effect, the crank-pin being enlarged to the point where it was keyed to the crankshaft directly, dispensing with the crank webs. The rod to the valve mechanism, called the eccentric rod, received its motion from a 'strap' passing round the eccentric, the whole becoming a basic piece of steam engine design.

Under an 1801 patent, Murray simplified the slide valve to the point where it became just a box sliding over three ports on the cylinder face. Called the 'locomotive D-valve', it was operated by the eccentric rod which, for starting, could be detached from the valve mechanism by 'gab' gear and thus allow the valve to be worked by hand. When the engine was going, the driver re-engaged the gab and the engine would continue running so long as sufficient steam was admitted to it. For backward running, an engine needed two eccentrics and two gabs and would run in accordance with whichever eccentric was engaged; to reverse the direction meant stopping, disengaging one gab and re-engaging the other. Early winding engines were worked in this way but it was a laborious task, even after the introduction of a lever device for changing over the gabs. The problem was solved in 1842, when William Howe of Chesterfield joined the ends of the eccentric rods by a curved link which both engaged the end of the valve rod or other part of the valve mechanism and could be slid over it. More precisely, the link was slotted and ran over a 'die-block'. Though the link was given an oscillating motion by both eccentrics all the time the engine was running, the motion imparted to the die-block corresponded to whichever end of the link, and therefore whichever eccentric rod, was in coincidence with it. Moving the link to bring the die-block to its mid-point provided a neutral position where the movement given to the die-block was insufficient to admit steam past the valve. Though modifications were later made to this link motion for various reasons, its basic simplicity and working principles never changed.

Up to this time, steam engine builders in other parts of the world were doing little more than copying Britain, but one of the few European contributors to early developments was Egide Walschaerts of Brussels. In 1844 he introduced a different form of reversing gear which needed only one eccentric. It was never used to any great extent in stationary engines, though it did become popular in locomotives and marine engines.

THE CORLISS ENGINE

A prime advantage of departing from the beam engine approach was that it enabled a higher degree of expansive working, and hence efficiency, for which a beam engine was structurally unsuitable. One man who realized this was an American, George Henry Corliss, who in 1849 patented an engine with rocking cylindrical valves, four to each cylinder. This event marked the beginning of the next phase of stationary steam engine development, which was to be one of refinement to secure greater economy and to enable higher power outputs to be obtained. The main advantage of Corliss valves was thermodynamic, for the inlet valves and ports were kept quite separate from the exhaust ports, and so they were not cooled on every stroke by the exhaust steam. In the conventional slide valve arrangement, the exhaust steam, expanded to a lower pressure and temperature compared with the live steam, passed through ports in close proximity to the inlet ports, thus cooling them, so that some of the heat in the incoming steam on the next stroke was wasted in raising the temperature of the inlet port.

In the Corliss engine, the valves were arranged at each end of the cylinder, the inlet at the top and the exhaust at the bottom. This provided another benefit, in that the position of the exhaust valves helped to drain the cylinder of condensate. This reduced the risk of damage inherent in any steam engine when starting from cold, because some of the steam is condensed when it comes into contact with cooler surfaces, and water is incompressible; and there have been numerous instances of cylinder ends being blown off accidentally.

Yet another advantage of the Corliss engine lay in the snap closing action of the steam inlet valves, the timing of which could be controlled by a governor, enabling close speed control to be maintained. All four valves were operated by short crank arms on their spindles, linked to four points on an oscillating 'wrist plate', or spider arm, which in turn received its motion from an eccentric on the crankshaft. On each inlet valve spindle was incorporated a trip mechanism which allowed the crank arm to 'let go' at a certain time (decided by the governor) after opening the valve. The valve would then be rotated smartly to the shut position by a spring.

Horizontal Corliss engines proved highly economical and their use spread rapidly in North America where the beam engine was not so well established as in Europe; American technology had been trying to catch up with British ever since the days of Oliver Evans. With the Corliss engine, US stationary engine practice made much better progress. Although Britain later adopted the American Worthington and inverted vertical triple expansion pumping engines, in 1850 she was not ready to accept ideas from the New World, and the Corliss engine remained little known in Europe until the Paris Exposition of 1867.

REFINEMENTS

By this time European engineers were also aware of the economies to be gained by linking the automatic governor with a device for varying the cut-off directly, instead of merely throttling the live steam on its way to the valve chest. In 1834 Zachariah Allen brought out an expansion valve fixed to the back of the normal slide valve, but much better known is the double slide valve introduced by a Frenchman, Jean Jacques Meyer, in 1842. The expansion valve, having a different motion, needed a separate eccentric and a device for adjusting its position relative to the main valve, controlled either manually or by the governor. Another solution was the use of drop valves, which were lifted bodily off their seats like the double-beat valves used in Cornish pumping engines and whereby the motions of the inlet and exhaust valves could be made independent. Henry Corliss was the first to use the governor to control the cut-off of drop valves in about 1840, a move which was also made by another American, Frederick E. Sickels, who in 1841 patented a drop-valve cut-off gear.

The largest textile-mill engines had four cylinders. This twin-tandem compound engine in Lancashire was built by John Petrie & Co., of Rochdale, in 1910 and drove the mill by gearing. The gear casing can be seen in the middle distance.

A typical two-crank horizontal reversing engine for driving rolling mill machinery. Such engines had to accelerate smartly and reverse instantly, so both cylinders took high-pressure boiler steam. This example features disc cranks incorporating balance weights to counteract the weight of the connecting rods and Gooch's link motion. It was built in 1869 by Thwaites & Carbutt for Landore steelworks near Swansea.

Following this approach, attempts were made to increase the sensitivity of the governors themselves. Watt's simple pendulum governor was effective enough as an anti-runaway device, but since a positive change in speed was necessary to give its response enough force to actually move a cut-off device, something better was required for drives which required a speed as near constant as possible to be maintained.

In 1858, in an effort to cut down the time lag between change in speed and governor response, an American, Charles T. Porter loaded the governor by a weight on the centre sleeve which tended to keep down the balls without increasing centrifugal action. He fitted this to a high-speed horizontal engine designed in conjunction with John T. Allen in which, by shortening the stroke and lightening the reciprocating parts, a speed of 125 rpm was attained. On trial the engine produced 125 bhp with steam at 177 kN per sq m ($25\frac{3}{4}$ lb per sq in) with a very low consumption of 1.25 kg (2.87 lb) of coal per indicated horsepower per hour. The steam inlet valves were driven through the medium of a curved, slotted link rocked by an eccentric, the point of take-off of the valve rod being controlled by the governor.

When this engine was shown at the 1862 International Exhibition in London, it caused a sensation because its speed was about double what was then usual. This caused British firms like Tangye of Birmingham and Robey of Lincoln, who were major builders of horizontal engines, to increase their speeds; and with this came the use of a trunk crosshead guide made integral with a girder-type frame joining the cylinder to the crankshaft pedestal. The advantage of speed was, of course, that it enabled more power to be obtained.

Subsequent improvements to governors included the provision of springs to supply the closing power, and balls deviating only a small amount from the horizontal plane to eliminate the effect of gravity. The Pickering, introduced in America in 1862 and discussed in chapter 7 in connection with agricultural traction engines, had the balls fixed to springs. Other designs of governor were the Buss in 1870, the Hartnell in 1876, the Hartung in 1893 and the Jahn in 1912. They show that the problem of sensitive governing was never entirely absent from the minds of steam engineers throughout the reign of the stationary reciprocating engine.

LATER DEVELOPMENTS

Developments of the stationary steam engine continued until construction all but ceased in the 1920s and 1930s. A few firms, however, in Britain and elsewhere, managed to build up a sufficient export business in sugar cane machinery to continue to manufacture, until the 1950s, mill engines of the smaller types, principally tandem compounds to provide the drive. With boilers fired on the waste product 'bagasse', these plants were highly economical and a few are still at work in the West Indies and other remote parts of the world.

However, at the time of the 1862 Exhibition and after, British engine-builders were producing an enormous variety of types of engine, from the more archaic forms such as beam engines, to the latest horizontal engines with drop valves and governor controlled cut-off gear. There was also the inverted vertical engine which James Nasmyth had introduced in 1851 in the likeness of his steam hammers (described later) and which, by 1862, was well established as a marine engine.

A typical 'second generation' beam engine is illustrated on page 161. It is a 60 horsepower Woolf compound made in 1863 by a firm who built a large number of beam engines, Easton & Amos of Southwark (later Easton & Anderson of Erith) and notable for its early use of a Porter–Allen governor. Beam engines continued to be produced for waterworks and sewage pumping duty long after they had been discarded for industrial drives. This was partly because of the conservatism of engineers in the public services. The very last to be built was a rotative compound Corliss beam engine built by Glenfield and Kennedy of Kilmarnock, for a Watford waterworks in 1919.

After the 1867 Paris Exposition, several British and European firms began to make engines with Corliss valves, at least on the high-pressure cylinder. Drop valves, piston valves like those used in locomotives, and D-valves all remained in common use but it was demand

A modern power station can be operated entirely from the control room. This one is at the Central Electricity Generating Board's Cottam coal-fired power station in Nottinghamshire.

for higher power and sensitive governing by the booming textile industry that really brought the Corliss valves into vogue in Britain. Power requirements in excess of 1,000 hp, principally of the Lancashire cotton mills, prompted the adoption of the triple expansion engine, first introduced by Daniel Adamson of Dukinfield in 1863 and later commonly employed in various forms. In 1878 he produced a quadruple expansion horizontal engine with two cranks, but this did not sell.

By the turn of the century, textile mill engines were being built in sizes up to 3,000 hp—one compound built by Hick Hargreaves in 1888 was of 4,000 hp—but they were often overloaded. These large machines were among the finest examples of the steam engine ever built. While running speeds stayed generally around 60–80 rpm, boiler pressures had gradually risen to 1,103 kN per sq m (160 lb per sq in), with superheat, and occasionally higher. Older engines were sometimes 'Corlissized' (just as beam engines had been McNaughted half a century earlier) by fitting a new high-pressure cylinder with Corliss valves and increasing the steam pressure. Superheating the steam, that is, heating it to a temperature higher than its saturation temperature before leaving the boiler, gave its own economy by cutting down condensation in engine cylinders. In the rest of Europe, it was found that, with higher degrees of superheat, Corliss valves were prone to lubrication problems in their tight-fitting bores. Because of this, from 1900 onwards the drop valve made a comeback, with the same double-seated type which was first developed for the Cornish pumping engine so that large valves would lift easily against high-pressure steam. Some sophisticated designs of trip gear emanated from Europe and a few engines were even imported into Britain.

A popular arrangement for large mill engines was the twin tandem compound with two pairs of cylinders arranged in line and with the rope or gear drive pulley in the middle of the crankshaft. Yates & Thom of Blackburn produced a number of mill engines of this type with Corliss valves on all cylinders, and the firm used a similar design for winding engines, two of which are still in service at a Yorkshire pit.

A development of the twin tandem compound was the four-cylinder triple expansion engine with the same cylinder arrangement, except that the high-pressure cylinder on one side was matched by the intermediate pressure on the other, there being two low-pressure cylinders, one on each side. Both these types were well balanced engines with all the parts readily accessible, though they did take up a lot of floor space. There did not seem to be any rule as to whether the low-pressure cylinders should be placed nearest the crankshaft or furthest from it. Typical dimensions for a Galloway 1,600 hp mill engine of 1907 were cylinders 660, 914, 1,016 and 1,016 mm (26, 36, 40 and 40 in) bore by 1.5 m (5 ft) stroke and a 8 m (26 ft) flywheel pulley with 36 ropes: the speed was 63 rpm.

A few cotton mills used the 'Manhattan' vertical-horizontal engine made by the Manchester firm of George Saxon. In it, one cylinder was vertical and the other horizontal, both driving on to the same crankpin. This type originated in the USA for power generation, and is discussed in more detail in chapter 10. Inverted vertical compound and triple expansion engines with the cylinders in line following marine practice were used to some extent for mill drives, particularly after the adoption of the triple for waterworks pumping.

Yorkshire textile mills generally required smaller power units in the 200–500 ihp range. For these engines and those up to 1,000 hp, the single tandem compound and cross-compound were the most popular and a few still remain at work. A tandem compound was convenient for fitting in a narrow engine-room but, to keep down the engine's overall length, the stroke had to be kept short. Normally there had to be a space between the cylinders for access to the two glands on the piston rod, but in Pollitt & Wigzell's popular three-rod engine, patented in 1870, the low-pressure cylinder abutted the high-pressure, immediately behind it, and there were two low-pressure piston rods passing along each side of the high-pressure cylinder. They and the high-pressure piston rod in the normal position were joined to a common crosshead. Smaller types of mill engine, down to single-cylinder examples, were by no means confined to the textile industry but were used in potteries, brick-

Corliss rotary valves originated in the USA and became widely used by stationary steam engine builders all over the world. This typical single-cylinder Corliss engine of about 1912 used to generate power at the US Tobacco Company's plant in New York. The cut-off was controlled by the governor on the tall stand near the middle of the picture.

works, sawmills, breweries and factories making a wide range of products. Until a few years ago, excellent examples of both the tandem and cross compound engine could be seen working at a Croydon brickworks.

Like the Cornish pumping engine, the textile mill engine enjoyed a long life, many fine examples giving good service for 50 to 60 years and more. Many British-built mill engines went abroad, mostly to the colonies, where they had similar records of long service. During the heyday of the mill engine, the number of different designs of trip gear used with Corliss and drop valves multiplied, as manufacturers strove to gain extra efficiency.

The famous Uniflow type of engine (known as Unaflow in the USA) came to Britain too late to be a major force in its industry, but on the Continent and in North America it was widely employed both in new plant and to replace older engines. The principle is a simple one: if the piston is made roughly the same length as the stroke and the cylinder about twice the length, the exhaust ports can be arranged in a belt around the middle of the cylinder and will be uncovered by the piston each time it nears the end of its travel. Thus no exhaust valves are needed but, more important, the ends of the cylinder where the inlet valves are situated stay hot and the middle of the cylinder stays at the lower exhaust temperature. In other words, the working temperature at any point along the cylinder remains constant and there is no heat loss due to alternately heating and cooling any part of it.

Though the idea went back to 1825, it lay dormant until 1908 when Dr Johann Stumpf of Charlottenburg designed and patented a valve gear to suit the engine. It was taken up by Erste Brünner Maschinenfabrik in Czechoslovakia and several Continental, American and British firms took out licences to manufacture. Among the most important were Gebrüder Sulzer of Winterthur, Switzerland, and the Skinner Engine Company of the USA.

The Uniflow's practical advantage was that the same efficiency as a triple expansion engine could be gained with a single cylinder, so it offered a big saving in space and upkeep. However, since it used an early cut-off to gain the degree of expansion normally requiring three cylinders, the initial blow on the piston was high which meant the reciprocating parts had to be heavily built and precision-balanced. Also, because the temperature gradient between ends and middle of the cylinder was so marked, the cylinder had to be bored with a slight taper so that under full working temperature it became truly parallel. With normal running speeds of the order of 120–160 rpm, Uniflow engines had the crankshaft and motion enclosed to retain the oil. Cylinder dimensions of a typical mill Uniflow were 559 mm (22 in) bore and 610 mm (2 ft) stroke, to produce 300 hp at 160 rpm.

Another use for Uniflows was in process industries which required large quantities of low-pressure steam for processing, as well as high-pressure steam for power, demands which varied independently. A so-called 'extraction' engine usually comprised a single drop-valve high-pressure cylinder and Uniflow cylinder with condenser in tandem with a common crank. The engine could run either as a normal tandem compound or as a high-pressure engine with the exhaust going to process before being condensed. The governor controlled the amount of high-pressure steam taken from the boiler in accordance with the demand for process steam. Extraction engines were built in some numbers in sizes up to 2,500 hp in the 1920s, coinciding with the Uniflow's brief but well-deserved spell of popularity. The truth is it could do little to stem the advance of the more compact and efficient steam turbine.

Large reciprocating engines seemed to be used for longer in Britain and the USA than they were in Europe, and even in the 1950s a large number were still at work. In America the widespread use of Corliss valves persisted, though a large number of Unaflow engines were also installed. But during the 1950s and early 1960s, their demise on both sides of the Atlantic was rapid, because of plant closures and mills being re-equipped with independently motorized machines which did away with the need for a central prime mover. Today there are only a few engines still working, but happily, the characteristic clickety-clack of Corliss trip gear in operation will not be lost. The Northern Mill Engines Society, based in Rochdale, the Brighton & Hove

These small tandem compound pumping engines provided hydraulic power to work elevators in a large New York building, until it was demolished.

This cross-compound, gear-drive mill engine, built by Ashton Frost of Blackburn in 1884, is still at work. It occupies an older building that used to house a beam engine. The high-pressure cylinder, of smaller diameter, is on the left and has Corliss valves with the cut-off controlled by the governor; the low-pressure cylinder on the right has a slide valve.

Steam navvies originated in the USA and were much used on the construction of Britain's last stretch of main line, the Great Central Railway at the turn of this century. This one is receiving attention from the 'black gang' (squad of fitters) who are using a steam crane to lift one of the drive shafts.

'Engineerium', and the Tokomaru Steam Engine Museum in New Zealand all have a preserved Corliss engine which runs under steam.

ENGINES IN THE IRON AND STEEL INDUSTRY

Reversing drives required for rolling mill operation called for no radical departures in design from the basic reciprocating engine as used for textile mill drives and other unidirectional applications. But there was a tremendous contrast in operation between the short-burst, maximum power efforts of a large rolling mill engine compared with the almost monotonous, closely governed motion of a textile mill engine against its load. In a rolling mill the maximum load came on when the red-hot metal ingot first entered the rolls. On each successive 'pass', the ingot became longer and so the engine had further to go before reversing, but against a lessening load. Finish rolling gave the engine the longest runs of all but required a negligible effort compared with that put out initially.

Some of the earliest rotative beam engines went to drive rolling mills but these only ran in one direction; the mass of metal being treated was relatively small and it was returned over or around the rolls for the next pass, much heat being lost in the process. As with other industrial drives, step-up gearing was used to convert the crankshaft speed to the higher speed needed at the rolls. To store enough energy for the first pass, it was common practice to put the flywheel on the high-speed roll shaft instead of the engine crankshaft.

The first horizontal engines always had two cylinders with the cranks at the usual 90 degrees and non-compounded, so there was no tendency to stop on dead centre. Use of link motion with two eccentrics enabled them to be reversed smartly between passes so the hot metal billet now went through the rolls in both directions. In smaller installations, this was sometimes achieved with a unidirectional engine and reversible gear drive. As steel rolling became an increasing requirement, together with the need to handle bigger and bigger sections, the two-crank rolling mill engine grew in size. When the cylinders reached 1,524 mm (60 in) bore and 1,524 mm (5 ft) stroke, it was at its practical limit. The ultimate development was the three-crank type with three high-pressure cylinders, and cranks at 120 degrees which were arranged in either horizontal or inverted vertical form.

The three-crank engine, usually exhausting to atmosphere, was widely employed in the 20th century and used steam pressures up to 1,100 kN per sq m (160 lb per sq in). Some rolling mill engines used locomotive-style piston valves arranged above the cylinders (or beside them in a vertical engine) and worked by Joy's valve gear instead of link motion which required no eccentrics on the crankshaft. This kept the engine as compact as possible, despite the large size of its main parts; the cylinders could be placed close together and the main bearings brought in close to the cranks. Joy's gear required little reversing effort and lent itself to controlling the engine by direct variation of the cut-off. Three-crank engines commonly exceeded 10,000 horsepower and could reach 20,000—some very large examples are believed to be still at work in the USA—but unfortunately, the sight and sound of one of these large engines at work are not likely to be preserved.

In any discussion of the use of the steam engine to fashion iron and steel, mention must be made of one of the basic tools of the ironfounder's art, the steam hammer. The need, particularly in the second half of the 19th century, was first to make large and more intricate forgings built up from welded wrought-iron, and later to make steel forgings in one piece, using first crucible steel, then Bessemer and finally open-hearth steel.

One of the most primitive types of industrial hammer was the trip or 'tilt' hammer driven by waterwheel and later by rotative beam engine. This consisted of a heavy timber rocking beam like a see-saw which carried a heavy metal head at one end. The other end was depressed and released by a cam on a shaft, causing the head to be raised and then dropped on an anvil.

The type of hammer which ultimately survived unchanged for more than a century was the direct-driven steam hammer in which steam, acting on a piston in a cylinder in the head of the machine, lifted the hammer block in a truly vertical line. James Nasmyth designed the first of this type in 1839 though Schneider of Le Creusot was just ahead of him in making one, which they each did in 1842. Nasmyth's had a hammer block

Later winding engines were normally of the horizontal duplex form with the two cylinders arranged to drive a crank on each end of the drum shaft. This splendidly kept, piston-valve example still works at a Scottish colliery and was built by Murray & Patterson of Coatbridge in 1924. The brake gear would have been steam-operated originally.

weighing 1.5 tonnes (30 cwt) with a fall of 1.2 m (4 ft). In the 20 years it took for use of such hammers to become universal, other ideas were tried. In 1852, for example, Krupps of Essen applied a direct-acting steam piston to lift the beam of a tilt hammer.

Later, Nasmyth improved the performance of his hammers by introducing 'top steam' in the cylinder, that is, his hammers became double-acting as steam was also used to assist gravity on the downstroke. The anvil then became much the heaviest part of the machine: a 25 tonne (25 ton) hammer built for the Bolton Iron & Steel Company, for instance, had a 213 tonne (210 ton) cast anvil block, a ratio approaching ten to one. Probably the biggest steam hammer ever built was a 126 tonne (125 ton) machine built at the Bethlehem Iron Company, USA, in 1891, but its vibration proved such a nuisance that is was replaced by a press.

Early steam hammers were all supported on two massive pillars, and the workpiece was handled within the arched frame formed by the pillars and the head. In later years much lighter examples were built using a single-pillar cast-iron frame incorporating the anvil and curved at the top to bring the head over it, a form still used in pneumatic hammers today.

The gravity principle of the tilt hammer was also used in 'Cornish stamps' which were used extensively in Cornish mines and elsewhere for crushing the ores down to a size which the concentration plant could handle. In this application the beam was dispensed with, and the hammer heads were carried instead on vertical shafts which were free to slide up and down in a frame. Each shaft carried a substantial round collar which was engaged by a cam projecting from a rotating shaft, and so designed so that it, too, could rotate about its vertical axis and thereby even out the wear. Cornish stamps were usually arranged in sets of eight with the cams staggered to even out the load on the prime mover. Drive was frequently provided by a double-acting beam engine with twin flywheels and as many as 112 heads of stamps driven off one engine have been recorded.

Blowing engines for supplying large quantities of air at low-pressure to blast furnaces were, both before and after 1800, non-rotative beam pumping engines, but having a large air cylinder and piston in place of the pump rod and pumps. A fine example of one built by Boulton & Watt in 1827 and which worked until the 1950s has been preserved as an industrial monument on an open site in Birmingham. Cornish beam engines were also used for furnace blowing, as were double-acting rotative beam engines.

In the last half of the 19th century, large single-cylinder inverted vertical engines became popular for this duty, the air cylinder with its self-acting valves being placed above the steam cylinder with the pistons working in tandem. These engines required a large flywheel for steady running but had the advantage of occupying the minimum ground area, a more important consideration than height on a congested steelworks site. A fine example of a single-cylinder Corliss valve blowing engine of this form, built in 1886 by the Lilleshall Company, has been preserved at the Ironbridge Gorge Museum in Shropshire, along with an earlier twin-beam rotative blowing engine by the same firm.

WINDING ENGINES

Mine winding was another application of the steam engine which did not, of itself, call for much variation of the steam engine's basic forms. Before the use of steam power, mineshafts had been comparatively shallow, and primitive means of getting coal and ore to the surface, such as the 'horse gin', were then adequate. Beam engines began to be employed early on, however, as the coming of the steam engine intensified the search for coal. The principal additions to the plain factory drive engine, apart from a winding drum, were link motion reversing gear and a brake. The winding drum was often fixed directly to the crankshaft next to a large flywheel; this was the normal practice in the Cornish mines where both single-acting engines, with a weighted connecting rod to effect the piston's up stroke and double-acting engines were employed for winding. In the former case, of course, the valves had to be

Facing page: *early winding engines for mines and collieries were beam engines, in which the drum and sometimes the front half of the beam were situated out in the open. This example is probably the last ever built, by Halman Brothers of Camborne in 1887, and is preserved on the site of East Pool Mine in Cornwall.*

Three common types of auxiliary steam engine, preserved at Kew Bridge. **Far right:** *a pair of vertical non-rotative boiler feed pumps by G & J Weir.* **Top right:** *a Worthington horizontal duplex non-rotative boiler feed pump.* **Bottom right:** *an economizer engine by E. Green of Wakefield with a Pickering governor. This engine appeared in Universal Pictures' 1979 production* Dracula.

worked by hand on starting to get the engine going in the right direction. Outside Cornwall it was sometimes the practice to gear down the winding drum to obtain more pulling power.

For working the deep coal seams in County Durham, a special form of single-cylinder condensing overcrank winding engine was introduced. It was in use for a very long time there, and was still being built long after double-cylinder engines had become generally accepted everywhere else. The design originated in 1800, when Phineas Crowther of Newcastle-on-Tyne took out a patent for a single-cylinder winding engine with the crankshaft above the cylinder and the crosshead guided by parallel motion. The advantage of it appeared to lie in dispensing with the beam, though, of course, it took the heavy winding drum to the top of the engine-house.

By 1830 Crowther had built several engines on this plan and they 'were found to succeed very well' according to a contemporary report. Though the patent drawing shows the winding drum geared down from the crankshaft, all known engines had the drum fixed direct to the crankshaft, support for which was gained from the engine-house wall at one end and a substantial A-frame or internal wall at the other end, close to the crank. During the 19th century these engines were a common feature of the district, with cylinders in sizes up to 1,727 mm (68 in) by 2 m (7 ft). All were built by local firms, with the exception of the final examples of 1890 which were made in Leeds.

These engines worked on low steam pressures of 207–345 kN per sq m (30–50 lb per sq in), the bigger ones being made condensing while the smaller ones exhausted to atmosphere. Valve gear was of the plug handle type, so the engines were started and stopped by hand. Because the engines only pulled a single cage, counterbalance was usually provided. Monkwearmouth B. Pit engine was one of the largest of a handful still at work in 1956; it was built by J. & G. Joicey in 1868, had a cylinder 1,651 mm (65 in) bore and 2 m (7 ft) stroke and in a 20-hour working day handled 1,218 tonnes (1,200 tons) of coal and 1,200 men. A characteristic of these Durham engines, which may help to explain their longevity, was that the headgear structure carrying the pulley wheel over the shaft was kept very simple, since it was braced directly to the tall engine-house. Happily, two of these engines have been preserved.

The vertical arrangement with the drum at the top of the house was also used for two-cylinder winding engines but, with the increasing use of horizontal cylinders in the 1850s, it was natural that winding engines should assume this form. From then on, the great majority that were built all over the world consisted of two high-pressure cylinders, widely spaced, with the winding drum occupying the central portion of the crankshaft. As time went on, steam brakes began to be fitted, the shoes generally acting on a single path or twin paths on the drum, and steam-powered reversing gear. In recent years, stricter colliery safety legislation has prompted the fitting of automatic stopping gear to guard against accidental overwinding, and 'slow banker' equipment to limit the speed in the shaft.

While duplex winding engines of the form just described continued to be built until the 1940s and are still at work, some of the deepest pits in Lancashire and Yorkshire, and in some other countries, required engines of abnormal size in order to cut the time taken in raising coal from the pit bottom. This brought into being winding drums of up to 10 m (40 ft) in diameter, cylinders as large as 1,092 mm (44 in) by 2 m (7 ft) stroke on duplex engines, the use of cross-compounding with low-pressure cylinders going up to 1,524 mm (60 in) in diameter and, in a few cases, the adoption of a four-cylinder arrangement consisting of a pair of tandem compound engines, one coupled to each crank. Both Corliss and drop valves were widely used in winding engines in combination with link motion valve gear, while some makers preferred piston valves.

Winding has always been a severe test of both man and machine, owing to the great variation in load, the

The pioneer Bellis & Morcom self-lubricating engine and generator of 1890, now preserved in the Birmingham Museum of Science and Industry.

speed and the need for accurate stopping. The driver usually gives the engine full steam for the first third of the wind, then either he or the trip gear cuts it off for the middle third, and the last third is used for stopping. On some engines, full steam is put against the engine to assist the brake. The cage finally has to land within 25 mm (1 in) of the right level, with the aid of a pointer and marks on the side of the drum. It was once common with a steam winder to make 58 journeys per hour through a 374 m (900 ft) shaft pulling double, one cage with 3 tonnes (3 tons) of coal ascending with the empty one descending. Seventy-five million such journeys used to be made with steam every year, and perhaps 25 million others with men and materials, with very few mishaps. Britain is one of the few places in the world where large winding engines may still be seen in daily operation, though they will all be gone by the middle 1980s.

At some very large American mines with several shafts, it became the practice not to have individual winding engines which were constantly reversing, but to install a large, constantly running engine driving a series of shafts to which the drum could be clutched in or out as required. The clutch would be engaged to hoist the cage and de-clutched to lower it under gravity, controlled by the brake. The wastefulness of not using two cages in counterbalance in each shaft was compensated by the higher efficiency of a well-governed mill-type condensing engine. Probably the finest example of this system was the 4,000 hp compound condensing Leavitt engine installed at a Michigan copper mine in 1881, which had cylinders of 1,016 and 1,778 mm (40 and 70 in) bore by 1.8 m (6 ft) stroke. Ten years later a two-crank, six-cylinder triple-expansion engine was installed at the same mine for drawing from a new 1,524 m (5,000 ft) shaft—it could raise 6 tonnes (6 tons) of ore from the bottom in $1\frac{1}{2}$ minutes.

The Leavitt engine, incidentally, used a bell crank to transmit motion from the piston in an inverted vertical cylinder to a horizontal connecting rod and crank. This was peculiar to North America and had the advantage of being lower in height than a normal inverted engine while being shorter than a horizontal; in other words, it made the engine roughly square as in the later Manhattan engine design.

All over the world, the humble portable or semi-portable engine, usually with overtype cylinders, was used for shaft sinking and other winding duties of a more temporary nature. Subsidiary winding duties for pit maintenance and the like were normally carried out by a small, geared two-cylinder winding engine or capstan sited in an annexe to the winding house. Such engines were sometimes installed in waterworks as part of the permanent installation for handling equipment down the well when maintenance was being carried out. Underground and surface haulage of coal tubs along a level track or up an incline shaft was another use for winding engines of the smaller variety. Some were put underground and ran on steam or compressed air fed from the surface; others performed underground haulage on the surface by driving an endless rope.

A type commonly used for winding and pumping at small mines was a derivative of the portable, the overtype or undertype stationary engine. This was essentially an enlarged portable engine without wheels and was used at various times for a wide variety of purposes: driving small factories, process plants, sawmills, quarry equipment and small pumping and generating plants, normally using belt or rope drive. The great majority were overtypes; two high-pressure or compound cylinders were contained in a common cast block, as in traction engine practice and which in an undertype formed the base of the smokebox. Most manufacturers were those who also built traction and portable engines. Some were supplied with small jet condensers; others had the engine unit placed alongside the boiler, instead of using the boiler as the frame. Horsepower generally did not much exceed 300.

The gas industry was another large user of steam engines for driving boosters, exhausters and small

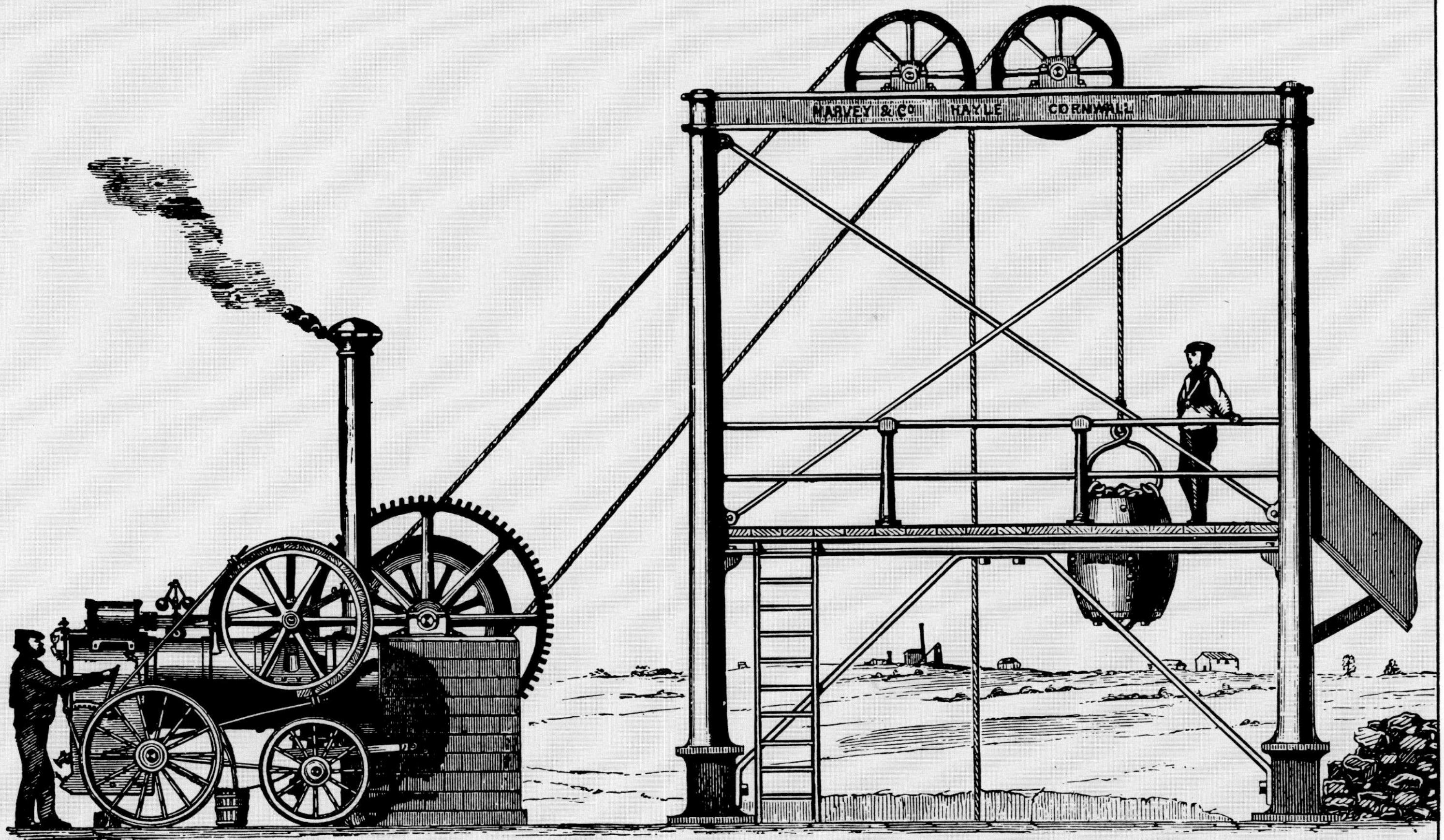
HARVEY & Co HAYLE CORNWALL

pumps. Small beam engines driven by belt soon gave way to medium-sized horizontal and inverted vertical slow-speed engines (usually single-cylinder), with which the firm Bryan Donkin of Chesterfield will always be associated. Small high-speed enclosed engines were used more and more after the mid-1920s.

During the first half of the 20th century, practical testing of steam engines came on the syllabus at many of the world's technical colleges and universities, and a number of small engines were specially built for this work. Usually having a single horizontal or vertical cylinder, but occasionally two for compound or duplex working, these engines had a water-cooled brake drum with an accurate load-measuring device as the main special feature.

STEAM IN CONSTRUCTION

The widespread adoption of horizontal cylinders from 1850 onwards, in place of the earlier vertical-cylinder beam engine and overcrank arrangements, did more than revolutionize the form of large stationary steam engines, it also brought into being the small steam power 'package'. If you scale down a two-cylinder simple horizontal engine with link motion and D-valves, provide it with solid flywheel-type disc cranks in which balance weights can be incorporated, fix a gear pinion on the crankshaft and connect the steam chests by pipe to a quick steam-raising vertical boiler, the result is a reversible power unit which can be attached to a wide range of machinery and equipment.

This is the way steam was applied to heavy construction equipment—excavators, grabs, hoists, cranes, dredgers, derricks, pile drivers and even concrete mixers. By the turn of the century, a visit to a major construction site in all but the most primitive countries would have found some steam equipment in use. For example, on a large dam site which would call for prodigious feats of earthmoving in a remote and hilly region, it was common to provide a temporary railway system worked by steam for the haulage of men and materials. Stone crushing and grading plants driven by portable engines, an electric light plant driven by high-speed steam engines and steamrollers for roadmaking and compacting fill material in the dam itself would also be found there. The civil engineer of those days had to be enthusiastic about steam power if he was to take advantage of the latest equipment technology could offer.

In railway construction, the coming of the American-inspired 'steam navvy' into this country revolutionized construction of the last great main line to be built, the Great Central Railway's extension from Nottingham to London. Gone were the hordes of men who used to tramp all over Britain and Europe, armed with picks, shovels and wheelbarrows, looking for railway work. Instead, on the Great Central, the crude-looking steam navvies with non-slewing, rail-mounted structure and jibs capable of swinging through a limited arc in front ate their way through massive cuttings, while fleets of narrow-gauge locomotives and tipper trucks took their spoil to form embankments.

Excavating is always the most laborious and costly operation in construction, and it was in this field that steam power was most widely employed. A typical excavator of the early 1900s was the 'No. 6' produced by Ruston & Hornsby of Lincoln, mentioned in chapter 6 in connection with crawler tracks. The firm's literature proclaimed that, 'fitted with a $\frac{7}{8}$ cubic yard bucket, the No. 6 is adaptable for use as a crane, navvy, dragline grabbing crane, trench excavator, pile driver or crane'. The machine consisted of a fully-revolving deck and superstructure carrying the boiler, jib and all crane motions, mounted on a four-wheel chassis which could have rail or road wheels or (at extra cost) crawler

tracks. The main engine operated the hoist and travel motions through a system of gears and clutches; it had cylinders of 140 mm ($5\frac{1}{2}$ in) diameter by 152 mm (6 in) stroke. A further engine with cylinders of 102 by 127 mm (4 by 5 in) was fitted for slewing, driving through bevel gears and, when fitted up as an excavator, a 102 by 127 mm (4 by 5 in) racking or 'crowd' engine was fitted halfway along the jib to operate the dipper arm and bucket. The vertical cross-tube boiler with injector and feed pump provided steam at 862 kN per sq m (125 lb per sq in) and, with the water tank, was placed in the customary position at the rear of the machinery deck where it helped to counterbalance the load on the jib.

Ruston & Hornsby was among the first firms to start building steam excavators in 1874, and also built some of the last. These, which again were based on American practice, were giant overburden strippers used in opencast mining. Ruston's 'No. 300' was one of the biggest: the firm continued to build them, principally for export, until the 1930s, but the one illustrated worked in a Northamptonshire ironstone mine until the early 1960s. These monsters had 4 cubic metre ($5\frac{1}{4}$ cu. yard) buckets, horizontal stoker-fired locomotive-type boilers and independent engines for each motion.

In more mundane situations, the twin-cylinder disc-crank horizontal was also to be found in winches and capstans of all types; on dredgers, pile-driving rigs, ships' decks, overhead and dockside cranes where the engine and boiler units were often located high in the air, derricks, hoists—indeed in every situation where at one time hand or hydraulic power were the sole alternatives. In low-speed hoisting applications, the engine unit was geared to the drum through single- or two-stage reduction gears. Some equipment used for heavy, infrequent lifts such as pit maintenance, employed worm drive, which had the advantage of a positive safeguard against accidentally letting go the load, in addition to the brake.

In mining particularly, small steam hoists were frequently built for running on steam or compressed air. The latter medium was, and still is, particularly useful for use underground where ample compressed air is available for other mining purposes. In a few instances, notably in South Wales, elderly steam winding engines have been kept in service as maintenance units, being operated on compressed air on occasions when the electric winder is out of service. This has resulted, in South Wales, in an 1875 engine being saved for preservation.

Space precludes description of all the many small steam engines used for auxiliary drives such as fans, economizers, feed and circulating pumps, mechanical stokers, coal hoists, small air compressors and so on.

Facing page, top: *one use of portable engines was to hoist the material when sinking a mineshaft or in shallow mining operations. This installation appeared in the catalogue of Harvey & Company of Hayle, dated 1884. The winding drum was geared to the engine crankshaft and the band brake was applied to the engine's flywheel.*

Facing page, bottom: *one of the first horizontal steam engines, a winding engine of 1803 designed by the Cornish originator of high-pressure steam, Richard Trevithick. Recessing the cylinder into the top of the boiler has the same effect as in his locomotive on page 162.*

This page, above: *early steam excavator produced by Bucyrus in the USA, in about 1850. Excavators remained substantially in this form, though increasing in size, for many years. They usually ran on rails but could also be mounted on crawler tracks.*

Major Events in the Development of Steam Power

c 50AD	Hero's Aeolipile and other devices	Egypt
1629	Branca's impulse turbine proposal	Italy
1642	Torricelli demonstrates a vacuum	Italy
1654	von Guericke makes an air pump	Germany
1690	Papin's piston-in-cylinder model	France
1698	Savery's pumping engine	England
1712	Newcomen's atmospheric engine and haystack boiler	England
1718	Newcomen engine attains duty of 3.6 millions	England
1719 (c)	Newcomen engine fitted with automatic valve gear	England
1725	Hammered iron plates and deadweight safety valve on boilers	England
1761	Watt's first experiments with steam	England
1765	Watt's first patent for separate condenser, air pump, closed cylinder top and steam jacket	England
1769	Cugnot's steam carriage	France
1774	Wilkinson's boring mill makes accurate cylinders	England
	Smeaton's improved Newcomen engine attains duty of 10.3 millions	England
1775	Boulton and Watt partnership	England
	Glass water gauge on boilers	England
1776	Watt's first pumping engine	England
1779	Watt pumping engine attains duty of 20 millions	England
1780	Pickard's patent crank used on steam engine	England
1781	Hornblower's single-acting compound pumping engine	England
1783	Watt's double-acting rotative engine	England
	Jouffroy d'Abbans' *Pyroscaphe* —first effective steamboat	France
1787	Watt applies centrifugal governor to steam engine	England
	Rumsey's jet-propelled boat	USA
1788	Symington's double paddle engine tried	Scotland
1789 (c)	Murdock's oscillating cylinder engine	England
1790	Fitch's oared steamboat	USA
1792	Watt engine attains duty of 32.8 millions	England
1799	Murdock's long slide valve	England
1800	Watt's restrictive patent expires	England
	Murdock devises the eccentric	England
	Crowther's patent overcrank winding engine	England
1801	Trevithick's high-pressure self-moving engine	England
	Symington engines the *Charlotte Dundas*	Scotland
1802	Trevithick's return-flue boiler and fusible plug	England
	Trevithick's London carriage and portable engine	England
1803	Oliver Evans' high-pressure engine	USA
	Trevithick's Coalbrookdale locomotive	England
1804	Trevithick's Pen-y-Darren locomotive hauls trains effectively	Wales
	Evans' grasshopper beam engine	USA
	Woolf reintroduces compounding	England
	Col. Stevens' twin screw vessel	USA
	Stevens' *Phoenix* makes first sea voyage under steam	USA
1805	Murray's side lever engine	England
1807	Maudslay's table engine	England
	Fulton's *Clermont*, first commercial steamer	USA
1812	Trevithick's high-pressure condensing pumping engine	England
	Blenkinsop and Murray's rack locomotives	England
	Trevithick's Cornish boiler	England
	Bell's *Comet*, first European commercial steamer	Scotland
1814	G. Stephenson's first locomotive, *Blücher*	England
1815	Murray's direct-loaded spring safety valve	England
1819	PS *Savannah* makes first Atlantic crossing with steam assistance	USA
1823	J. Perkin's flash boiler	England
1825	Opening of Stockton and Darlington Railway	England
(c)	Elephant boiler	France
1827	Maudslay's improved oscillating cylinder	England
	Hancock's patent boiler for road vehicles	England
	Hancock's patent boiler for steam buses	England

1829	M. Seguin's multi-firetube boiler for locomotives	France
	G. and R. Stephenson's 'Rocket'	England
	Rainhill locomotive trials	England
1833	Hall's surface condenser	England
	Stephenson's 'Patentee' six-wheeled locomotive	England
1834	West's Cornish engine at Fowey Consols mine attains duty of 125 millions	England
1835 (c)	Otis' steam excavator	USA
1838	*Sirius* and *Great Western* cross Atlantic entirely under steam	England
1839	SS *Propeller*, later *Archimedes*, demonstrates effectiveness of screw propulsion	England
1841	Stephenson's long boiler locomotive	England
	Meyer expansion valve gear	France
	Jucke's mechanical stoker	England
1842	Stephenson link motion invented by Williams and Howe	England
	Ransome's traction engine	England
1843	Brunel's *Great Britain*, iron screw steamer	England
1844	Walschaerts single eccentric valve gear	Belgium
1845	Tug of war between HMS *Rattler* and *Alecto*	England
1846	Boydell's patent 'endless railway' for traction and portable engine wheels	England
1848	Galloway's Lancashire boiler with large water tubes	England
1849	Corliss valves and valve gear	USA
	Bourdon pressure gauge	France
1850	Worthington's duplex non-rotative pumping engine	USA
1852	Semmering locomotive trials	Austria
1854	Bach's gear drive traction engine (Boydell system)	England
	Hackworth's radial valve gear	England
1858	Ramsbottom's double safety valve	England
	Fowler's balance plough for steam cultivation	England
	Giffard's injector	France
	Brunel's *Great Eastern*	England
	Porter's centrally loaded governor	USA
1860 (c)	Double expansion (compounding) used at sea	England
1861	Aveling's steam-jacketed cylinder traction engine	England
1863	Riggenbach's rack railway patent	Switzerland
	Clark and Batho's steamroller	England, India
1866	Vicar's mechanical stoker	England
1867	Babcock and Wilcox water-tube boiler	England
	Aveling and Batho's three-wheeled steamroller	England
1870	Aveling's improved traction engine with extended hornplates	England
1871	Normand introduces triple-expansion marine engine	France
1875	Crosby's pop safety valve	England
1876	Mallet introduces the practical compound locomotive	France
1879	Joy's radial valve gear	England
1881	Fowler's double-crank compound traction engine	England
1882	Webb's first three-cylinder compound locomotive	England
1884	Parsons' axial-flow turbine and generator	England
	Willans' high-speed central valve engine	England
1886	First four-cylinder compound locomotive	France
1887	Sauvage's three-cylinder compound locomotive	France
1887	Mallet articulated locomotive	France
1889	De Laval's steam turbine	Sweden
1889	First Vauclain compound locomotive	USA
1893	Bellis and Morcom high-speed self-lubricating engine	England
1894	Edwards' independent air pump	England
1896	Thornycroft's prototype steam commercial vehicle	England
1897	Serpollet's flash boiler for steam vehicles	France
	Parsons' *Turbinia* at Queen Victoria's Diamond Jubilee Naval Review	England
1900	Schmidt's fire-tube superheater	Germany
	Triple-expansion pumping engine attains duty of 165 millions	England
1902	Parson's augmentor condenser	England
	Serpollet steam car gained world land-speed record of 121 kph (75.06 mph)	France
1905	HMS *Dreadnought* revolutionizes capital ships	England
1907	RMS *Mauretania* gains Blue Riband with turbines and holds it for 22 years	England
1908	Garratt locomotive invented	England
1910	Ljungstrom duplex radial-flow turbine	Sweden
1923	Sentinel's two-cylinder undertype steam lorry	England
1929	Chapelon's first rebuilt Pacific revolutionizes the possibilities of steam locomotives	France
1932	Sentinel's S4 steam lorry, undertype engine totally enclosed and shaft drive	England
1936	*Queen Mary* and *Normandie* rivals on North Atlantic	England, France
1938	*Mallard* attains world speed record for steam locomotives at 202 km (126 miles) per hour	England
1940	Chapelon's later 4-8-0 locomotives set records for power-to-weight ratio and thermal efficiency of 40 hp per ton, and 12.8%	France
1941	Union Pacific *Big Boy*, largest and most powerful steam locomotive ever built	USA
1948	Lenz and Butenuth produce steam automotive engine converted from internal combustion	Germany
1953	SS *United States* becomes final Blue Riband holder	USA
1956	World's first commercial nuclear power station opened at Calder Hall	England
1959	World's first power-producing fast reactor at Dounreay	Scotland
1962	Hinkley Point B accelerated gas-cooled reactor nuclear power station	England

Preserved Steam Engine Locations

This list is confined to established museums and sites, and is not exhaustive. There are hundreds of other preserved sites, preserved operating railway lines, and enthusiast organizations which hold open days or steam rallies. For details of these it is necessary to consult annually produced guides. There are also very many places where steam can be seen in action, which are not preserved sites or museums. Main line steam railways are active in many places in Africa, India, China, South America and elsewhere. Steam ships in service include all the largest vessels afloat, but small steamers are also still to be found all over the world. Because the preservation movement is so active, there are new schemes opening every week, and it is not possible to give an account of these here.

Argentina	Quilmes	Museum of the History of Transport
Australia	Adelaide	Railway Museum
	Brisbane	Redbank Railway Museum
	Fremantle	Maritime and Historical Museum
	Goulbourne	Museum of Historic Engines
	Mannum	Marion Paddlewheel Museum
	Melbourne	Railway Museum
	Melbourne	Science Museum of Victoria
	Parramatta	Steam Tram and Railway Museum
	Sydney	Museum of Applied Arts and Sciences
	Sydney	Rail Transport Museum
Austria	Bad Wimsbach	Transport Museum
	Vienna	Austrian Railway Museum
	Vienna	Museum of Industry and Technology
Belgium	Antwerp	National Maritime Museum
	Antwerp	Open Air Maritime Museum
	Brussels	Belgian Railway Museum
	Brussels	Tramway Museum
Brazil	Campinas	Railway Museum
	Porto Velho	Madeira-Mamore Railway Museum
Canada	Dawson City	SS *Keno* Museum
	Kingston	Pump House and Steam Museum
	Montreal	Canada Railways Museum
	Ottawa	National Museum of Science and Technology
	Quebec	St Constant Railway Museum
	Toronto	Ontario Science Centre
	Whitehouse	SS *Klondyke* Museum
China	Tientsin	People's Hall of Science
Czechoslovakia	Kosice	Technical Museum
	Prague	National Technical Museum
Denmark	Copenhagen	Railway Museum
	Helsingor	Danish Technical Museum
Egypt	Cairo	State Railways Museum
France	Compiegne	Museum of Road Transport
	Mulhouse	Railway Museum
	Paris	National Technical Museum (Museum of Arts and Trades)
	Pithiviers	Transport Museum
	Uzès	Museon di Rodo
E. Germany	Dresden	Transport Museum
W. Germany	Berlin	Transport Museum
	Frankfurt	Model Railway Museum
	Koblenz	Rhine Museum
	Munich	German Museum
	Nuremberg	Transport Museum
Gt. Britain	Aviemore	Spey Valley Railway
	Aylesbury	Quainton Railway Society Collection
	Beamish	Open Air Museum
	Beaulieu	National Motor Museum
	Belfast	Transport Museum
	Birmingham	Museum of Science and Industry
	Birmingham	Tyseley Steam Centre
	Bradford	Industrial Museum
	Bridgnorth	Severn Valley Railway
	Brighton	Brighton and Hove Engineerium
	Bristol	SS *Great Britain*
	Cardiff	National Museum of Wales
	Cardiff	Welsh Industrial and Maritime Museum
	Carnforth	Steamtown
	Darlington	North Road Railway Museum
	Diss	Bressingham Steam Museum
	Edinburgh	Royal Scottish Museum
		Prestongrange Mining Museum

	Falkirk	Scottish Railway Preservation Collection
	Glasgow	Museum of Transport
		PS *Waveley*
	Great Bedwyn	Crofton Pumping Station
	Leicester	Abbey Pumping Station
	Liverpool	City Museums
	London	HMS *Belfast*
	London	Kew Bridge Living Steam Museum
	London	National Maritime Museum
	London	Science Museum
	Lough-borough	Great Central Railway
	Manchester	Museum of Science and Technology
	Newcastle	Museum of Science and Engineering
	Nottingham	Papplewick Pumping Station
		Wallaton Hall Museum
	Portmadoc	Festiniog Railway and Museum
	Portsmouth	Eastney New Beam Engine House
	St Austell	Wheal Martyn China Clay Industry Museum
	Sheffield	Abbeydale Industrial Hamlet
	Shrewsbury	Coleham Pumping Station
	Sunderland	Monkwearmouth Station Museum
	Sunderland	Ryhope Pumping Station
	Swindon	Great Western Railway Museum
	Telford	Ironbridge Gorge Museum
	Towyn	Narrow Gauge Museum
		Tal-y-Lyn Railway
	Wansford	Nene Valley Railway
	Worksop	National Coal Board's Lound Hall Museum
	York	National Railway Museum
Hungary	Budapest	Hungarian Transport Museum
India	Delhi	Railway Museum
Ireland	Stradbally	Steam Museum
Italy	Milan	Leonardo da Vinci Museum of Science and Industry
Japan	Tokyo	Museum of Transport and Technology
Netherlands	Kamperzee-dijk	Mastenbroek Steam Engine Museum
	Utrecht	Netherlands Railways Museum
	Vijfhuizen	Cruquius Museum
	Weesp	Netherlands Tramways Museum
New Zealand	Auckland	Museum of Transport and Technology
	Christchurch	Ferrymead Trust
Norway	Hamar	Railway Museum
	Oslo	Norwegian Science and Industry Museum
		SS *Boroysund*
Poland	Warsaw	Railway Museum
Rhodesia	Bulawayo	Rhodesian Railways Museum
Romania	Bucharest	Railway Museum
South Africa	Johannesburg	South African Railways Museum
Spain	Madrid	Railway Museum
Sweden	Stockholm	Technical Museum
	Gavle	Swedish Railways Museum
Switzerland	Lucerne	Swiss Museum of Transport
	Winterthur	Swiss Technorama
USA	Ashland	Santa Fe Museum
	Baltimore	Baltimore and Ohio Transportation Museum
	Bellows Falls	Steamtown
	Chattanooga	Tennessee Valley Railroad Museum
	Corinne	Railroad Museum
	Dallas	'Age of Steam' Museum
	Dearborn	Henry Ford Museum
	Exton	Thomas Newcomen Memorial Library and Museum
	Fall River	Marine Museum
	Fort Worth	Pate Museum of Transportation
	Green Bay	National Railroad Museum
	Houston	Battleship *Texas*
	Long Beach	RMS *Queen Mary*
	Louisville	Kentucky Railway Museum
	Marietta	Steamer *WP Snyder Jnr*
	New York	Brownville Railroad Museum
	New York	South St. Seaport Museum
	Oneonta	National Railroad Museum
	St. Louis	National Museum of Transport
	Sandy Creek	Rail City
	Seattle	Puget Sound Railway Museum
	Strasburg	Strasburg Railway and Railroad Museum of Pennsylvania
	Truckee	Emigrant Trail Museum
	Union	Illinois Railway Museum
	Vicksburg	Steamer *Sprague* Museum
	Washington	National Museum of History and Technology (Smithsonian Institution)
	Wilmington	USS *North Carolina*
	Winona	Steamboat Museum
	Worthington	Ohio Railway Museum
USSR	Leningrad	Museum of Railway History
Venezuela	Caracas	Museum of Transport
Yugoslavia	Belgrade	Museum of Yugoslav State Railways

Glossary

Adams blastpipe Blastpipe (q.v.) with an annular orifice, the inside being arranged to draw on the lower tubes of the boiler.

Adams bogie Bogie (q.v.) capable of bodily sideways movement, controlled by springs.

Air pump A device usually driven off the main engine for scavenging air and water out of the condenser (q.v.).

Allan gear Link motion with a straight link. (*See* index.)

Annular compound Arrangement of compound cylinders in which the low-pressure cylinder surrounds the high-pressure one, giving rise to a low-pressure piston of annular form.

Anti-carbonizer Provision for a slight leak of steam into valves and cylinders to prevent carbonization of oil.

Arch head Curved end of the beam in early beam engines, over which chains passed to allow straight line motion of the piston and pump rods.

Arch tube Water tube in a locomotive boiler firebox, arranged to assist in supporting the brick arch (q.v.).

Atmosphere Pressure of approx. 1 bar, 14.7 lb per sq in, 100 kN per sq metre, or one Hectopieze. *See also* Gauge.

Backhead That part of a boiler with the firedoor(s) and usually the gauges and controls.

Baltic Locomotive with the wheel arrangement 4-6-4 (2-3-2 or 2-C-2).

Bauer Wach System of applying an exhaust turbine to a marine engine.

Beam Large lever transmitting the motion imparted by the piston to either a connecting rod or pump rod, or both. *See also* Grasshopper.

Belly tank Tank slung beneath the boiler of a traction engine, road locomotive etc., to increase the water-carrying capacity.

Belpaire Type of outer firebox on a locomotive boiler, having a predominantly flat top and sides.

Beugniot bogie System which allows driven axles to move sideways in opposite directions on curved track.

Bissel truck Frame with carrying wheels, pivoted outside its wheelbase.

Blading The system of fixed and moving blades in a turbine, through which the steam passes.

Blastpipe The exhaust pipe in a locomotive boiler smokebox which, in connection with the chimney, is arranged to induce a draught.

Blow-down cock Valve fitted to a low point on a boiler for blowing out accumulated sediment and sludge.

Blower Steam jet in a smokebox, arranged to induce draught when there is no exhaust from the engine.

Bogie Four-wheeled truck of carrying wheels, able to rotate about its approximate centre. (*See also* Adams bogie.)

Boiler manhole Larger version of a mud door (q.v.) provided for internal inspection and cleaning of a boiler.

Bolster Cross piece within a bogie, carrying the pivot.

Brake horsepower (bhp) Measured power delivered by an engine, i.e. power available for actual work after deduction of frictional losses in the engine.

Brick arch Brick-built arch within a locomotive firebox, to assist combustion. (*See* index.)

Brotan boiler Boiler with a water-tube firebox, named after the Hungarian engineer who designed it.

Bull engine Type of Cornish engine (q.v.) used mostly in waterworks, in which the beam is dispensed with and the cylinder inverted over the well.

Bunker Fuel container at the rear of a tank locomotive; also an important part of the structure of the rear of a traction engine, steamroller, or similar.

Butterfly valve Regulating valve in a steam pipe or steam chest consisting of a circular disc, rotatable about its axis to provide a throttling effect; usually associated with a centrifugal governor.

Bypass Arrangement for putting the two ends of a cylinder in communication, to avoid pumping action when running without steam. Also an arrangement to allow a pump to return water or oil to a reservoir.

Calliope Organ composed of steam whistles used in North America to advertise the arrival of a travelling circus or show.

Camelback Type of American locomotive, with the driver's cab on top of the boiler and the fireman's at the rear.

Caprotti gear Poppet valve gear, operated by rotating cams, invented in Italy.

Cartazzi axle Two-wheeled carrying axle, capable of sideways movement, and controlled by diagonal slides to give a radial action, with spring control.

Cartwright's rabbit gear Valve gear used to control the movement of steam swing centre engines.

Cataract Dashpot device for regulating the stroking rate of a non-rotative beam pumping engine.

Centre engine Special form of portable engine used to drive a steam roundabout, switchback or swings.

Chain grate Continuously moving grate, in the form of an endless chain, to provide mechanical stoking. It is usually fed with coal from a hopper.

Check valve Non-return valve used for water or oil feeding.

Circulating pump Auxiliary pump for maintaining a flow of cooling water through a condenser (q.v.).

Clack valve Another name for a check valve (q.v.).

Cladding Lagging insulation (q.v.).

Cleading The thin covering, usually of metal, over lagging.

Columnar engine Beam engine in which the beam is centrally pivoted on trunnions supported on a single column.

Combination lever That part of Walschaerts valve gear (q.v.) which combines the effect of the link with that of the crosshead (q.v.).

Compensating gear Another name for a differential gear when fitted to the hind axle or intermediate drive shaft of a steam traction engine or road locomotive. It enables the driven wheels to rotate at different speeds when negotiating corners.

Compound Engine in which expansion takes place in two cylinders successively.

Compression point Point at which the exhaust valve closes, leaving residual steam to be compressed in the cylinder.

Condense Turn steam into water by cooling. Saturated steam (q.v.) can be condensed by compression, but this was only applied experimentally to steam engines.

Condenser Vessel attached to fixed types of steam engine in which the exhaust is condensed, to boost efficiency.

Connecting rod Rod connecting a piston or a crosshead to a crank.

Consolidation Locomotive with the wheel arrangement 2-8-0 (1-4-0, 1 D).

Corliss gear Valve gear with semi-rotating plug valves and separate inlet and exhaust valves, providing a rapid cut-off action (q.v.).

Cornish engine Single-acting non-rotative beam pumping engine working on the Watt cycle, as modified by Trevithick and contemporaries, to use high-pressure steam expansively.

Cossart gear Locomotive valve gear, with four piston valves per cylinder, operated by rotating cams. French.

Coupling rod Rod joining the cranks of the coupled and driving wheels of a locomotive.

Crampton Locomotive firebox, the outer wrapper of which conforms in its upper part to the shape of the boiler barrel, and is not raised or square. Also, in 19th-century France, synonymous with a fast train.

Crosshead Block provided at the joint of piston and connecting rods, equipped with lubricated surfaces for sliding along guide bars, or pin joints, for parallel motion linkage.

Cross compound Engine in which two compounded cylinders work on separate cranks, so the steam crosses the engine.

Crown sheet The roughly horizontal sheet forming the top of the inner firebox of a locomotive or other fire-tube boiler.

Cut-off Point at which the supply of fresh steam to a cylinder is cut off, measured as a percentage of stroke.

Cylinder That part of a steam engine in which the piston is acted upon by live steam; or, in a turbine, the part of the casing containing blading.

Cylinder cover Stiffened plate closing the end of the cylinder, with or without a gland (q.v.) for the piston rod to pass through.

Cylinder lubricator Device, usually of polished brass, to maintain flow of heavy 'steam oil' to lubricate the valves and piston(s). Superheated engines usually have a mechanical pump to perform this function (*see also* Roscoe lubricator).

Dabeg French firm specializing in poppet valves (q.v.), feed-water heaters and other equipment.

Damper Air control device in a furnace; an air door in a locomotive ashpan; or a flap valve in a furnace uptake.

Darby digger Type of steam cultivator with the digging mechanism carried on the main body of the machine.

Dart Handle used to secure the smokebox door centrally on a locomotive-type boiler.

Dome Raised piece on top of a boiler, used to collect dry steam and sometimes to hold the main steam valve.

Downcomer Part of a water-tube boiler, depending from the top or steam drum, and connecting it to water tubes arranged beneath it.

Dragbox Strong frame stretcher at the rear of a locomotive, to which the drawbar is attached.

Drain cock Valve, usually manually operated, to release trapped water from cylinders.

Drop valve Vertically moving valve. Two or four of these sometimes replace a single slide or piston valve.

Double acting Engine in which both inward and outward strokes of the piston provide power.

Double-crank compound Compound cylinders arranged side-by-side. The most common arrangement in a traction engine or road locomotive.

Duty Numerical expression for the comparative efficiency of a pumping engine and its boilers: the number of lb of water lifted 1 ft high by the consumption of each bushel of coal. In Cornwall the bushel was usually taken as equivalent to 94 lb and is so used in this book.

Dynamometer Device for measuring work done and the rate at which it is done, and therefore power.

Eccentric A 'sheave' or thick circular disc mounted eccentrically on a shaft with a 'strap' encircling it, in which it rotates. The eccentric rod is fixed to the strap, and so moves back and forth. It is equivalent to a crank, but can only be used to convert rotary to linear motion and can be fixed to an uninterrupted shaft.

Economizer Heat exchanger using flue gases to preheat boiler feed water. Sometimes also used to extract heat from exhaust steam.

Ejector Device for producing a vacuum by passing steam through a series of nozzles. Large and small ejectors are required on locomotives fitted with a vacuum brake. A type of condenser depends on an ejector (which can also work with water instead of steam). *See also* Giesl.

Element A steam tube in a super-heater.

Engerth Type of locomotive in which a large overhanging firebox is partly supported on the tender. Originally the tender wheels were coupled to those of the locomotive.

Equalized suspension Means of linking the springs of a locomotive by levers to ensure that the intended weights are carried on the axles, regardless of track irregularities or spring settlement.

Equilibrium valve The valve in a Watt or Cornish pumping engine which allows the expanded steam to pass from the upper to the lower part of the cylinder on the 'outdoor' stroke (q.v.).

Expansion link In a valve gear, the slotted link (or, in marine practice, sometimes a solid bar) from which a drive to the valve can be taken anywhere along its effective length, to vary valve travel and effect reversing.

Fairlie locomotive Type of locomotive with a swivelling powered truck under one or, more commonly, each end. It may have a normal boiler, or a double-ended one with a central, divided firebox, and a smokebox and chimney at each end.

Feed pump Term always applied to the feeding of water into a boiler.

Fireless locomotive Steam locomotive in which a steam reservoir, charged periodically from a stationary boiler, replaces the boiler.

Flexible stay Jointed boiler stay, usually used at the ends of the firebox, where differential expansion between inner and outer plates is at its greatest.

Footplate That part of a self-moving engine or locomotive where the driver and fireman or steersman stand.

Forced draught Boiler draught increased by introducing air under pressure beneath the grate, or by pressurizing the stokehold.

Forecarriage Steerable front axle of a portable engine or steam road vehicle.

Foundation ring The bottom of the water spaces around the firebox of a locomotive type or similar boiler. It joins the inner and outer boxes, and is formed of bar or channel steel.

Franco Crosti Locomotive boiler of Italian design, fitted with a second, smaller, barrel acting as an economizer (q.v.).

Fusible plug Screwed-in plug placed in a boiler just below the normal water level. It is bored and filled with metal of low melting point, which melts and discharges steam into the fire if the water level falls too low.

Gaines wall Low brick wall separating a fire-grate from a combustion chamber leading to tubes, flues etc. It serves a similar purpose to a brick arch (q.v.) but permits a higher grate.

Garratt locomotive Locomotive with two powered units and the boiler slung on a cradle between them, so usually making it short and of large diameter. Fuel and water supplies are carried on the powered units, which are pivoted. A speciality of Beyer, Peacock and Co., Manchester.

Giesl ejector System of multiple blastpipes for locomotives, in which 7 small nozzles are arranged in line, and discharge into a narrow, elongated chimney with petticoat (q.v.).

Gland Point where a rod passes out of a steam space but is kept steam tight by packing. If the packing is soft it is sometimes called a stuffing box.

Gooch gear Form of link motion in which the link is 'stationary' and drive is taken from it to the valve rod by a radius link, which can be shifted up and down. The curvature of the link is concave towards the valve, and it is rocked by two eccentrics (q.v.).

Governor Device for maintaining constant speed, originally by regulating the steam supply but latterly by altering the cut-off.

Grasshopper engine An engine in which the beam is pivoted at one end instead of centrally. The end connected to the piston rod usually moves in a vertical straight line, so the pivot has to allow movement, and is on a vertical link pivoted at the bottom.

Gresley (conjugated) gear A simple arrangement of levers allowing three piston valves to be operated by two sets of valve gear. The cranks of the three cylinders are phased at 120°. Used on the London and North Eastern Railway, and also in locomotives in Australia, the USA, Malaya, etc.

Haystack boiler Boiler with a greatly raised firebox top, usually domed in vertical section and square in horizontal section.

Header A casting or forging into which a number of tubes are inserted, especially the saturated and superheated manifolds of a superheater.

Hectopieze Unit of pressure commonly used in Europe: one kilogramme per cm^2, approximately one atmosphere or one bar, or 100 kN per sq metre.

Heisler Type of American logging locomotive. The V-twin-cylinder steam engine under the boiler barrel drives the axles of two swivelling trucks via Cardan shafts.

Hornblock A steel casting within which an axlebox is able to move vertically, controlled by the spring of the suspension, but not horizontally. Found in locomotives, traction engines, etc. If divided into two portions, one ahead of and one behind the axlebox, they are known as horncheeks. In either case, their lower ends are joined by a removable horn stay, known in America as a pedestal tie.

Hornplates Upward and rearward extension of the firebox side plates to provide a rigid mounting for the crankshaft, intermediate shaft and hind axle bearings. Used in European-designed traction engines, road locomotives, etc.

Horsepower The power required to raise 33,000 lb by one foot in one minute, originally calculated in those terms by James Watt. Equivalent to 746 Watts in electrical units. (*See also* Brake Horsepower and Indicator).

Houlet superheater Used in French locomotives, it has an annular outer element used for flow in one direction, and a tubular inner element for the flow in the other direction; the whole is within a flue tube as in other types of fire-tube superheater.

Hudson The usual American name for a Baltic locomotive.

Husband's safety governor Dashpot device fitted to some Cornish pumping engines to prevent damage in the event of the breakage of a pump rod.

Ice locomotive Type of road locomotive used for passenger and goods haulage on frozen rivers and lakes.

Indicator Device for recording on paper the pressure fluctuations within a cylinder, thus enabling calculation of the 'indicated horsepower' (ihp), i.e. the power available before the deduction of frictional losses.

Indoor stroke The steam stroke of a Watt or a Cornish pumping engine which depresses the piston, raising the nose of the beam and pump(s) in the well or shaft.

Induced draught Draught produced by the creation of a vacuum in the smokebox, as in road and railway locomotives.

Injection cock Small valve which controls the cooling water inlet to a jet condenser.

Injector Invented by Henri Giffard, it provides the commonest way of feeding water into locomotive boilers.

Intensifore lubricator One in which boiler pressure is increased by an arrangement of two cylinders with pistons of slightly different diameter, to enable the oil to be delivered into the steam pipe at a higher pressure.

Jack link In Joy's valve gear, the link which pushes the dieblock up and down the slide.

Jet condenser Condenser in which the steam is cooled by a spray of cold water, as in a Newcomen engine. The separate jet condenser, outside the cylinder, was Watt's invention.

Joy gear Valve gear used in locomotives and marine engines, in which the valve motion is derived from the angularity of the connecting rod, which is used to move a die block along a slide of variable inclination.

Junk ring A removable ring on a piston or piston valve, which enables soft packing to be compressed (in a piston) or piston rings to be placed or removed.

Keep Lower part of an axlebox; it keeps the sides in place and contains the oil reservoir.

Knot Measure of speed at sea, equal to one nautical mile per hour, i.e. approx. 1.76 kilometres per hour. Now largely superseded by metric measurement, it derives from the knots on the rope attached to a stationary 'log' cast into the sea, the passage of these knots through a seaman's hand was counted during the running of a sand glass.

Krauss-Helmholtz truck Type of locomotive bogie, incorporating a single pair of carrying wheels and the leading coupled wheels, to increase flexibility on curves. *See also* Zara truck.

Kylchap Type of locomotive blastpipe, with the jet split into four streams and passing into the chimney via a four-part petticoat and a one-part petticoat. Usually made in double form and, as such, the most efficient pumping device known for its purpose. The name derives from Messrs. Kylala and Chapelon.

Lap Amount by which the valve of a cylinder overlaps the steam or exhaust ports when in central position; or its theoretical equivalent in other types of valve. Steam lap is considerable, exhaust lap very small or non-existent, or even negative, in which case it is known as clearance. Lap has a significant effect on timing and the word is used to describe the effect.

Lagging insulation Non-heat-conducting material, traditionally asbestos, applied to the external surfaces of boilers, steam pipes, cylinders and valve chests to conserve heat.

Lead The amount of valve opening to steam at dead centre, measured in the direction of travel.

Lemaître exhaust Locomotive smokebox arrangement, in which a number of blast nozzles (normally 5) surround a central one with a variable aperture, and discharge into a chimney of large diameter. On the British Southern Railway a similar arrangement omitted the central variable nozzle.

Lentz gear Valve gear, with poppet instead of slide or piston valves, of Austrian origin. Oscillating cam Lentz gear depended on Walschaerts gear (q.v.) for the primary motions, and rotary cam Lentz gear had a camshaft with variable setting of the cut-off control. *See also* Caprotti, Cossart and Dabeg.

Lifting link Link in a valve gear which lifts the expansion link or the die block of a fixed link gear.

Living van Large caravan towed behind a steamroller, traction or ploughing engine in which the crew live.

Luffing Vertical movement of the jib or arm of a crane.

Luttermoller axle Method of coupling the extreme axles of a locomotive with a long coupled wheelbase. The use of gears in the middle of the axles, instead of coupling rods, allows a radial swing of the extreme axles.

Mallet Articulated locomotive with a rear powered wheel set rigid with the boiler, and a leading powered set able to swing laterally on curves. (*See* index.)

Manifold Cast box provided with boiler steam, to which valves are attached controlling supplies to auxiliaries such as injectors, lubricators, air pumps, etc. Usually placed on top of a locomotive boiler.

Meyer valve A second slide valve placed on the back of the main valve and independently driven. Used mainly in stationary engines, it controls the rate of expansive working.

Michell bearing Bearing in which the journal or thrust collar is surrounded by pivoted pads instead of rigid brasses or liners. The pads take up a position determined by the laws of hydrodynamics, to provide an effective, wedge-shaped oil film. Michell thrust bearings for ships' propeller shafts ran cool and saved much space as they needed only a single collar. Previously, a dozen collars might have been necessary.

Mikado Locomotive with the 2-8-2 (1-4-1 or 1-D-1) wheel arrangement.

Mogul Locomotive with the 2-6-0 (1-3-0, 1-C) wheel arrangement.

Mountain Locomotive with the 4-8-2 (2-4-1, 2-D-1) wheel arrangement.

Mud door Oval cover secured by a nut and bridgepiece over an opening in a boiler for internal cleaning and inspection, an alternative to a washout plug (q.v.).

Notch up To adjust a valve gear away from the full gear position, to obtain a greater degree of expansion.

Oil separator Device for removing lubricating oil from exhaust steam or condensate, before the condensate is returned to the boiler as feed water.

Oscillating cylinder Cylinder in which the piston rod also serves as connecting rod, and therefore swings on trunnions to accommodate itself to the resulting angularity.

Outdoor stroke The return stroke of a Watt or Cornish engine in which the weight of the pump(s) and pump rod take charge and cause the piston to be raised under the control of the equilibrium valve.

Overhead gallopers Type of roundabout in which the up and down movement of the galloping horses in derived from the top frame by cranks and gearing.

Pacific Locomotive with the 4-6-2 (2-3-1, 2 C 1) wheel arrangement.

Palm stay Boiler stay used for joining surfaces more or less at right angles, as for tying tube plates to barrels. One end (attached to the barrel) is connected to a palm-shaped plate in the same plane as the axis of the stay. The plate is attached to the barrel by several bolts.

Parallel motion A linkage designed to guide the outer end of a piston rod in a straight line, without the use of crosshead guides. Originally devised by Watt, whose linkage gives a close approximation to a straight line, but only the much later and seldom used linkage of Paucellier gives a true straight line.

Pedestal American term for a hornblock (q.v.).

Petticoat Downward, and usually flared, extension of a chimney within a smokebox.

Piston valve Type of cylinder steam distribution valve in which cylindrical pistons work within cylindrical liners which have ports cut all round.

Platform gallopers Type of galloping horse roundabout in which the up and down movement of the horses is derived from wheels driven round a fixed track beneath the platform.

Ploughing engine Type of traction engine with a special large winding drum for towing the plough.

Pony truck Two-wheeled truck attached beyond the coupled wheelbase of a locomotive, arranged to swing radially on curves by means of a radius bar, and provided with control of its sideways movement by spring or swing links.

Pop safety valve Valve which opens rapidly at a predetermined pressure, and does not close until the pressure has dropped by a specific percentage.

Poppet valve Valve which moves on to and off a seating, in a direction at right angles to the plane of the seating, as distinct from the sliding action of slide and piston valves.

Portable engine Steam engine of relatively low power, complete with boiler (which usually forms the frame) and mounted on wheels so it can be towed.

Ports The openings in a valve chest which communicate with the ends of the cylinder and are covered and uncovered by piston or slide valves.

Prairie Locomotive with the 2-6-2 (1 C 1, 1-3-1) wheel arrangement.

Precipitator Device in a chimney uptake for removing solid particles from the flue gases.

Pressure The quoted figures for pressure are usually 'gauge' pressures, i.e. measured taking the pressure of the atmosphere as zero. However, absolute pressure (gauge pressure plus atmospheric pressure) has to provide the basis for calculations involving expansion, especially where condensing machinery is involved.

Priming An engine is said to 'prime' when water is carried into the cylinders with the steam: usually caused by the water in the boiler being dirty.

Radius rod The rod, in Walschaerts valve gear (q.v.), which takes the drive from the expansion link.

Reach rod The rod which operates the valve gear, and is connected to the driver's or engineer's control

Reducing valve Valve for reducing steam pressure, designed to give a constant low pressure despite fluctuations of the higher pressure.

Regulator A locomotive's main steam valve.

Release The point, in the operating cycle of a steam cylinder, at which expanded steam trapped in the cylinder is released to exhaust.

Relief valve Spring-loaded valve for automatically releasing water trapped in a cylinder.

Road locomotive Refined form of traction engine designed primarily for road haulage work.

Roscoe lubricator Type of cylinder lubricator in which a small supply of steam condenses, sinks below the oil, thereby lifting it, so it feeds it into the steam pipe against steam pressure. It is then carried to valves and cylinders with the main steam supply.

Running plate The horizontal plating over the wheels of a railway locomotive to provide a platform for cleaners and to prevent dirt from being thrown up from the wheels when running.

Saddle tank Locomotive with its water reservoir fitted over the boiler.

Saturated The condition of steam within a boiler, in contact with water, when temperature and pressure are precisely related. If saturated steam is expanded outside the boiler, without doing work, it becomes superheated; if compressed, it condenses.

Serve tube Type of boiler tube with longitudinal fins on the inside (fire side).

Shay Type of American logging locomotive, with two or three powered bogies driven by Cardan shafts along one side and a vertical three-cylinder engine upright beside the boiler, which is slightly offset to accommodate it.

Showman's engine Road locomotive (q.v.) equipped with a dynamo and special adornments for fairground work.

Side rod Colloquial term for a locomotive coupling rod.

Side tank Locomotive carrying its reserve water supplies in tanks beside the boiler (and usually in one beneath the bunker as well).

Sight feed lubricator Displacement lubricator, basically on the Roscoe principle, in which drops of oil pass upwards through a glass tube full of water, thereby giving visible evidence of the rate of feed, which is adjustable.

Simple Steam engine or locomotive in which the expansion of steam takes place within one cylinder only, or several cylinders in parallel, but not in series.

Simpling valve Cock worked from the footplate of a traction engine or road locomotive with double-crank compound cylinder arrangement to admit live steam to the low-pressure cylinder to assist starting. Also used on marine engines and called a 'starting valve'.

Single Locomotive with one driven axle.

Single-acting Cylinder arrangement in which only one direction of piston movement is powered.

Single-crank compound Arrangement of compound cylinders used in traction engines and a few road locomotives in which the piston rods from both cylinders drive a common connecting rod and crank.

Siphon Suction device for liquids, in which the vacuum is created by the action of gravity on a liquid column in an inverted U-tube. Used in some lubricators. A thermic siphon is a stayed water passage in a firebox, in which water rises by convection, to promote circulation.

Slide (Joy) The component of Joy's valve gear which takes the place of the expansion link. In mid gear, the slide is set vertically, and the die block works up and down, driven by the connecting rod, without any fore and aft movement. Tilting the slide one way or the other gives a fore and aft component its movement, which operates the valve.

Slide bars The bars which guide a crosshead.

Slide valve Cylinder valve in the form of a flat block, with a rectangular recess (the exhaust cavity) underneath. Its reciprocating motion admits steam to each end of the cylinder in turn, and connects the opposite end to the exhaust pipe via the exhaust cavity.

Snifting valve Valve which allows cylinders to 'breathe' when the engine is running with steam turned off. In a locomotive it may be arranged to allow the cylinders to draw air through the superheater elements.

Spinning top Roundabout or steam switchback in which the centre engine rotates the top frame through an upright shaft and gearing.

Standpipe tower A structure, sometimes combined with the chimney, used to house standpipes which 'ironed out' pressure fluctuations in water mains caused by slow speed reciprocating pumping engines.

Standing top Type of switchback in which the cars or animals are driven round an undulating track by a spinning frame geared to the centre engine.

Stationary engine Term covering all types of steam engine solidly mounted on a foundation to provide a fixed power source for pumping, winding, factory and mill drives, etc.

Stay Tension bar within a boiler to prevent swelling under pressure. Occasionally, stiffening girders across the firebox crown are also called stays.

Steam dryer Device to prevent the entraining of drops of water in the steam supply.

Steam jacket An outer cylinder fitted round an engine cylinder, for the purpose of applying steam to the intervening space to maintain temperature in the cylinder.

Steamroller Derived from the traction engine, it has smooth roller treads instead of normal wheels, and is used for roadmaking.

Steam tractor Small form of the road locomotive designed to overcome problems due to taxation by weight.

Steam wagon (or waggon) Another term for a steam lorry.

Stephenson locomotive Term applied to any locomotive having the main characteristics of the *Rocket*, i.e. fire-tube boiler, direct drive from cylinders to wheels, and induced draught.

Stephenson's link motion The first valve gear giving control of expansive working by means of a slotted link. The two ends of the link are connected to two eccentrics, one for forward and one for reverse running, and the valve drive can be taken from any position of the link between the ends.

Stern tube Tube through which a ship's propeller shaft passes from the inside to the outside of the hull.

Stroke The movement of a piston between front and back dead centres.

Strum box Box containing strainers, arranged in the feed water supply of a boiler.

Stuffing box Gland (q.v.) with soft packing.

Suction hose Armoured hose used with the water lifter on a steam vehicle.

Superheated Steam at a higher temperature than that equivalent to its pressure when saturated.

Surface condenser Condenser in which the steam is cooled by contact with a cold surface, such as is commonly provided by tubes of cold water.

Swing link Device for providing lateral control of bogies, (q.v.) pony trucks (q.v.) etc. They are provided in pairs, usually double, and in the central position their upper ends (attached to the bogie) are closer than their lower ones (attached to the frame). Lateral movement causes a tilting action which compresses the springs on one side more than on

Tandem compound Cylinders arranged one behind the other so that they can share a piston rod, crosshead and crank.

Tender Vehicle coupled to a locomotive to carry fuel and water supplies. Also, the rear portion of a traction engine, road roller, etc., which serves the same purpose but is an integral part of the engine.

Ten wheel American term for a locomotive with the 4-6-0 (2-3-0, 2 C) wheel arrangement.

Ten wheel tank Tank locomotive with the 4-4-2 (2-2-1, 2 B 1) wheel arrangement.

Throat plate Front part of the downward extension of a boiler firebox, immediately below the barrel.

Throw The eccentricity of an eccentric (q.v.) or crank. The throw of a crank equals half the piston stroke.

Thrustblock In a screw steamer, the point at which the thrust of the propeller is taken by the hull. Usually it takes the form of bearings in which collars, fixed to the propeller shaft, bear against stationary collars.

Top feed The introduction of feed water into a boiler through the steam space, usually via a cascade of water trays.

Traction engine A derivative of the portable engine in which the engine is able by chain or gear drive to propel itself along a road or track and haul a load.

Trunk engine Type of screw engine.

Trunk piston A piston carrying the gudgeon pin for the small end of the connecting rod within its length. The piston itself therefore acts as a crosshead.

Try cocks Fitted to the boiler backhead to enable the water level to be verified independently of the reading of the water level gauge.

Tube plate A plate with holes in it, into which tubes are inserted. There are tube plates in locomotive fireboxes and smokeboxes, and also in condensers and some water-tube boilers.

Union link In Walschaerts valve gear (q.v.), the link joining the crosshead (q.v.) and the combination lever (q.v.).

Uptake In stationary and marine plant, the connection between the boiler and the chimney or funnel.

Vauclain compound Compound without a receiver, in which high- and low-pressure piston rods operate upon the same crosshead (q.v.).

Von Borries Inventor of features associated with compound locomotives: a starting arrangement with direct admission of high-pressure steam to the receiver (without direct exhaust of high-pressure cylinders); a valve gear in which the low-pressure cylinders have their own combination levers (q.v.) but also derive movement from the high-pressure expansion link.

Walschaerts valve gear Single eccentric valve gear, with a centrally pivoted expansion link for control of travel and direction.

Washout plugs Removable plugs in a boiler to permit internal examination and to enable loose scale, etc. (sometimes called mud) to be dislodged.

Water lifter Device worked by a steam jet and fitted to steam road vehicles for sucking water from ponds etc. to replenish the tank(s).

Weighshaft Shaft from which the lifting links of a number of valve gears are suspended, and operated by the driver's control.

Wheel splasher Casing over the upper part of a railway locomotive driving wheel.

Woolf compound Compound engine with no receiver, the pistons being in phase or 180° out of phase.

Wootten firebox Very wide firebox.

Wrapper Outer shell of a locomotive firebox.

Zara truck Italian equivalent of the Krauss-Helmholtz truck (q.v.).

Index

References in italics are to illustrations.

Acknowledgements

The publishers would like to thank the following individuals and organisations for their kind permission to reproduce the photographs in this book:

J. Adams 121 above; W. J. V. Anderson 53 below left; I. Belcher 16 above and below, 110, 121 below, 124 right, 125 above; Birmingham Museums and Art Gallery 160, 175; C. Bowden 15 above, 35, 37, 166, 170 below, 173; The British Tourist Authority 40; K. Brown 15 below, 128, 133, 171, 172; F. K. Burton 50 above; J. I. Case (P. Myers) 2–3; C. E. G. B. 168; Circus World Museum 124 left; R. Clark 123 below, (Merryweather Museum) 137; Compagnie General Maritime 107 above; Cunard Leisure 108–109; L. de Selva 120; A. M. Erlish 43; Fotocolor Fiore (Fiore Lazio) 7; V. Goldberg 62, 63; H. Hopfinger 161; A. Hornak 10, 34 above, 34 below, 36, 38–39, 44, 45 above, 112, 113, 116 above, 162 left, 163, 164, 174 above left, 174 below left, 174 right; J. M. Jarvis 47, 58 above left, 58 above right, 61 above, 103; London Transport 115; J. H. Meredith 118 below; C. Milster 4–5, 165, 169, 170 above; Ministry of Defence 107 below; National Maritime Museum 106 above, (M. Holford) 97 above, 100, (E. Tweedy) 97 below, 101; National Railway Museum 48 below, 45 below, 46, 50 below, 51; L. A. Nixon 58–59, 60–61; Nord Preservation Society and Nene Valley Railway 55; C. A. Parsons 156; Picturepoint 122 below; W. D. Sawyer 39 above left; Scala 12–13, 48 above; Science Museum 22, 41, 42, 56 above, 57 above, 104 above and below, 106 below, 114, 162 right; S.N.C.F. (Lamarche) 54, 56–57; Spectrum 8–9, 11; J. Topham 14 below, 102, 119, (G. M. Wilkins) 125 below, 126; Van der Pols 39 above right; J. Van Reimsdijk (Science Museum) 52, 53 above, 61 below, 105; T. Varley 116 below; La Vie Du Rail 53 below right, 64 below; ZEFA (W. F. Davidson) 14 above, (E. Hummel) 6, (F. Walther) 109 inset.

PDO 83-1213